Protein/Peptide Sequence Analysis: Current Methodologies

Editor

Ajit S. Bhown, Ph.D.
Associate Professor
Division of Clinical Immunology and Rheumatology
Department of Medicine
University of Alabama at Birmingham
Birmingham, Alabama

CRC Press
Taylor & Francis Group
Boca Raton London New York

CRC Press is an imprint of the
Taylor & Francis Group, an **informa** business

First published 1988 by CRC Press
Taylor & Francis Group
6000 Broken Sound Parkway NW, Suite 300
Boca Raton, FL 33487-2742

Reissued 2018 by CRC Press

Library of Congress Cataloging-in-Publication Data

Protein/peptide sequence analysis.

Bibliography: p.
Includes index.
1. Amino acid sequence. I. Bhown, Ajit S..
1934-
QP551.P69745 1987 574.19'245 87-718
ISBN 0-8493-6583-X

A Library of Congress record exists under LC control number: 87000718

ISBN 13: 978-1-315-89698-4 (hbk)
ISBN 13: 978-1-351-07608-1 (ebk)

Visit the Taylor & Francis Web site at http://www.taylorandfrancis.com and the
CRC Press Web site at http://www.crcpress.com

PREFACE

The functional properties of a protein molecule have long been known to depend upon its primary structure. It is, therefore, knowledge of the amino acid sequence that allows an investigator to understand: the interaction between an antibody molecule and its antigen; the ability of a hormone to bind to a receptor; the ability of an enzyme to bind and cleave its substrate with the ability for a variety of molecules to form aggregates and, hence, develop a structural framework upon which other biological activities occur. The fascination of protein chemistry, therefore, depends very much upon our ability to gain information regarding the primary structure of protein molecules and how that amino sequence directs the molecular function. Even with the new excitement generated by the very rapid methods allowing DNA sequencing to be transformed into protein structure, we still have an enormous need to determine amino acid sequences on ever-decreasing amounts of available polypeptide samples. Since structure and function are so intimately involved in proteins, even our knowledge of the DNA sequence does not always allow us to make a leap to functional relevance without first knowing and determining at least some parts of the amino acid sequence of a given protein molecule. This book is an attempt to provide in a single source current state-of-the-art methodologies for protein sequence analysis. It is hoped that these various chapters are presented in such a way that both the newcomer and the established protein chemist will find useful information and directions to new techniques.

The contributors to this volume have been carefully selected by the editor from among the world leaders in this field. Each chapter is written in a way that establishes the individual author's particular perspective on the field of his research. Chapter 1 outlines experimental details for the purification of proteins and peptides suitable for sequence analysis by high pressure liquid chromatography. Chapters 2 through 5 describe in detail how the purified protein-peptide can be subjected to sequence analysis. Use of homologous coupling reagents has been described in Chapter 6, while amino acid sequence analysis from the carboxyterminal end by chemical and enzymatic procedures has been discussed in Chapters 7 and 8, respectively. Mass spectrometry as a tool for sequence analysis is discussed in Chapter 9, and Chapter 10 covers the novel methods of amino acid sequence analysis. The last two chapters (11 and 12) deal with the identification procedures of modified (methylated) and derivatized (phenylthiohydantoin) amino acids.

This book offers a rich array of techniques and methods for sequencing proteins and peptides. It should meet the expectations of investigators in protein chemistry who wish to update their knowledge of sequencing techniques, and of those who wish to reacquaint themselves with the best available current technologies.

J. Claude Bennett

THE EDITOR

Ajit S. Bhown, Ph.D. is a Research Associate Professor in the Division of Clinical Immunology and Rheumatology, Department of Medicine, the University of Alabama at Birmingham, Birmingham, Alabama where he heads the Protein Sequencing Core Facility.

Dr. Bhown obtained his Ph.D. degree in 1965 from the University of Rajasthan, Jaipur, India, and his post-doctoral training at the University of Tennessee Medical Center, Knoxville, from 1965 to 1967 and at the University of Alabama at Birmingham from 1974 to 1975. He was on the Teaching Faculty in the College of Medicine in Jodhpur, India from 1967 to 1974. His major research interest has been in the development of techniques for the structural studies of proteins and peptides. Dr. Bhown has presented numerous lectures at international and national scientific meetings and seminars and has published more than 80 scientific papers and chapters.

CONTRIBUTORS

J. Claude Bennett, M.D.
Professor and Chairman
Department of Medicine
University of Alabama at Birmingham
Birmingham, Alabama

Russell F. Doolittle, Ph.D.
Professor
Department of Chemistry
University of California, San Diego
La Jolla, California

Rikimaru Hayashi, Ph.D.
Associate Professor
Research Institute for Food Science
University of Kyoto
Uji, Kyoto, Japan

Michael W. Hunkapiller, Ph.D.
Scientific Director
Applied Biosystems
Foster City, California

Rodney A. Jue, Ph.D.
Senior Research Scientist
Immunology Department
Hybritech, Inc.
San Diego, California

Sangduk Kim, Ph.D.
Associate Professor
Fels Research Institute
Temple University School of Medicine
Philadelphia, Pennsylvania

Henry C. Krutzsch, Ph.D.
Cancer Expert
Laboratory of Pathology
National Cancer Institute
National Institutes of Health
Bethesda, Maryland

Hyang Woo Lee, Ph.D.
Associate Professor
Department of Biochemistry
Sung Kyun Kwan University
Suwon, Kyunggi-do, Korea

James J. L'Italien, Ph.D.
Senior Scientist
Molecular Genetics, Inc.
Minnetonka, Minnesota

Woon Ki Paik, M.D.
Professor
Fels Research Institute
Temple University School of Medicine
Philadelphia, Pennsylvania

Minnie Rangarajan, Ph.D.
Research Officer
Centre for Biotechnology
Department of Biochemistry
Imperial College
London, England

Johann Salnikow, Ph.D.
Professor
Institute of Biochemistry and Molecular Biology
Technical University Berlin
Berlin, West Germany

Jin Shan-Wei
Senior Scientist
Shanghai Institute of Organic Chemistry
Shanghai, China

George E. Tarr, Ph.D.
Lecturer
Department of Biochemistry
University of Michigan
Ann Arbor, Michigan

Kenneth J. Wilson, Ph.D.
Research and Development Scientist
Applied Biosystems, Inc.
Foster City, California

B. Wittmann-Liebold, Ph.D.
Professor
Molecular Genetics
Max Planck Institute
Berlin, West Germany

ACKNOWLEDGMENTS

I would like to express my sincere gratitude to the authors who have helped me in producing this book. I appreciate the support given to me by Dr. J. Claude Bennett and his help in the completion of this book. I am grateful for the critique of the Advisory Board which was helpful in the selection of the authors and subject matter. I would like to thank Ms. Sandra Reid for the typing responsibilities. I am also grateful to CRC Press for publishing this book.

In memory of my Parents

This book is affectionately dedicated to my wife Meera, daughter Sonchita, and son Abhoyjit.

TABLE OF CONTENTS

Chapter 1

PURIFICATION OF PROTEINS/PEPTIDES FOR STRUCTURAL STUDIES

K. J. Wilson

TABLE OF CONTENTS

I. INTRODUCTION

Since the introduction of the gas phase sequencer,[1] there have been changes and improvements in isolation methodologies used for sample preparation. The manipulations required for any isolation often involve numerous methods which differ in their basic operating principles. These methods frequently introduce contaminates that chemically modify the desired end product and/or further compromise the purity of the sample. In addition, the time required for an isolation, its cost, overall yields, and the availability of laboratory instrumentation are collectively important factors which must be considered prior to the start of an isolation.

Given that there are a number of variables which require consideration, it is often helpful to consider the sequencing results from a random selection of samples which, for the most part, have *not* been "optimally" isolated.

Table 1 summarizes such results from samples submitted over a 12-month period to our Custom Sequencing Facility. The percentage of each group that was successfully sequenced (defined by a major-interpretable sequence) differed considerably: 51% of the proteins and 74% of the peptides.

Such results suggest that the analytical methods used for determining protein homogeneity and/or quantity are, at best, wanting. The high peptide heterogeneity (16%) supports a similar conclusion. The lower "contamination" (12%) noted for proteins speaks well of the isolation methods used for this sample set.

Insufficient amount and/or the absence of an available α-amino group could not be differentiated and the quantity estimates supplied with the samples were assumed to be correct. However, the obviously low initial yields from what could be considered a "typical" set of proteins suggest that either quantity estimates were inflated or significant amino-terminal blockage had occurred during isolation, or both. Since peptide amounts were, for the most part, unknown, the appropriate calculations could not be done.

As indicated in Table 1, the final steps for purifying proteins were based predominately (57%) on conventional chromatographic methods. This is not unexpected, since the majority of samples came from laboratories experienced in protein preparation. Although the isolation methods used were not necessarily among the more "modern" ones, including polyacrylamide gels and the various modes of HPLC, they appear to have been adequate.

Conversely, the almost exclusive use of reverse-phase chromatography (RPC) for peptide isolation (despite considerable contamination problems, 26% being unsequenceable for one reason or another) suggests that this method has found wide acceptance.

A survey of the literature has indicated that gas phase sequencer users are also employing conventional methodologies quite effectively (Table 2). It was interesting to see that affinity chromatography was being used almost as frequently as the ion exchange methods, and that size exclusion was being used for both isolation and desalting. The preference for an electrophoretic method is similar to the ratios noted in Table 1. The frequent use of various forms of HPLC for characterization (fragment isolation) was also supported in this survey.

The aim of this review is to indicate those methodologies found to be convenient sample preparation schemes by us and other gas phase sequencer users. Some of the techniques reviewed are designed so that sample volumes and sizes are compatible with the cartridge design of gas phase instruments. Since there are only a few laboratories that are employing all of the techniques, every attempt is made to give sufficient detail so that the appropriate literature can be located. Note that the topics covered and a person's familiarity with each are considered to be of *absolute* importance when sequencing at high sensitivities of 25 pmol or less.

Table 1
CUSTOM SEQUENCING SUMMARY

	Proteins	Peptides
Total number samples	75(100)	69(100)
Number successfully sequenced	38(51)	51(74)
Insufficient amount/blocked	28(37)	7(10)
Heterogeneous	9(12)	11(16)
% Initial yield	30 ± 20[a]	ND
Last step in purification		
Conventional	43(57)	2(3)
SDS-PAGE	24(32)	—
HPLC	8(11)	67(97)

Note: Values in parentheses represent percentages of total. ND, not determined, insufficient data.

[a] Calculated only on those samples where sufficient data were available.

Table 2
ISOLATION/CHARACTERIZATION METHODS EMPLOYED IN PREPARING PROTEINS/PEPTIDES FOR HIGH-SENSITIVITY STRUCTURAL INVESTIGATIONS

Isolation		Characterization	
Chromatography, liquid		High-performance liquid chromatography	
Affinity	28	Ion-exchange	19
Chromatofocusing	1	Reverse-phase	86[a]
Hydrophobic	1	Size-exclusion	14
Ion-exchange	37		
Reverse-phase	[a]		
Size-exclusion	52		
Electrophoresis, PAGE			
Isoelectric focusing	4		
Preparative SDS-PAGE	27		

Note: A survey of 109 papers which have employed the Applied Biosystems 470A Gas Phase Sequencer from 1982 through mid-1984 (6 in 1982, 23 in 1983, and 80 in 1984).

[a] Values under Characterization and Isolation are combined.

II. CHROMATOGRAPHIC METHODS

A. Reverse-Phase Chromatography (RPC)

There are numerous reasons why RPC is suited for sample preparation. Among these are the short separation times that range between 30 and 120 min and the high sensitivities that can be achieved using either UV or fluorescence detectors for preparing microgram or smaller amounts of material. In addition, sample work-up is usually simple and the general methodology is suitable for most peptides and many proteins.

The disadvantages of RPC are few, but extremely important. For example, column-to-

Table 3
CHROMATOGRAPHIC VARIABLES IMPORTANT IN REVERSE-PHASE SEPARATIONS OF PROTEINS AND PEPTIDES

	Proteins	Peptides
Support		
Phase	C_3—C_{18}, CN, phenyl	C_4—C_{18}
Pore size (Å)	300—500	100—300
Particle size (μm)	3—20	Same
Column		
Length (cm)	3—25	Same
Diameter (mm)	1—4.6	Same
Conditions		
Temp (°C)	5—35	Ambient
pH	2—8	Same
Buffers	Varied depending on protein activity/ stability	Normally including a volatile component (TFA, etc.) and low pH
Organic modifier	Acetonitrile, 1- or 2-propanol, mixtures with butanol or others	Acetonitrile or propanol
Flow rate (mℓ/ min)	0.1—2.0	Same

column reproducibility, from the same vendor or from different vendors, is frequently poor. Of equal importance is the initial equipment cost for mid- to high-sensitivity applications, and the fact that many proteins/enzymes are not compatible with the chromatographic conditions often used.

There are also problems associated with sample preparation which directly affect the overall yield. These include possible introduction of contaminates by the chromatographic step and the probability of chemically modifying the sample of interest. Sample losses on surfaces (glass, polypropylene, etc.) by either adsorption from a liquid or following a drying/ lyophilization step often reduce overall recoveries. As will be discussed later (Table 8), there are general guidelines that allow one to minimize such interactions or losses for many samples.

1. Chromatographic Variables

RPC is based on the hydrophobic differences between the sample being chromatographed and the bonded phase on the support material. There are a number of variables which effect these associations. The most important are listed in Table 3.

The utility of an RPC separation, under a defined set of experimental conditions, is dependent on the factors:

1. Column retention and selectivity
2. Material or mass recovery
3. How the chromatographic conditions affect recovery of biological or enzymatic activity

Since there are a number of variables which must be considered both prior to and following chromatography (see Table 3), it is frequently necessary to look at a series of well-defined experiments which optimize at least one of these three factors. One can often achieve high mass and activity recoveries, but not increase the overall purification (specific activity, biological or enzymatic); alternatively, retention and selectivity might be optimized at the expense of sample recovery.

Sample recoveries are usually not a serious problem given that "correct" chromatographic parameters can be found. For example, many of the earlier RPC isolations were carried out on samples that were either minimally affected by the chromatographic conditions or denatured during chromatography, with renaturation occurring during sample recovery by lyophilization, dialysis, or neutralization.

Comparisons of columns from various vendors have shown that differences do exist and the methods used by such vendors for column characterizations do not yield useful information as far as protein chromatography is concerned.[2] In fact, the product descriptions (particle size, pore size, carbon coverages, etc.) are often obscure or even incorrect, and it is virtually impossible to determine a support's particular efficacy. This information is disconcerting to the protein chemist that simply wants something that "will work" and not be forced into carrying out testing on each newly purchased column to determine its utility. It is even more of a concern when he realizes that even the same column from a given vendor can generate quite different results. There are other reports on nonspecific factors thought to contribute to the mass loss of sample: adsorption by stainless steel tubing,[3] frits,[4] or by the packing material itself.[5] These studies have been carried out on only a limited number of samples and it is not clear how generally applicable the results are to peptide/protein RP-HPLC.

2. Protein Separation by RPC

More definitive factors that affect both recovery and selectivity are the pore size of the support[5-10] (100 to 300 Å being optimal) and the choice of the organic modifier used for sample elution.[5,7,10] The most popular modifiers have been either acetonitrile or propanol, with several recent reports suggesting that mixed organic phases have distinct advantages. For example, combinations of propanol-butanol, acetonitrile-propanol, or even propanol-2-methylbutanol have proven effective.[11,12]

The actual pH during chromatography is of equal importance and has been illustrated in a number of publications.[13,14] The effects of parameters, including the pH, aqueous buffer, or organic component(s), on activity recovery can often be determined by simple incubation experiments. It is imperative that one not only perform these experiments prior to chromatography, but that the recovery yields from each be compared. Only through such evaluations can the factors, specific to a given support material, be identified and, perhaps, understood.

Other factors long considered unimportant, such as column temperature and sample-bonded phase contact time, have recently been investigated. Using soybean trypsin inhibitor,[15] papain,[16] lysozyme,[17,18] and ribonucleae A[19] as model proteins, it was demonstrated that both irreversible and reversible denaturation can occur during chromatography. Again, the denaturation was a function of the chromatographic conditions.

Native and denaturated species that create two or more peaks can often be separated, and the kinetics of unfolding can be determined.[14-18] The ability to detect both the native and denatured forms is a function of the refolding kinetics. The slower the refolding the greater the separation. Conversely, the faster the refolding the higher the probability of eluting a sharp peak. Distorted peaks are observed when the half-life of refolding was comparable to the elution time of chromatography.[18]

Another investigation[20] studied the optical changes which occurred in a number of proteins as a function of propanol concentration and pH. By comparing UV, circular dichroism, and/or fluorescence emission spectral changes, it was possible to account for greater than 80% renaturation. These results were consistent with the induction of an altered, though highly ordered, conformation.

Since RPC can detect very slight hydrophobic alterations in a protein, the presence of multiple peaks concerns the protein chemist interested in determining a primary structure. One of the most obvious detectable modifications that occurs is the substitution of an amino acid at one or more positions in a protein.[21,22] Others are chemical modifications such as glycosylation,[23] or oxidation of amino acid side chains.[24] The presence of disulfides can also be detected and has been employed to determine the disulfide bridging in proteins, including interferon,[25] T4 lysozyme,[26] and human growth hormone.[27] Similarly, various mixed disulfide forms of the high mobility group proteins[28] and recombinant α-interferons[29] have been isolated and characterized.

As previously discussed, denaturation caused by the chromatographic conditions can lead to peak multiplicity. For example, Ca^{2+}-binding proteins such as calmodulins, parvalbumins, etc. can chromatograph as broad, ill-defined peaks.[30] These proteins elute as sharp, well-defined peaks when either $CaCl_2$ or its specific chelator, EGTA (ethyleneglycoltetracetic acid), is added to the buffers. The elution positions for this particular class of proteins can be modulated through changes in the pH of the chromatography buffer. A decrease in the buffer pH from 7 to 2.5 causes increased retention. Since it is known that bound Ca^{2+} is lost at low pH, it stands to reason that the molecular structure is altered and the "open" configuration (lacking Ca^{2+}) interacts more strongly with the support.

Therefore, when a particular cofactor is required by the protein being isolated, it should be either removed from the sample or added to the buffers prior to chromatography. An example of the latter case was the need to add $CaCl_2$ to the buffers when chromatographing trypsin and α-chymotrypsin.[31] For both enzymes the recovery yields were significantly increased following the addition.

3. Peptide Separation by RPC

The application of RPC for the separation of peptides is now the method of choice in most labs for either mapping or isolation for sequence analysis. The chromatographic conditions are listed in Table 2.

The ligands most frequently used are either C_8 or C_{18} on 100- or 300-Å pore-size supports. Various authors have noted slight improvements of one particle size over another for specific applications, i.e., the use of a 3-μm particle for hydrophilic peptides. In general, 5- and 10-μm supports work equally well. Figure 1 illustrates how the reduction in column length affects peptide separation. With the exception of the most hydrophilic components, there are basically no differences in selectivities as a function of column length.

The eluting "strength" of the organic solvent needs to be sufficiently high to cause even the very hydrophobic fragments to elute. For this reason the most commonly used buffers contain either acetonitrile or propanol and, less frequently, methanol.[7,10] It is not, however, possible to substitute one organic for the other and expect the same resolution from a given column.[7] The most commonly utilized solvent system contains one of the perfluoralkanoic acids, usually 0.05 to 0.5% TFA or even HFBA (heptafluorobuteric acid). The volatile nature of these buffers makes them especially convenient to use.[32-34]

Ambient temperature and flow rates in the range of 1 mℓ/min are most frequently employed. One of the least recognized variables available for significantly altering the elution characteristics is the wide pH range over which separations can be performed. Often a simple pH adjustment from 2 to 6 or 7 will allow one to separate peptides or detect contamination without having to use other modes of HPLC or changing to another buffer system.[35]

4. Increased Sensitivities with RPC

Figure 2 illustrates the separation of peptides generated by exposing a 150,000-dalton protein to trypsin for 2 hr prior to RPC. Since less than 100 pmol of sample was available, a 1-mm microbore column was used for preliminary mapping and determining gradient

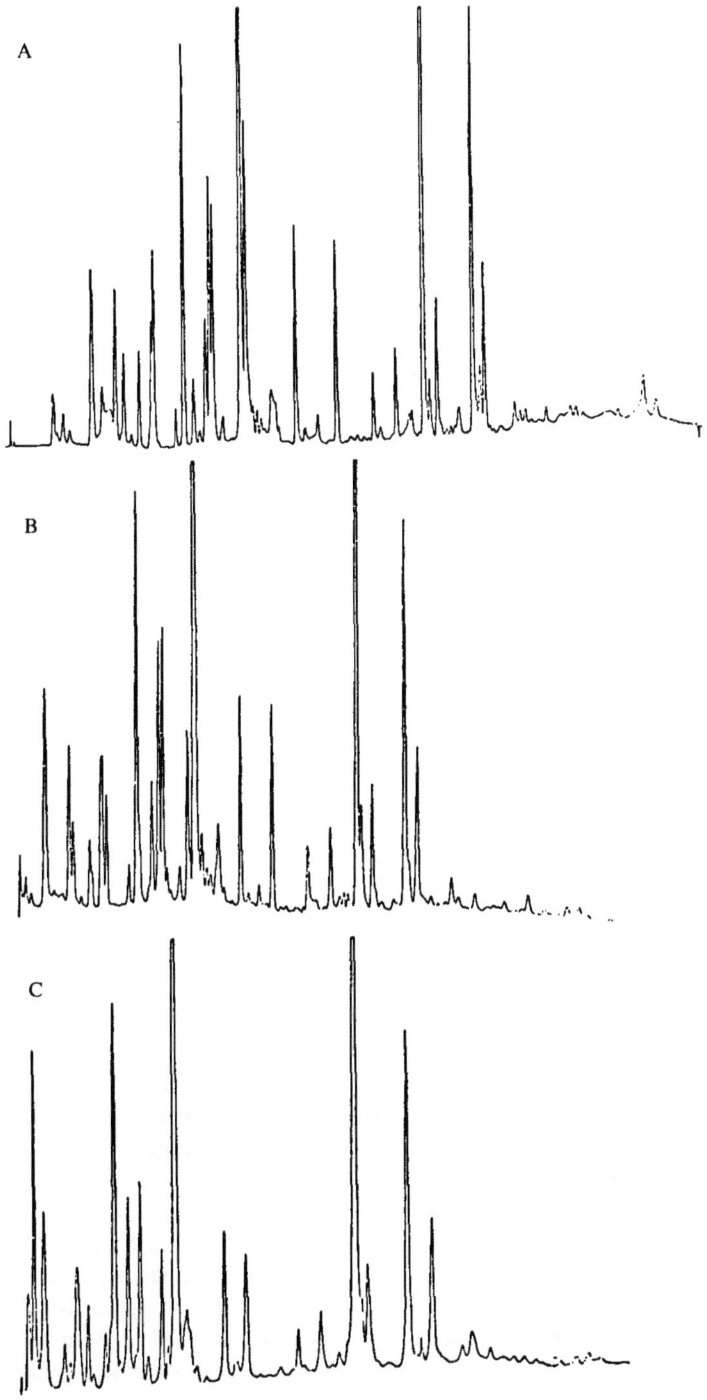

FIGURE 1. Peptide separations as function of column length. Apomyoglobin was tryptically digested and then 500 pmol amounts chromatographed on a series of Aquapore RP-300 cartiridge columns. All columns had internal diameters of 2.1 mm, but lengths varied from 220 mm (A) to 100 mm (B) to 30 mm (C). The solvent systems used were A buffer, 0.1% TFA; B buffer, 60% acetonitrile in 0.1% TFA. A gradient from 0 to 100% B was developed at 200 μℓ/min over 45 min.

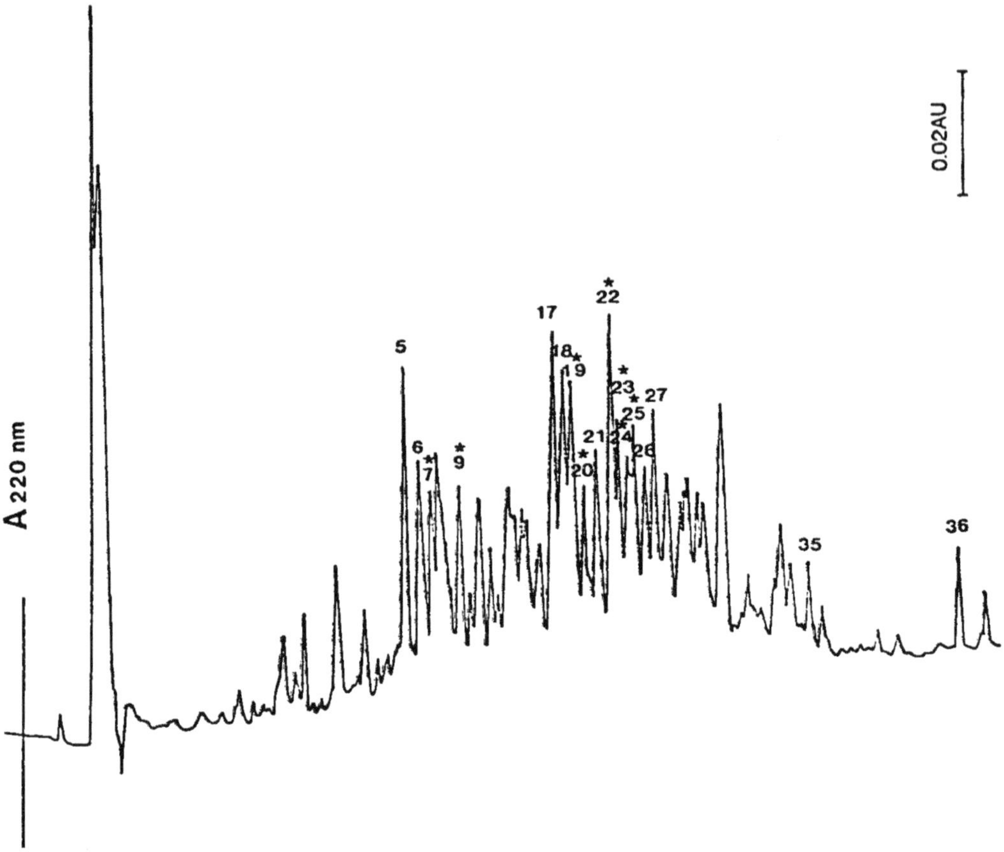

FIGURE 2. Preparative isolation of peptides. The tryptic fragments from a 150,000 dalton protein (approximately 100 pmol) were chromatographed on a 2.1 × 220-mm Aquapore RP-3000 column at 200 μℓ/min. Buffers were A, 0.1% TFA and B, 60% acetonitrile in 0.1% TFA.

profiles. Subsequently, a narrow-bore column was used for the indicated separation. Each peak was collected and those numbered were sequenced in a gas phase instrument. With the exception of the peaks denoted by an asterisk, all were found to contain multiple peptides. Given the size of the protein one would expect more fragments than were detected. Such results suggest that either the digestion was incomplete or the majority of the peaks contain multiple peptides, or both.

Approximately 100 residues were identified and most peptides were sequenced to the carboxy termini, starting with initial yields ranging from 5 to 50 pmol. This information, subsequently used for cloning, could only be derived by sequencing smaller peptides and not the intact protein. In the latter case only a few residues are generally identified due to the quick build-up of PTH background caused by nonspecific acid cleavage of the polypeptide backbone. The reasoning behind not attempting some type of either limited chemical or enzymatic cleavage was that aggregation frequently causes problems with subsequent separations.

In the above example and as Figure 2 illustrates, increased sensitivities were achieved by reducing the column internal diameter. Several possibilities are available to the protein chemist when higher detection sensitivities are required:

1. Decrease to lower UV wavelengths
2. Increase the detector sensitivity

3. Implement one of the stream-splitting, post-column reaction systems such as fluorescamine[36,37]
4. Reduce the internal diameter of the column

The first two possibilities are often used, but limited by either the absorbance and/or refractive index of the buffers commonly employed. Stream-splitting is less frequently implemented due to the expense and problems associated with maintaining the additional equipment. However, most of the HPLC units presently available can carry out gradient elutions on 2-mm ID columns without much difficulty. They are usually not optimized, however, to perform at full-scale expansions of greater than 0.05 AUFS at flow rates of 100 to 200 μℓ/min with proper mixing. When the column diameter is further reduced to 1.0 mm (microbore), problems increase. Pumping accurately and mixing at flows of 25 to 50 μℓ/min are very difficult. Also, the dead volumes must be sufficiently small so that the desired gradient reaches the separating column.

The series of chromatograms in Figure 3 illustrates the usefulness of reducing column ID and flow rate proportionately, while maintaining the sensitivity of the detector constant at 0.1 AUFS and reducing the amounts injected. The quantities chromatographed were 1000 and 100 pmol on the 4.6-mm ID support; 200 pmol on a 2.1-mm ID support, and 100 pmol on a 1-mm ID support.

It is clear that the sensitivity has been drastically increased (compare Figure 3B with 3D). Although not as obvious, the elution volumes of the individual peaks have decreased by a factor of approximately 20-fold, from approximately 500 μℓ with the 4.6-mm ID column to 30 μℓ with the 1.0-mm ID column. This results in simplified sample collection, handling, and concentration.

This significant reduction in elution volume provides an advantage to the gas phase sequencer user in that the glass fiber support disk can be used directly, either before or after polybrene precycling, as a fraction collector. This is accomplished by collecting the two to four drops of liquid containing the eluted peak onto the disk supported in an appropriate container, such as a 1.5-mℓ Eppendorf tube. The results presented in Table 4 indicate how sequencing yields can vary as a function of the collection method.

Similar increases in sensitivity are possible during protein HPLC on reduced ID columns (Figure 4A). Under these chromatographic conditions sample loss does not appear to be a problem (Figure 4B), and as Nice et al.[40] have shown, these smaller columns can be conveniently used for sample trace enrichment. Also, when appropriately excited, the native fluorescence of tryptophan in samples can provide a detection method with a higher sensitivity than direct absorbance measurements at low UV wavelengths. This latter point made by the above authors[40] is also one that we have substantiated.

One application for such a microbore system is illustrated in Figure 5 for the chromatography of a protein which was available in only very limited amounts. The sample, recombinant interleukin 2, was chromatographed at the 6-pmol level (approximately 120 ng). Two major components are clearly discernible as are a number of smaller "background" peaks which, due to their absence in the blank, must represent contaminates of the protein sample.

B. Ion Exchange Chromatography (IEC) and Chromatofocusing

1. IEC Chromatographic Variables

The recent availability of suitable anion and cation supports has provided both IEC and chromatofocusing as alternatives to RPC. The conditions required to elute proteins from an IEC matrix are usually compatible with maintaining native biological or enzymatic activities. Variables to be considered when performing such elutions are pH, the nature of the anion and cation of the particular salt being employed as the eluant, and the gradient rate.[41-46]

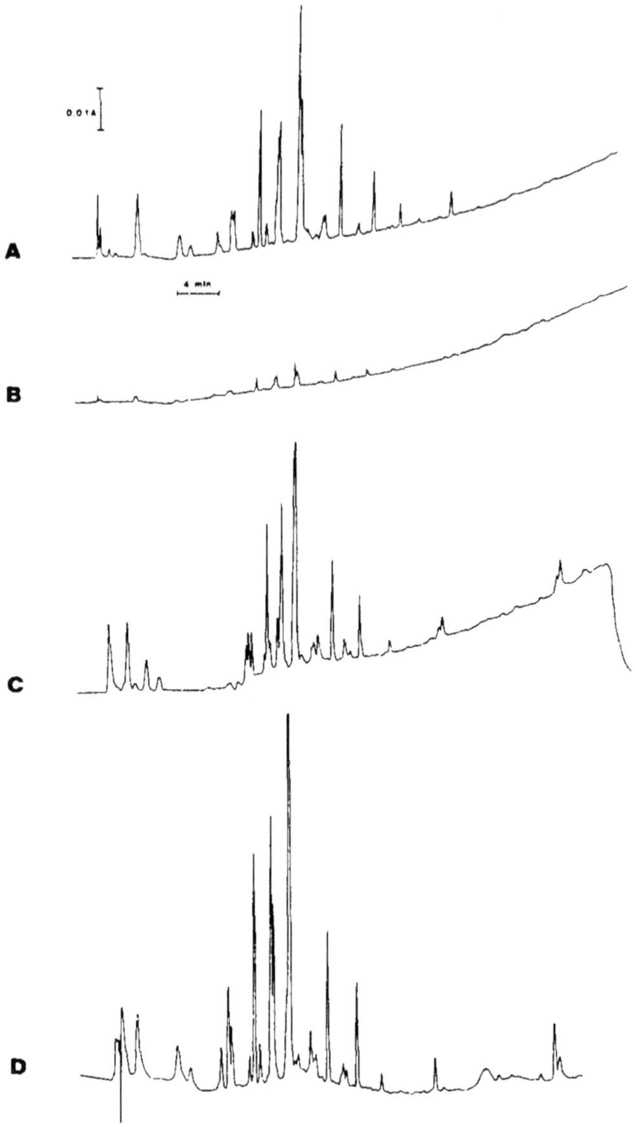

FIGURE 3. Peptide mapping with narrow and microbore columns. A tryptic digest of apomyoglobin was chromatographed on the following series of Aquapore RP-300 columns: (A) 4.6 × 250 mm at 1 mℓ/min, 1000 pmol; (B) as in A, 100 pmol; (C) 2.1 × 220 mm at 200 μℓ/min, 200 pmol; (D) 1.0 × 250 mm at 50 μℓ/min, 100 pmol. Detector sensitivity (0.1 AUFS at 220 mm) and chart speed (0.5 cm/min) were maintained constant. The linear gradient over 45 min was between 0.1% TFA (Buffer A) and 60% acetonitrile in 0.1% TFA (Buffer B).

2. IEC Applications

As Figure 6 illustrates, this particular form of chromatography is often complementary to the others. The RPC of diphtheria toxin A fragments failed to separate the three polypeptides known to be present (Figure 6A). The polypeptides, which differ only at their carboxytermini, end in -Arg-Val-Arg-Arg, -Arg-Val-Arg, and -Arg, and derive from the treatment of diphtheria toxin with trypsin.

Chromatography of the same mixture on an anion exchanger and elution with a NaCl

Table 4
SEQUENCING YIELDS AS A FUNCTION OF COLLECTION METHOD

Collection method	Peptide, initial yields (%) T-3		T-4	T-5	T-6
Onto filter	89(E), 21(H)		31(H)	33(V)	38(A)
Into tube	111(E), 46(H)		50(H)	43(V)	71(A)
Percentages filter ÷ tube	80	46	62	77	54

Note: A tryptic mixture (300 pmol) of apomyoglobin was chromatographed on a 1 × 250-mm Aquapore RP-300 column under conditions identical to those in Figure 3D. The peaks were collected either onto filters or into Eppendorf tubes and, subsequently, sequenced. See Wilson et al.[38] and Yuan et al.[39] for further details.

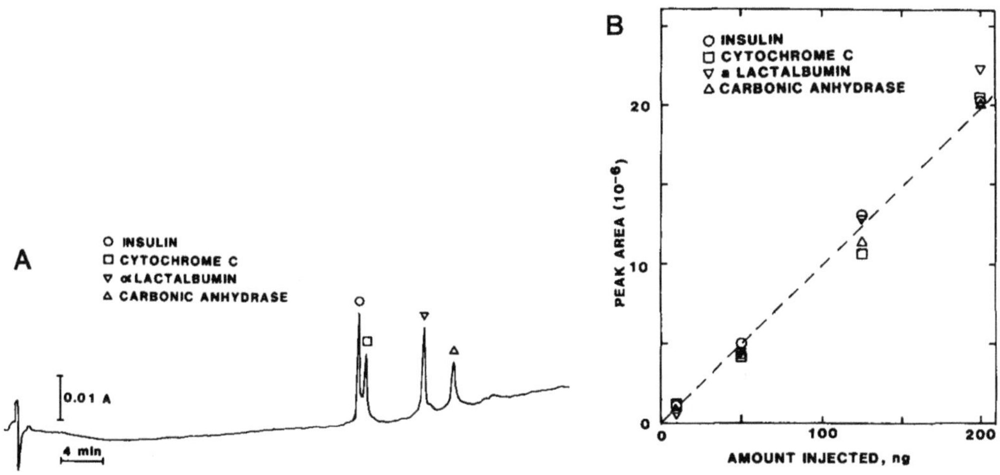

FIGURE 4. Microbore RP-HPLC of proteins. (A) Chromatography was performed as described in Figure 2; 50 ng of each protein were injected; (B) correlation between integrated peak areas and nanogram quantities injected.

gradient resulted in baseline separation of all components (Figure 6B). This was performed on intermediate amounts of material and provided a rapid, easy means of preparing sufficient quantities for further enzymatic and structural investigations.

Chromatofocusing of the identical mixture on another anion exchanger and eluting with a decreasing pH gradient of ampholites also resolved the components (Figure 6C). From this type of HPLC application, it is possible to obtain information about the isoelectric points (pI) of the chromatographed polypeptides. The average pH from three separate elutions of the diphtheria proteins were within 0.05 pI units of the IEP values determined by isoelectric focusing in polyacrylamide gels.

Chromatographic results have indicated that glycosylated and nonglycosylated proteins, such as ribonucleases B and A, can be separated by RPC (results not shown).[23] The only known difference between these two enzymes is the presence of a single, high-mannose-type oligosaccharide chain at an Asparagine residue in position 35. As Figure 7A shows, they can be separated by cation exchange chromatography. Here the main peak is ribonuclease B and the smaller peak (indicated by the asterisk) is the elution position of ribonuclease A.

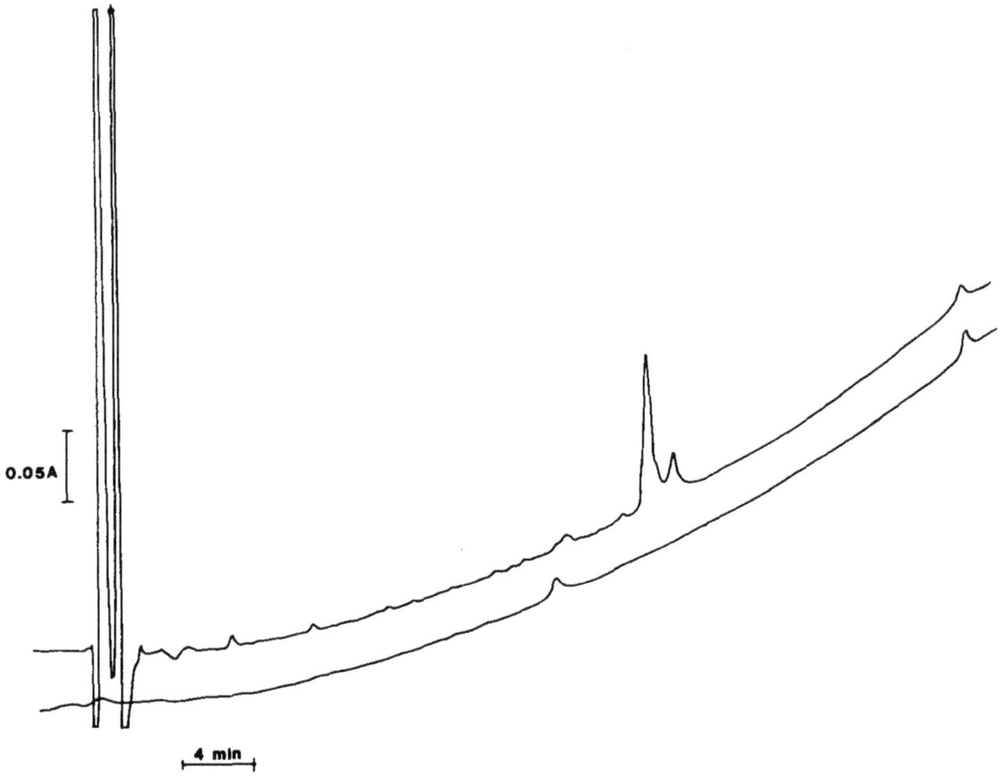

FIGURE 5. Microbore RP-HPLC of recombinant IL-2. Chromatography was performed on 6 pmol of material using a 1 × 250 mm Aquapore RP-300 column eluted at 50 µℓ/min. The gradient was from 50 to 100% using TFA/acetonitrile buffers (see Figure 1).

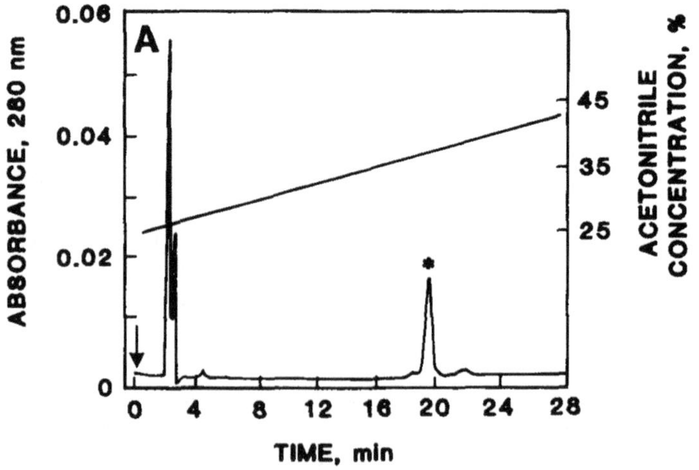

FIGURE 6. Chromatography of diptheria toxin A (DTA) fragments. (A) RP-HPLC of 12.5 µg DTA on C_4 column (4.6 × 250 mm) at 1 mℓ/min employing a linear gradient using 0.1% TFA (A buffer) to 60% acetonitrile in 0.1% TFA (B buffer) over 30 min; (B) IE-HPLC at 1 mℓ/min of 120 µg DTA on an anion-exchange (Mono Q) column using 50 m*M* Tris-HCl, pH 8.5 (A buffer) and 2 M in NaCl in A buffer (B buffer); (C) chromatofocusing of 235 µg DTA on an anion-exchange (Mono P) column equilibrated in 25 m*M* bis-Tris, pH 6.25, and eluting at 1 mℓ/min with a 1:10 dilution of Polybuffer 74, pH 4.0.

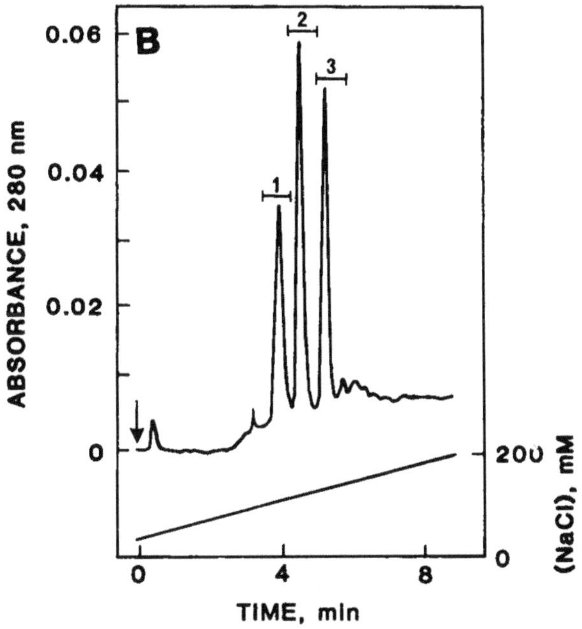

FIGURE 6B

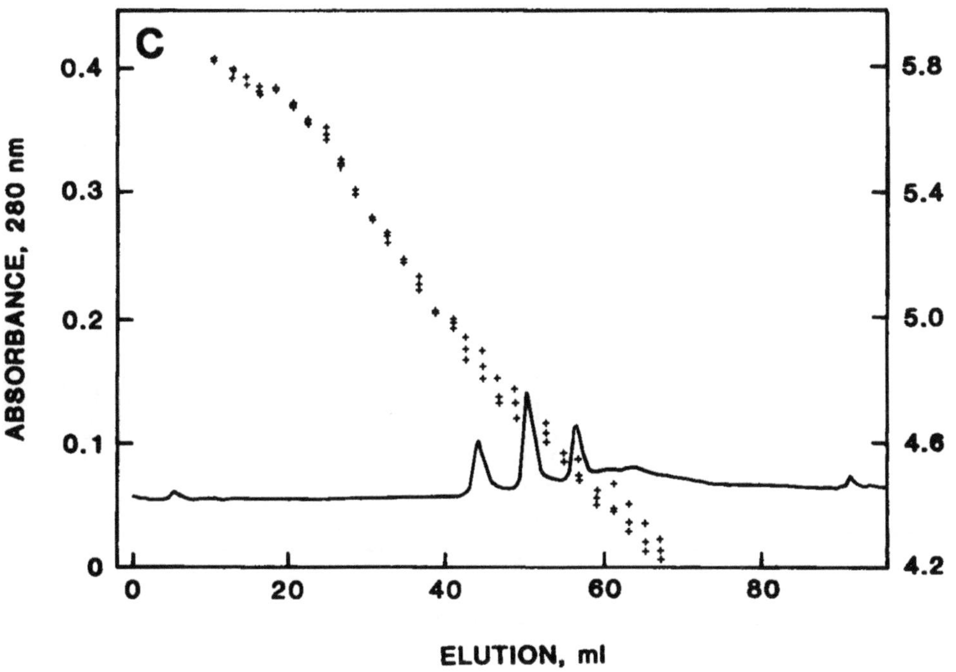

FIGURE 6C

This form of HPLC can also be used as a method for following enzymatic deglycosylation. For example, exposure of ribonuclease B to α-mannosidase (Figure 7B), and either endoglycosidase H (Figure 7C) or D (Figure 7D), produced discernible changes in the elution positions of the resulting products. The "fine" detail provided by IEC is a significant improvement over that observed by following molecular weight changes on SDS polyacrylamide gels (Figure 7E).

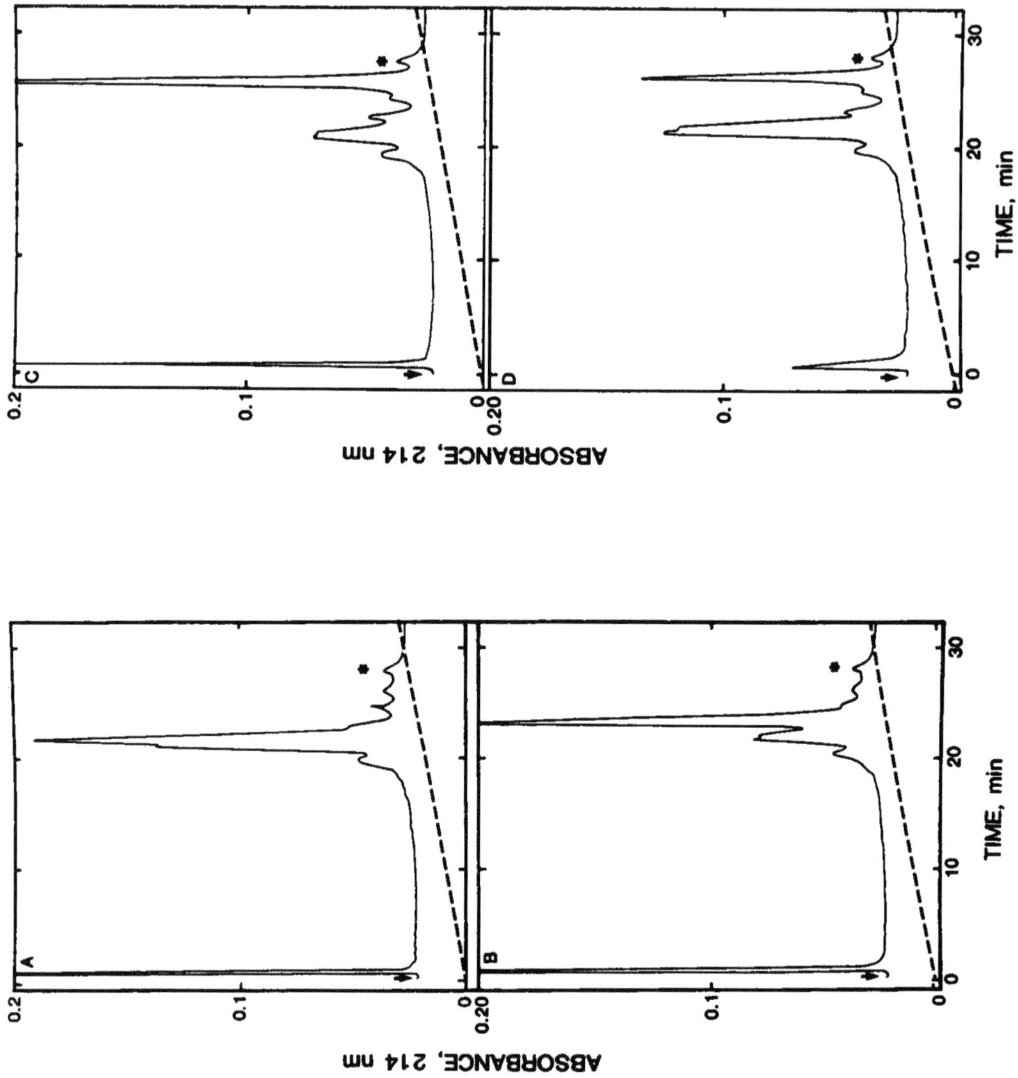
A
B
C
D
ABSORBANCE, 214 nm
ABSORBANCE, 214 nm
0.2
0.1
0
0.20
0.1
0
TIME, min
0
10
20
30
*

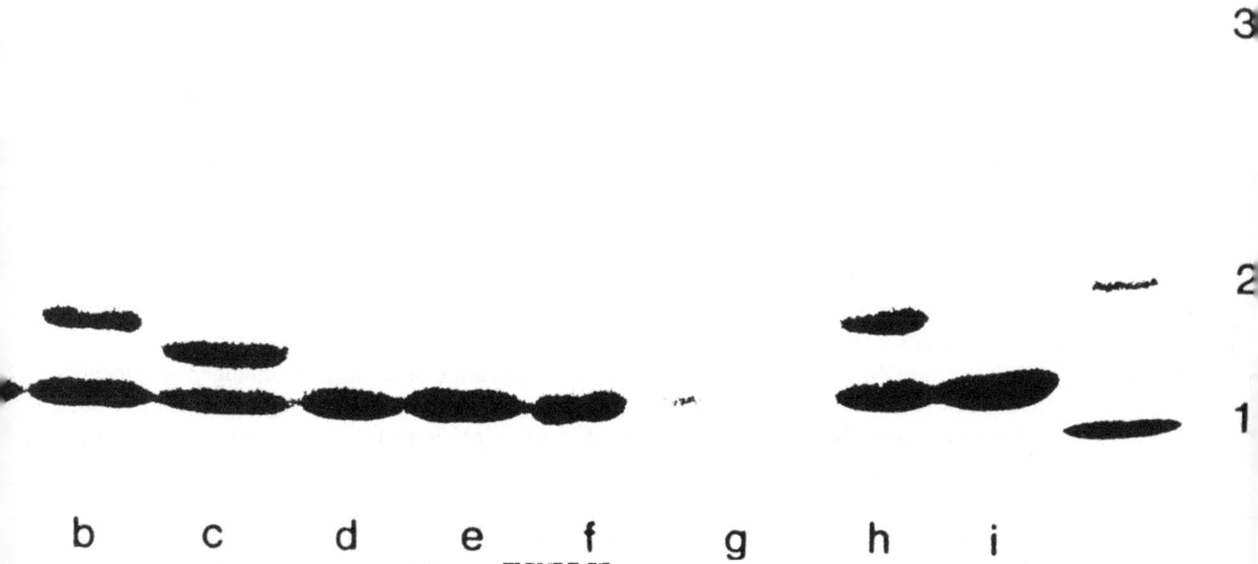

FIGURE 7E.

: HPLC of bovine RNase B. (A to D) IE-HPLC on Mono S column using a linear gradient formed between 25 mM sodium phosphate, pH 6.0 (A
Cl (B buffer). Samples (10 to 25 μg) were chromatographed at room temperature using a flow rate of 1 mℓ/min. (A) RNase B; (B) RNase B
cosidase H; (D) with endoglycosidase D. The asterisk (*) on each chromatogram designates the elution position of RNase A. (E) SDS-PAGE of R
Gradient gels (12 to 20%) were used, 1 to 5 μg per protein applied, and Coomassie blue staining employed for detection. Sample identification: (
s A and B; (c) α-mannosidase-treated RNase B; (d) RNase B treated with endoglycosidases D, (e) F, and (f) H; (g) peptide: *N*-glycosidase or *N*

IEC has been compared with gel isoelectric focusing in a recent publication by Kato et al.[47] Using a commercially available form of superoxide dimutase, they studied the resolution of IEC as a function of pH and composition of the eluant, and the steepness and flow rate of the gradient. Similarly, the resolution of polyacrylamide gel isoelectric focusing was studied as a function of the pH gradient. They concluded that the two techniques provide comparable resolution, but that IEC is more advantageous because of speed, simple sample recovery, reproducibility, quantitation, and ease of scaling up.

The application of either type of ion-exchange support to peptide isolation has not received much attention. This is primarily due to the availability of RP supports for such purposes and the necessity of sample desalting prior to sequencing.

The use of a cation exchanger in the first separation dimension, and RP in the second, has, however, proven effective in analytical peptide mapping of hemoglobin variants.[48] A similar approach has proven effective for Takahashi et al.[49] for separating complex peptide mixtures. They have developed an automated tandem chromatography system which is actually composed of two HPLC units. The first elutes the sample mixture isocratically, i.e., stepwise, from an anion exchanger onto a RP column. The salts are not retained on this support and the peptides are eluted with an acetonitrile gradient, followed by column reequilibration. The next group of peptides are then transferred onto the hydrophobic support by increasing the salt concentration in the eluant for the anion exchanger and the process repeated. Using this methodology it was possible to resolve/detect several hundred peaks within a relatively short time period (16 hr).

Another study[50] found that the elution times of peptides on a cation exchanger are linear with respect to net charge at pH 3.0 and 6.5. Such an application might find utility in correlating the overall charge of a peptide at a given pH with its amino acid composition. This would then be similar to the relationship between peptide molecular weight and charge during paper high-voltage electrophoresis.[51]

Both cation and anion exchangers have proven effective in the purification of bovine pituitary peptides.[52] Following an acidic pH extraction, supernatants were first enriched on C18 cartridges. The subsequent eluates were further fractionated into basic, acidic, and neutral pools on the two types of ion-exchange supports. This combination provided a method that circumvented the overloading problems of distortion, peak trailing, and large eluate volumes that arise when purifying minor components.

C. Hydrophobic Interaction Chromatography (HIC)

The tendency of many proteins to denature under conditions used for RPC has led to the development of support materials possessing both lower surface hydrophobicities and ligand densities. Such supports, either polymer or silica based, can be utilized for protein chromatography. The use of a more hydrophilic surface minimizes denaturation upon adsorption. It also allows one to elute using milder conditions such as high ionic strength and neutral pH rather than low pH, low ionic strength, and organic eluants required for RPC.

1. HIC Variables

Table 5 lists those chromatographic variables investigated and found important to HIC. The most commonly used supports, phenyl or butyl, are normally eluted around neutral pH at ambient temperature using a decreasing gradient of a salt. The most commonly used salt is ammonium sulfate at 2 to 3 *M* in the A buffer which is appropriately diluted with the same A buffer to form the gradient. Other salts have also been used, including Na_3 citrate, Na_2SO_4, NaCl, NH_4SCN, NH_4 acetate, $NaClO_4$, and Li_2SO_4.

2. HIC Applications

The recoveries given in Table 6 serve to illustrate the potential utility of this method for

Table 5
CHROMATOGRAPHIC VARIABLES IMPORTANT IN HIC OF PROTEINS

Support
  Ligand chain length
    Density
    Capacities
Column length vs. resolution
  Stability vs. pH
  Effect of surface tension on retention
  Sample size vs. retention volume
Conditions
  Flow rates
  Gradient times
  pH
  Temp and denaturation
  Eluant salts

Note: See References 53—61 for details.

Table 6
TYPICAL PROTEIN RECOVERIES FROM HIC

	Recovery (%)		
Protein	**Butyl-**	**Phenyl-**	**Ref.**
Cytochrome *c*	95	93	53
Ribonuclease	96	108	53
Lysozyme	101	99	53
Bovine serum albumin	105	100	53
Ovalbumin	99	101	53
α-Chymotrypsin	94	102	53
		92	58
		90	57
α-Chymotrypsinogen	88	95	53
Myoglobin	94	98	53
Lipoxidase		89	58
Ferrodoxin NADP reductase		100	58
α-Amylase		100	56
β-Amylase		80	58
β-Glucosidase		100	56
Lactic dehydrogenase		100	57

protein isolation. Since these values are generally quite high and certainly higher than would be expected from the RPC of the same set of samples, the inclusion of this method in any isolation scheme should be seriously considered.

The two chromatographic modes, HIC and RPC, in fact, compliment each other in a number of ways. For example, the selectivities of each are often sufficiently different toward a given sample that they can be used consecutively in an isolation. This is clearly illustrated in the selected comparisons given in Table 7. The obvious similarities in selectivities of two closely eluting proteins, such as cytochrome *c* and myoglobin, on HIC might be quite different on a RP support; conversely, proteins having similar elution properties on RPC, e.g., lysozyme and myoglobin, can be conveniently separated by HIC.

Since samples from an HIC separation are heavily contaminated with salts, a subsequent

Table 7
ELUTION PROPERTIES OF PROTEIN STANDARDS FROM RP AND HIC SUPPORTS[57]

HIC protein	RP t_r(min)	RP protein	t_r(min)
Cytochrome *c*	0.6	Ribonuclease A	10.7
Myoglobin	0.8	Cytochrome *c*	12.6
Ribonuclease A	1.6	Lysozyme	14.3
Lysozyme	8.5	Myoglobin	14.6
α-Chymotrypsinogen	18.1	α-Chymotrypsinogen	16.8
Bovine serum albumin	20.8	Bovine serum albumin	17.1

RPC step can both desalt as well as further purify the component of interest, i.e., remove buffers and/or salts which do interfere with microsequencing. As will be discussed later, it is the combination of chromatographic modes that can eliminate certain steps, such as dialysis, and allow one to carry out isolations while maintaining both high sensitivities, concentrations, and recoveries.

D. Size Exclusion Chromatography (SEC)

Of the four different chromatographic modes available, SEC is probably the least useful in isolating either peptides or proteins for structural determination. The primary reason for this is that the column has a very limited separation volume as a function of the molecular weight range being separated. Normally, one finds that there is only a maximum of 10 to 15 mℓ available for separating components differing three or more orders of magnitude in molecular size. Because of the poor efficiencies inherent to these supports, there is significant cross-contamination of the components in any given molecular weight range. Additionally, many of these supports still contain gel surfaces which interact electrostatically with the charged amino acid side chains of the polypeptides being chromatographed.

SEC may be useful for high-sensitivity applications, such as at an early step of isolation or as a substitute for desalting prior to sequencing. Again, due to their low selectivities, the latter use implies that the relatively large eluate volumes of 0.5 to 2.0 mℓ would have to be recovered by either drying or lyophilization prior to sequencing.

Although not directly related to sequencing, these columns have proven useful in the study of protein folding and stability as a function of urea and thermal denaturation.[62] Also, dissociation constants of oligomeric proteins have been investigated.[63] Separations of monomer and dimer fractions of cytochrome *c*[64] and albumin[65] have also been successfully performed.

E. Recovery Factors Controlled by the Investigator

The role that the investigator plays in sample recovery and modification is also of prime importance. When working with small amounts of material, serious losses occur through surface adsorption. It is necessary to recognize that such losses can be minimized by careful selection of materials in which samples are either transferred or stored. Alternatively, detergents, stabilizers, or even proteins can be added to minimize sample loss.

The method to improve sample recovery can often be easily and quickly established (Table 8). This is best done by determining the extent of sample loss through a series of incubation experiments under the appropriate chromatographic conditions of pH, buffer, and temperature, to which the sample will be exposed. The simplist way to limit loss is to minimize sample handling by reducing the number of steps required for an isolation. Drying, evaporation, lyophilization, and/or dialysis can often be circumvented in HPLC by diluting the sample with the appropriate buffer to reduce organic or salt concentrations, and pumping or injecting it onto the next column for further purification.

Table 8
REDUCTION OF SAMPLE LOSS

Attempt to eliminate steps involving either drying, lyophilization, and/or dialysis
When performing a series of chromatographic steps, reduce organic or salt concentrations by dilution and pump onto the next column
Arrange chromatographic separations intelligently, i.e., IEC, HIC, and/or SEC carried out prior to RPC
When necessary, utilize a cold methanol or ethanol precipitation method for sample recovery

Steps in a chromatographic series need to be arranged intelligently. For example, IE, HIC, and/or SEC steps should precede RPC. When necessary, samples can be recovered by a cold ethanol precipitation (see Section III.A) or by lyophilization from a volatile buffer. Currently, many labs remove the buffer used in the final isolation step by loading the entire sample onto a reverse-phase support equilibrated in a volatile buffer, often 0.1% TFA, and then elute with either acetonitrile or propanol. A preliminary experiment is necessary to establish recoveries prior to committing the entire sample to this desalting step. If high recovery of activity is desired, then chromatography in a low pH system might not be appropriate. Using ammonium bicarbonate as the buffer is often a reasonable alternative.

It is essential to realize that solvents as well as surfaces which come in contact with the sample might lead to a chemical modification. The oxidation of all methionyl side chains in rat brain parvalbumin was strongly suspected when isolations were performed under acidic conditions using 0.1% H_3PO_4-10 m $NaClO_4$ at pH 2.2, rather than 50 m*M* Tris.HCl at pH 7.5.[66]

The chemicals which one employs in preparative SDS (sodium dodecyl sulfate) PAGE need to be pure, otherwise any number of modifications are possible, including amino group blockage, oxidation, and polypeptide bond cleavage.[67]

High concentrations of some acids can also modify samples. For example, the exposure of proteins to formic acid, under conditions frequently used to get hard-to-solubilize proteins into solution,[68,69] can cause aspartic acid-proline bond cleavages as well as formyl esterification at serine and threonine hydroxyl groups.[12]

III. ELECTROPHORETIC RECOVERY TECHNIQUES

The electrophoretic separations illustrated in Figure 8 are examples of the resolving power inherent to this technique. In a single dimension, separation is based on molecular weight differences (Figure 8A). The component of interest is clearly separated from surrounding proteins and can be easily isolated, provided a suitable recovery method is available. The two-dimensional separation (Figure 8B) illustrates what is possible when combining isoelectric focusing in the first dimension with electrophoresis in the presence of SDS in the second dimension. For preparative purposes, either of these techniques can be used with the latest methods for recovering proteins from gels.

There are a number of advantages inherent to gel electrophoretic techniques. Perhaps the most important is that the method, as well as the equipment required, is simple. Thus, the expense and time required for installation are relatively little. Since there are simple staining methods suited to low-scale, high-sensitivity applications, one does not have to invest heavily in detection equipment.

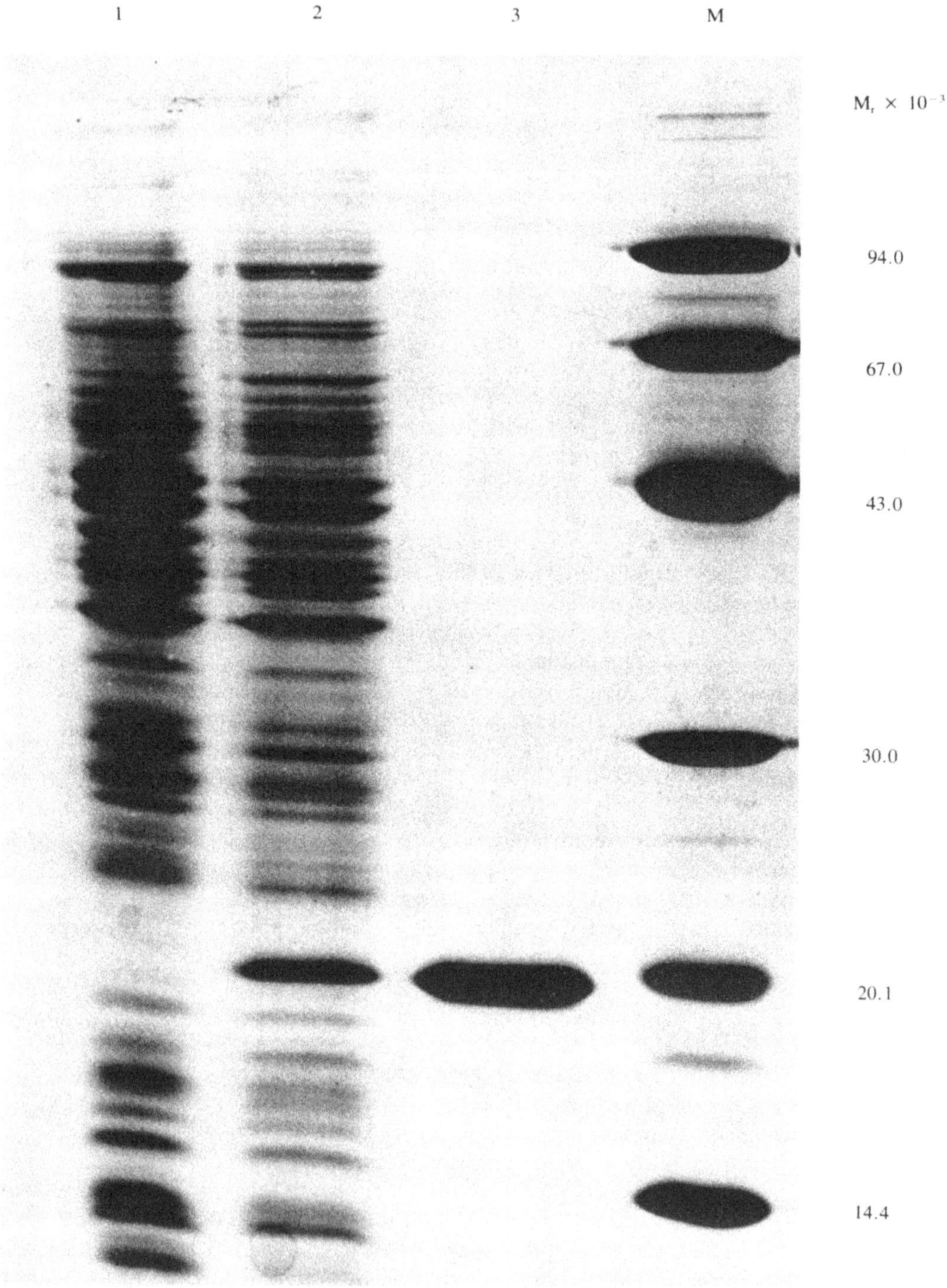

A

FIGURE 8. Protein electrophoretic separations. (A) SDS-PAGE: (1) an *E. coli* extract; (2) an extract of *E. coli* containing the gene for a human protein; (3) the purified protein; (M) a molecular weight standard. (B) A two-dimensional separation of a cell extract with the first dimension using isoelectric focusing and the second SDS-PAGE.

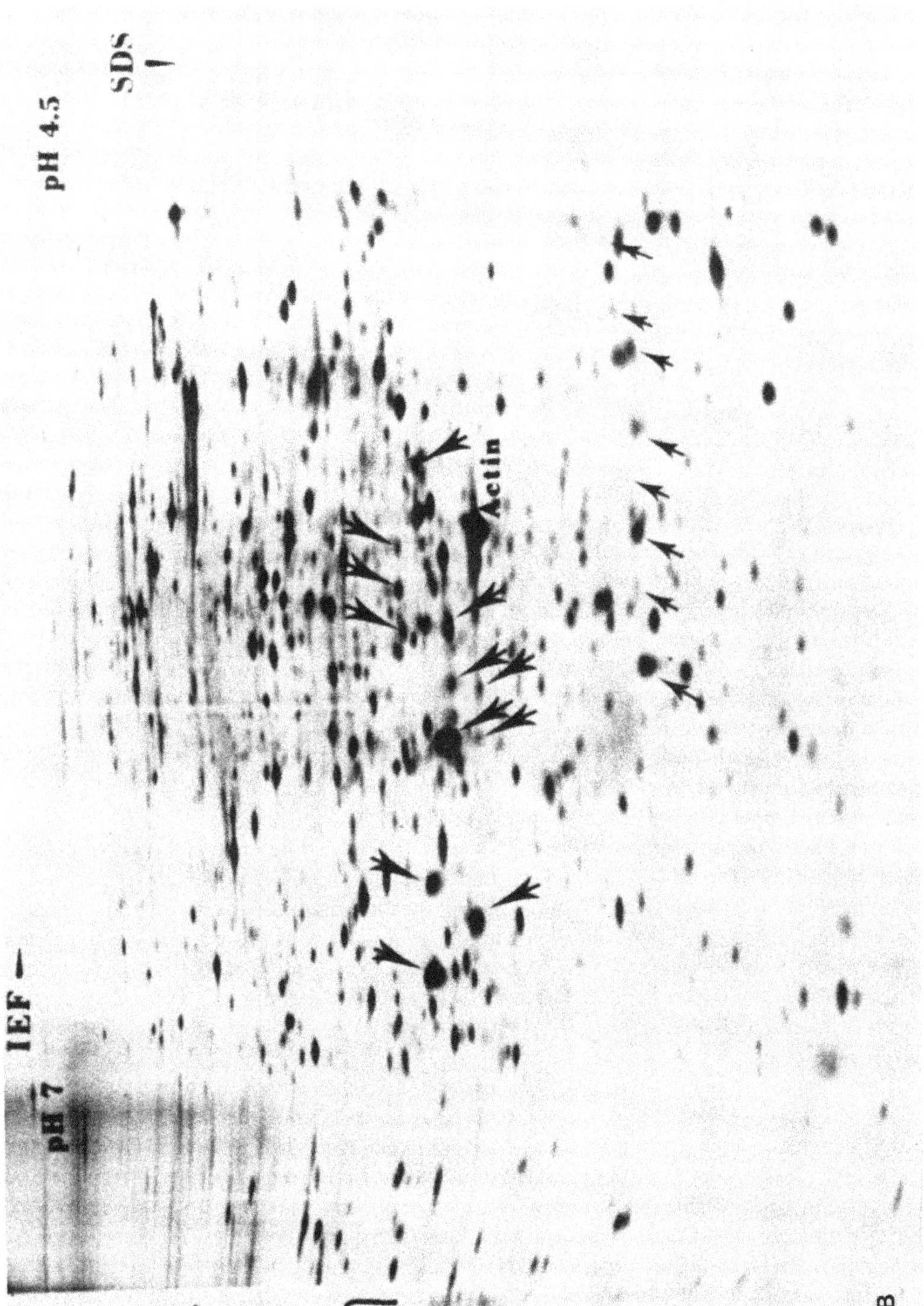

FIGURE 8B

As with any method, there are problems inherent in using gel electrophoresis for isolation. Perhaps the most significant is the slowness of both separation and recovery steps. There can often be a loss of biological activity, and fraction work-up is somewhat tedious (at least with electroelution and dialysis). In addition, protein fragmentation is possible during the staining and destaining steps, as well as any number of chemical modifications that can potentially occur. These modifications are of obvious concern, since a method that gives high recoveries must also yield samples which have remained unmodified.

The potential pitfalls can be circumvented if a few precautions are taken. These include the need to minimize amino-terminal blockage, modification of certain amino acid side chains, cleavage of the polypeptide chain, and introduction of contaminates which could interfere with either the Edman chemistry or PTH-residue identification.

A. Electroelution and Electrodialysis

Of the two sample methods which will be covered, this was the first to be perfected and has been utilized in the isolation of hundreds of proteins. It found general use, for example, in preparing samples of cloned and expressed polypeptides where possible inefficient amino-terminal processing can be readily detected. This is often seen in bacterially expressed products. To the molecular biologist and those responsible for carrying out various induction/expression protocols, it is necessary to know this as soon as possible so that the appropriate corrective steps can be attempted.

The technique has been adequately described by Hunkapiller et al.[67] and researchers are encouraged to follow the directions *exactly*. The quality and source of chemicals are extremely important so that chemical modifications are minimized. One also needs to further purify the commercially available SDS, add an antioxidant such as sodium thioglycolate to the buffer reservoir for electrophoresis, and use "mini" gels so that a high protein-to-gel ratio is maintained.

One of the drawbacks to this method is the relatively long time (24 to 36 hr) required to isolate a sample following electrophoretic separation. The time required to electroelute the sample and collect it on a dialysis membrane is usually done overnight in 0.1% SDS/50 m*M* ammonium bicarbonate. The buffer is subsequently changed to 0.02% SDS/10 m*M* ammonium bicarbonate, for the electrodialysis which requires another 12 to 20 hr.

At the end of this step the sample is removed from the electroelution cell and lyophilized. What remains is a "plug", usually dark blue in color, which has to be further purified by precipitation before sequencing. The two suggested techniques for this step are given in Table 9, and of the two the ethanol method has been most widely used in our laboratories.

A primary problem with any precipitation technique is *knowing* that the sample is soluble at a given step or, if precipitated, where it is located in the container. For either of these points it is essential that a magnification of about 10 to 20× be used, so that even a microgram or less of material can be seen.

If the samples are not "cleaned up" by precipitation, then a number of artifact peaks will be found in the first few Edman cycles. These can severely interfere with PTH amino acid identification and arise, for the most part, as degradation products from the Coomassie blue.

Table 10 lists a selected number of proteins which have been prepared by this method. Note that in cases where recoveries were determined, the values ranged between 52 and 85% for protein loads of 0.6 to 120 μg. Additionally, the recovery of a wide range of sample molecular weights indicates general utility of this method.

One can also conveniently use the SDS gel as either a separation medium for previously cleaved material or, alternatively, as a support medium in which to carry out cleavages. Some chemical cleavage methods are given in Table 11. Since gels of either the correct acrylamide percentage or gradient range are capable of separating proteins differing in molecular weight by only 2%, the utility of these fragmentation methods in combination with PAGE can be very useful.

Table 9
SAMPLE PRECIPITATION

Method 1	Dissolve lyophilized material in 50 μℓ water, add 450 μℓ −20°C ethanol, let stand for 4 to 18 hr at −20°C, and recover by centrifugation[a]
Method 2	Dissolve lyophilized material in 100 μℓ water, add 400 μℓ methanol, vortex, and centrifuge Add 100—200 μℓ chloroform, vortex, and centrifuge Add 300 μℓ water, vortex, and centrifuge: discard upper phase (carefully) Add 300 μℓ methanol, vortex, and centrifuge Discard supernatant and dry pellet with nitrogen *Caution*: chloroform contains stabilizers that can block amino termini of polypeptides (see Wessel and Fluegge[70] for further details)

[a] Note that minimal amounts (1 μg or less) are distributed over a rather large area and many proteins are, subsequently, insoluble in buffers lacking SDS.

B. Electroblotting

As discussed in the preceding section, one of the major limitations has been the time between electrophoretic separation and obtaining sequence information. Although this factor can be decreased somewhat by handling multiple samples at any given time, they still must be electroeluted and dialyzed separately, followed by individual lyophilization and precipitation. Most of these steps can be eliminated if the separated proteins could be transferred onto an appropriately prepared surface that can be directly used in the gas phase sequencer, and detected with a system that does not interfere with the subsequent Edman chemistry. Two recently developed procedures[78,79] address each of these points. They appear to be techniques that will be extremely useful for quickly recovering small amounts of protein, in the 1- to 10-μg range, for either direct amino-terminal degradation or further fragmentation and separation prior to sequencing.

Both of these methods are based on an electroblotting step which transfers the separated proteins onto an appropriately activated, glass fiber filter sheet. This step accomplishes the important sample immobilization which increases overall yields by reducing the number of steps required in the electroelution-dialysis technique. Following sample blotting of all the proteins on a gel, they are detected on the filter by a staining method, individually cut out, and directly sequenced.

In general, the technique involves blotting onto a glass fiber filter that has been prepared by one of four different methods (Table 12). The first involves transferring the separated proteins onto a glass fiber sheet that has been impregnated with polybrene. Here the gel can be first stained with Coomassie blue so that the protein(s) of interest are visible or, alternatively, the separated proteins are blotted and then detected by staining with fluorescamine. Not only can the blotted samples be used for gas phase sequencing, but also amino acid analyses. In the latter case quite reasonable compositions are observed, with only minimal background problems arising from either the gels, polybrene, or buffers.

The other three electroblotting procedures start with the same "activation" step of the glass fiber. It involves soaking either Whatman GF/C or GF/F glass fiber filters in TFA to introduce negative charges onto the glass surface. The second blotting method involves the removal of the SDS by soaking in Nonidet™ p-40 (octylphenoxypolyethoxyethanol), followed by transferring at low pH the proteins that have been separated by electrophoresis. The adsorbed components are then detected by Coomassie blue staining and destaining.

The third and fourth procedures involve preparing a positively charged filter, by deriv-

Table 10
SELECTED EXAMPLES OF PROTEINS RECOVERED FROM SDS-PAGE BY ELECTROPHORETIC ELUTION OF COOMASSIE-STAINED GEL SLICES

Protein/peptide	Mol wt (daltons)	Amount recovered (μg)	% recovery
Water-soluble proteins			
Human serum albumin	68,000	8	80
Human β-interferon	18,000	0.5	85
Mouse interleukin 2	22,000	2	—
^{3}H-basic pancreatic trypsin inhibitor	6,400	(1—5)	70
^{3}H-methylated myoglobin	18,000	(1)	71
^{3}H-methylated lysozyme	16,000	(1)	85
^{3}H-methylated cytochrome *c*	14,000	(1)	53
^{3}H-methylated carbonic anhydrase	31,000	(1)	74
Integral membrane proteins			
Mouse I_A antigen A^x	25,000	0.5	50
Acetylcholine receptor			
Torpedo subunit	65,000	4	71
Eel subunit	40,000	20	—
Calf subunit	42,000	2	—
Diol dehydratase subunit	29,000	100	80
Rat liver gap junction protein	28,000	20	60
Photosynthetic reaction center subunits of *Rhodopseudomonas capsulata*	25—30,000	(1—5)	75
Protein fragments			
V8 protease-derived fragments of *Torpedo* acetylcholine	9,000	20	—
Receptor subunit	2,500	2	—
Proteinase K-derived fragment of MM-creatine kinase	41,000	(1—5)	—
Diptheria toxin A subunit fragment	21,000	(1—5)	—

Note: Values in parentheses indicate approximations of amount recovered.

Table 11
PEPTIDE MAPPING BY SDS-PAGE

Cleavage/labeling method	Bond cleaved	Ref.
CNBr	Met-X	71
Acid	Asp-Pro	72, 73
N-Chlorosuccinimide	Trp-X	74
Hydroxylamine	Asn-Gly	75
Bolten-Hunter reagent	H_2N-X	76
Numerous	Above plus Arg-X	77

Table 12
ELECTROBLOTTING PROCEDURES

Method	Modified form of glass fiber filter	Surface charge	Blotting conditions	Type of association	Staining conditions	Ref.
1	Polybrene impregnated	Positive	Borate buffer (pH 8)	Ionic	Coomassie blue or fluoresamine[a]	78
2	TFA "activated"	Negative	Acetic acid (pH 2)	Ionic	Coomassie blue	79
3	Aminopropyl or quarternary ammonium	Positive	Tris buffer (pH 8)	Ionic	3,3′-Dipentyl oxa-carbocyanine iodide (fluorescence)	79
4	DITC — glass	NA	Tris buffer (pH 8)	Covalent	NA	79

[a] Staining can be done with fluorescamine after blotting or with Coomassie blue prior to transfer.

atizing the TFA-activated filter to produce either the aminopropyl or quarternary ammonium form. The filters are, subsequently, used for blotting at pH 8 in Tris buffers where the highly negative-charged SDS-containing proteins are ionically bound to the glass support. The blotted samples are then detected by staining with 3,3′-dipentyloxacarbocyanine iodide and observing protein-bound fluorescence at 254 nm. The proteins recovered by either of these methods are cut out of the filter blot and sequenced directly.

DITC filters can be prepared by reacting the aminopropyl derivative with phenylene isothiocyanate. This particular covalent blotting method might be very useful if the appropriate conditions can be found that provide high coupling yields.

The techniques of transferring onto either negatively or positively charged filters require approximately the same amount of time to perform. As seen by the recoveries of the protein standards in Table 13, the use of the high-pH blotting method is generally preferred. Differences between initial and repetitive yields on samples blotted by either procedure are minimal. A list of proteins sequenced using these techniques can be found in Table 14.

The chromatograms in Figure 9 illustrate the results from an Edman degradation of a 150,000-dalton protein which was blotted at pH 8.3 (see Table 14) and, subsequently, sequenced. Observation of the first six cycles shows that the PTH analyses have minimal background contamination from by-product peaks, and the repetitive yields were 91% over the first 24 residues. Thirty micrograms of this sample was used to carry out SDS PAGE and electroblotting. The resulting 33-pmol initial yield clearly illustrates the utility of this method.

IV. SUMMARY

The techniques described in the preceding sections cover methods found convenient for isolating small amounts of material. Some of these, microbore HPLC and electroblotting, directly utilize the cartridge of the gas phase sequencer. Other techniques, due to amounts or volumes, require further sample handling, such as lyophilization which increases the possibility of sample loss.

As discussed, there are distinct advantages and disadvantages inherent to each method. These need to be considered prior to initiating any isolation. The final choice is often based on factors not directly related to sequencing and might include the need to determine biological activity, an amino acid composition, or using physical characterizations as optical measurements (ORD, CD) where denaturation would obscure the results.

Table 13
BLOTTING AND SEQUENCING RESULTS FROM SDS-POLYACRYLAMIDE GELS[79]

					Sequencing yields				
			Blotting efficiency		Low pH			High pH	
Protein	**$M_r \times 10^{-3}$**	**pI**	**Low pH**	**High pH**	**Load (pmol)**	**IY**	**RY**	**Load (pmol)**	**IY**
Bovine α-lactalbumin	14	4.6	68	94	48	61	93	45	61
Sperm whale myoglobin	17	8.3	78	75	140	67	96	41	76
Soybean trypsin inhibitor	21	4.5	42	99	24	71	92	33	62
Bovine carbonic anhydrase	29	5.8	96	74	—	—	—	—	—
Bovine serum albumin	68	4.9	77	76	38	74	94	29	66
Rabbit phosphorylase B	97	5.5—6.3	33	n.d.	—	—	—	—	—
E. coli β-galactosidase	116	4.5	9	81	—	—	—	—	—

Note: IY, initial yield; RY, repetitive yield.

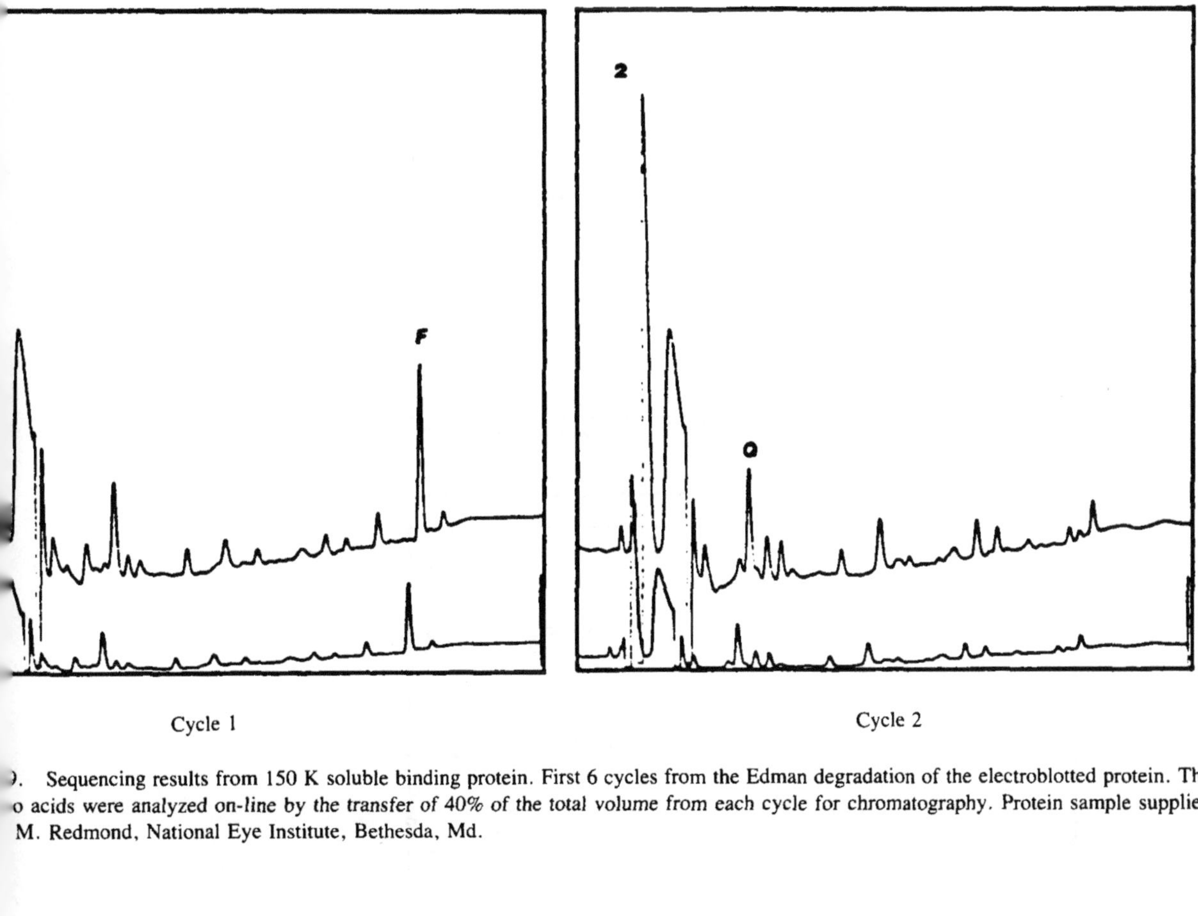

Cycle 1

Cycle 2

). Sequencing results from 150 K soluble binding protein. First 6 cycles from the Edman degradation of the electroblotted protein. The o acids were analyzed on-line by the transfer of 40% of the total volume from each cycle for chromatography. Protein sample supplied M. Redmond, National Eye Institute, Bethesda, Md.

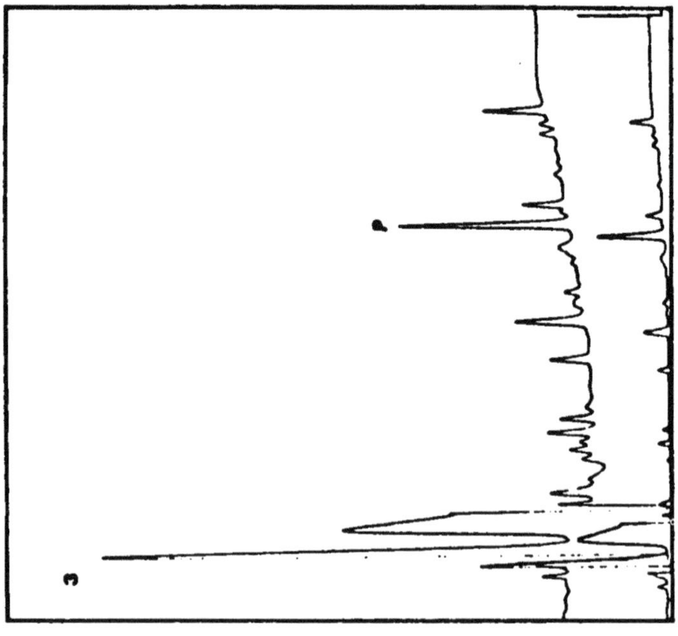

FIGURE 9: Cycle 3

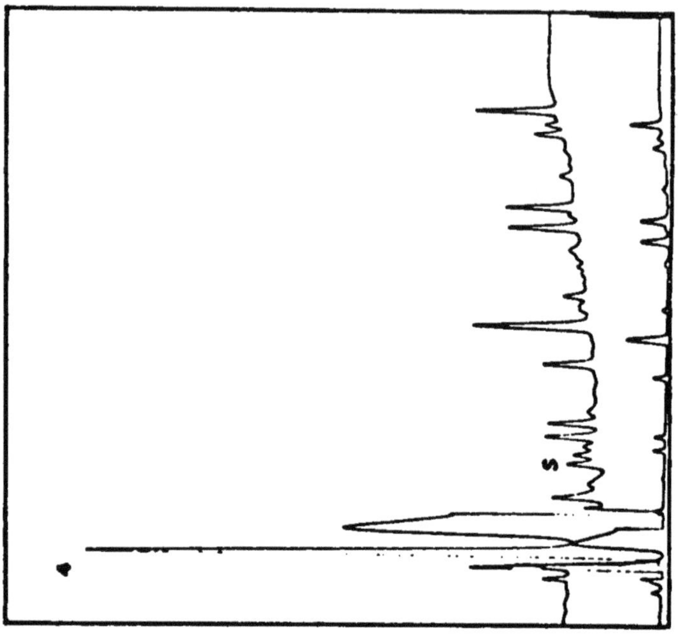

FIGURE 9: Cycle 4

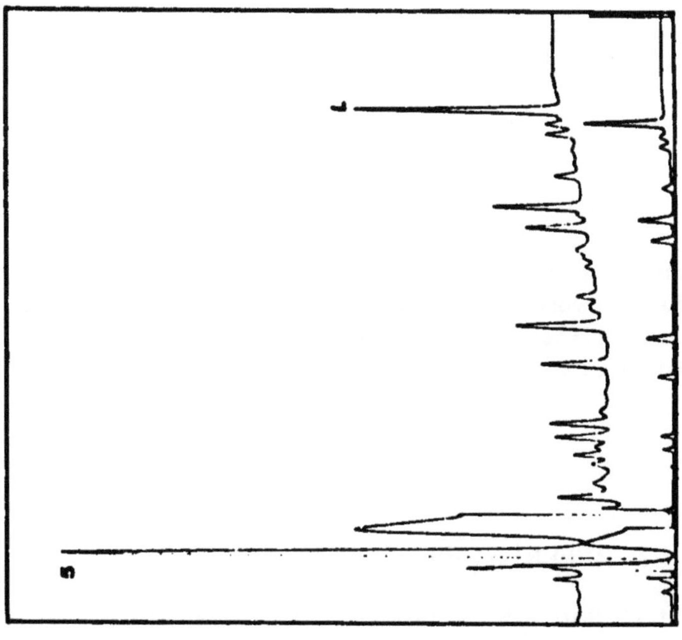

FIGURE 9: Cycle 5

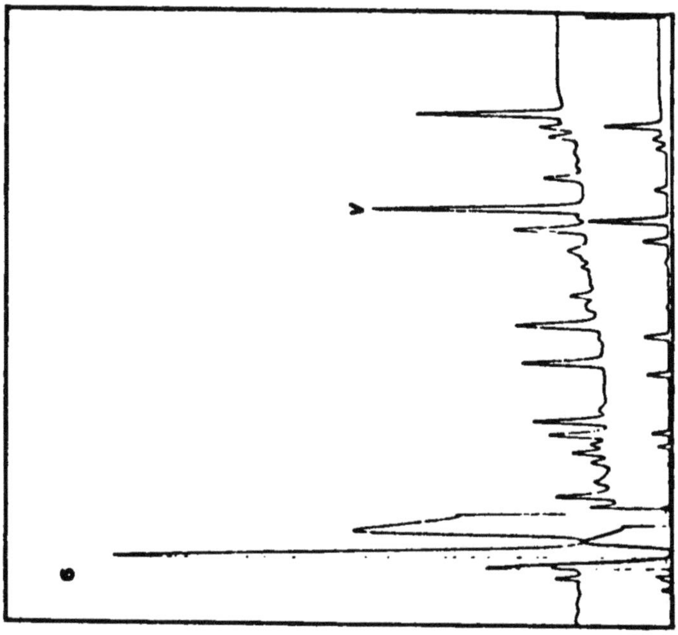

FIGURE 9: Cycle 6

Table 14
PROTEINS SEQUENCED USING THE ELECTROBLOTTING RECOVERY TECHNIQUE

Protein	$M_r \times 10^{-3}$	Sample preparation	Sequenceable amount (pmol)	No. of residues determined
Rat liver gap junction protein, chymotryptic fragment	10	Low pH blot	7	9/14
Rat liver gap junction protein, chymotryptic fragment	14	Low pH blot	15	10/15
IgG H-chain	58	Low pH blot	8	14/20
Mouse liver gap junction protein	28	High pH blot	12	16/18
Mouse liver membrane protein (gap junction associated)	21	High pH blot	5	10/16
IgG heavy chain	58	High pH blot	110	20
IgG light chain	28	High pH blot	12	13/17
Cholinergic differentiation factor	48	High pH blot	10	10/13
HF-treated cholinergic differentiation factor	28	High pH blot	12	9/12
Soluble binding protein	150	High pH blot	33	21/24
Bovine serum albumin	68	High pH blot	12	13/17
Soybean trypsin inhibitor	21	High pH blot	11	12/13
Sperm whale myoglobin	17	High pH blot	12	10/11
Bovine α-lactalbumin	14	High pH blot	12	11/13
Yellow fever virus protein	45	Spotted	400	20
Mu-phage transposase	70	Spotted	5	8/11
Scrapie-associated protein cyanogen bromide fragment	1.8	Spotted	50	12/14
(Arg^8) — vasopressin	1	Spotted	60	8/9
Synthetic peptide	4.5	Spotted	400	32/40
DNA binding protein tryptic peptide	—	Spotted	25	12/13

Modified from Aebersold, R. H., Teplow, D. B., Hood, L. E., and Kent, S. B. H., *J. Biol. Chem.*, 261, 4229, 1986. With permission.

ACKNOWLEDGMENTS

The author would like to thank Dr. Pau Yuan for contributing Figures 2 and 9; Table 1 was kindly supplied by Mr. Russ Blacher and Dr. Yuan. Dr. Carl March of Immunex contributed the recombinant IL-2 sample used in Figure 5. Similarly, gratitude is expressed to Dr. Steve Kent and colleagues at CalTech for allowing me access to the electroblotting data prior to publication.

REFERENCES

1. **Hewick, R. M., Hunkapiller, M. W., Hood, L. E., and Dryer, W. J.,** A gas-liquid solid phase peptide and protein sequenator, *J. Biol. Chem.*, 256, 7990, 1981.
2. **Cohen, K. A., Grillo, S. A., and Dolan, J. W.,** A comparison of protein retention and selectivity on large-pore reversed-phased HPLC columns, *LC* 3, 1, 37, 1985.
3. **Trumnore, C. N., Tremnlay, R. D., Penrose, J. T., Mercer, M., and Kelleher, F. M.,** Unusual flow behavior and losses of proteins in high-performance liquid chromatography capillary tubing, *J. Chromatogr.*, 280, 43, 1983.

4. **Sadek, P. C., Carr, P. W., Bowers, L. D., and Haddad, L. C.,** A radiochemical study of irreversible protein loss on high-performance liquid chromatography column frits, *Anal. Biochem.*, 144, 128, 1985.
5. **O'Hare, M. J., Capp, M. W., Nice, E. C., Cooke, N. H. C., and Archer, B. G.,** Factors influencing chromatography of proteins on short alkylsilan-bonded large pore-size silicas, *Anal. Biochem.*, 126, 17, 1982.
6. **Nice, E. C., Capp, M. W., Cooke, N., and O'Hare, M. J.,** Comparison of short and ultrashort chain alkysillane-bonded silicas for the high performance liquid chromatography of proteins by hydrophobic interaction methods, *J. Chromatogr.*, 218, 569, 1981.
7. **Wilson, K. J., Honegger, A., and Hughes, G. J.,** Comparison of buffers and detection systems for high pressure liquid chromatography of peptide mixtures, *Biochem. J.*, 199, 43, 1981.
8. **Wilson, K. J., Honegger, A., and Hughes, G. J.,** The behaviour of peptides on reverse-phase supports during high-pressure liquid chromatography, *Biochem. J.*, 199, 31, 1981.
9. **Pearson, J. D., Lin, N. T., and Regnier, F. E.,** The importance of silica type for reverse-phase protein separations, *Anal. Biochem.*, 124, 217, 1982.
10. **Wilson, K. J., Van Wieringen, E., Klauser, S., and Berchtold, M. W.,** Comparison of the high-performance liquid chromatography of peptides and proteins on 100- and 300-Angstrom reversed-phase supports, *J. Chromatogr.*, 237, 407, 1982.
11. **Chang, J. P., Melander, W. R., and Horvath, C.,** Reversed-phase chromatography of proteins with diphenyl-silica column and hydro-organic eluents containing two organic solvents, *J. Chromatogr.*, 318, 11, 1985.
12. **Tarr, G. E. and Crabb, J. W.,** Reverse-phase high-performance liquid chromatography of hydrophobic proteins and fragments thereof, *Anal. Biochem.*, 131, 99, 1983.
13. **Hughes, G. J. and Wilson, K. J.,** High-performance liquid chromatography; analytic and preparative applications in protein-structure determination, in *Methods Biochemistry,* Glick, D., Ed., John Wiley & Sons, New York, 1983, chap. 3.
14. **Bristow, A. F., Wilson, C., and Sutcliffe, N.,** Reversed-phase high-performance liquid chromatography of human thyroid-stimulating hormone, *J. Chromatogr.*, 270, 285, 1983.
15. **Cohen, S. A., Dong, S., Benedek, K., and Karger, B. L.,** Multiple peak formation in the reversed phase liquid chromatographic separation of soybean trypsin inhibitor, in *Symp. Proc. 5th Int. Symp. on Affinity Chromatography and Biological Recognition,* Chaiken, I. M., Wilchek, M., and Parikh, I., Eds., Academic Press, New York, 1983, 479.
16. **Cohen, S. A., Benedek, K. P., Dong, S., Tapuhi, Y., and Karger, B. L.,** Multiple peak formation in reversed-phase liquid chromatography of papain, *Anal. Chem.*, 56, 217, 1984.
17. **Cohen, K. A., Schellenberg, K., Benedek, K., Karger, B. L., Grego, B., and Hearn, T. W.,** Mobile-phase and temperature effects in the reversed phase chromatographic separation of proteins, *Anal. Biochem.*, 140, 223, 1984.
18. **Benedek, K., Dong, S., and Karger, B. L.,** Kinetics of unfolding of proteins on hydrophobic surfaces in reversed-phase liquid chromatography, *J. Chromatogr.*, 317, 227, 1984.
19. **Cohen, S. A., Benedek, K., Tapuhi, Y., Ford, J. C., and Karger, B. L.,** Conformational effects in the reversed-phase liquid chromatography of ribonuclease a, *Anal. Biochem.*, 144, 275, 1985.
20. **Sadler, A. J., Micanovic, R., Katzenstein, G. E., Lewis, R. V., and Middaugh, C. R.,** Protein conformation and reversed-phase high-performance liquid chromatography, *J. Chromatogr.*, 317, 93, 1984.
21. **Petrides, P. E., Jones, R. T., and Bohlen, P.,** Reverse-phase high-performance liquid chromatography of proteins: the separation of hemoglobin chain variants, *Anal. Biochem.*, 105, 383, 1980.
22. **Terabe, S., Nishi, H., and Ando, T.,** Separation of cytochromes c by reversed-phase high-performance liquid chromatography, *J. Chromatogr.*, 212, 295, 1981.
23. **Wilson, K. J., Csejtey, J., Desmond, J. T., and Strickler, J. E.,** High-performance liquid chromatography of glycoproteins, in preparation.
24. **Zull, J. E. and Chuang, J.,** Preparation of characterization of radioactive monoiodotryrosine and diiodotryrosine derivatives of parathyroid hormone, *Anal. Biochem.*, 140, 214, 1984.
25. **Wetzel, R.,** Assignment of the disulphide bonds of leukocyte interferon, *Nature,* 289, 606, 1981.
26. **Perry, J. L. and Wetzel, R.,** Disulfide bond engineered in T4 lysozyme: stabilization of the protein toward thermal inactivation, *Science,* 226, 555, 1984.
27, **Kohr, W. J., Keck, R., and Harkins, R. N.,** Characterization of intact and trypsin-digested biosynthetic human growth hormone by high-pressure liquid chromatography, *Anal. Biochem.*, 122, 348, 1982.
28. **Elton, T. S. and Reeves, R.,** The effects of oxidation on the reverse-phase high-performance liquid chromatography characteristics of the high mobility groups 1 and 2 proteins, *Anal. Biochem.*, 149, 316, 1985.
29. **Felix, A. M., Heimer, E. P., Lambros, T. J., Swistock, J., Tarnowski, S. J., and Wang, C.-T.,** Analysis of different forms of recombinant human leukocyte interferons and synthetic fragments by high-performance liquid chromatography, *J. Chromatogr.*, 327, 359, 1985.

30. **Berchtold, M. W., Heismann, C. W., and Wilson, K. J.,** Ca^{2+}-binding proteins: a comparative study of their behavior during high-performance liquid chromatography using gradient elution of reverse-phase supports, *Anal. Biochem.*, 129, 120, 1983.
31. **Titani, K., Sasagawa, T., Resing, K., and Walsh, K. A.,** A simple and rapid purification of commercial trypsin and chymotrypsin by reverse-phase high-performance liquid chromatography, *Anal. Biochem.*, 123, 408, 1982.
32. **Schaaper, W. M. M., Voskamp, D., and Olieman, C.,** Perfluoroalkanoic acids as lipophilic ion-pairing reagents in reversed-phase liquid chromatography of peptides including secretin, *J. Chromatogr.*, 195, 181, 1980.
33. **Harding, D. R. K., Bishop, C. A., Tarttelin, M. E., and Hancock, W. S.,** Use of perfluoroalkanoic acids as volatile ion pairing reagents in preparative HPLC, *Int. J. Pept. Protein Res.*, 18, 214, 1981.
34. **Mahoney, W. C. and Hermodson, M. A.,** Separation of large denatured peptides by reverse phase high performance liquid chromatography, *J. Biol. Chem.*, 255, 11199, 1980.
35. **Bennett, H. P. J.,** Isolation of pituitary peptides by reversed-phase high-performance liquid chromatography. Expansion of the resolving power of reversed-phase columns by manipulating pH and the nature of the ion-pairing reagent, *J. Chromatogr.*, 266, 501, 1983.
36. **Rubinstein, S., Familetti, P. C., and Pestka, S.,** High-performance liquid chromatography and picomole-level detection of peptides and proteins, *Methods Enzymol.*, 79, 7, 1981.
37. **Meek, J. L.,** Derivatizing reagents for high-performance liquid chromatography detection of peptides at the picomole level, *J. Chromatogr.*, 266, 401, 1983.
38. **Wilson, K. J., Dupont, D. R., Yuan, P. M., Hunkapiller, M. W., and Schlabach, T. D.,** Practical and theoretical aspects of microbore HPLC of proteins, peptides and amino acid derivatives, in *Proteins: Structure and Function. Selected Proceedings from the First Symposium of American Protein Chemists,* L'Italien, J., Ed., Plenum Press, New York, 1987, 21.
39. **Yuan, P. M., Uyen, S., Bergot, J., Hunkapiller, M. W., and Wilson, K. J.,** Progress toward polybrene purification and utilityin micro-peptide/protein sequencing, in *Proteins: Structure and Function. Selected Proceedings from the First Symposium of American Protein Chemists,* L'Italien, J., Ed., Plenum Press, New York, 1987, 325.
40. **Nice, E. C., Lloyd, C. J., and Burgess, A. W.,** The role of short microbore high-performance liquid chromatography columns for protein separation and trace enrichment, *J. Chromatogr.*, 296, 153, 1984.
41. **Gooding, K. M. and Schmuck, M. N.,** Comparison of weak and strong high-performance anion-exchange chromatography, *J. Chromatogr.*, 327, 139, 1985.
42. **Kopaciewicz, W., Rounds, M. A., and Regneir, F. E.,** Stationary phase contributions to retention in high-performance and anion-exchange protein chromatography: ligrand density and mixed mode effects, *J. Chromatogr.*, 318, 157, 1985.
43. **Gooding, K. M. and Schmuck, M. N.,** Ion selectivity in the high-performance cation-exchange chromatography of proteins, *J. Chromatogr.*, 296, 321, 1984.
44. **Rounds, M. A. and Regnier, F. E.,** Evaluation of a retention model for high-performance ion-exchange chromatography using two different displaying salts, *J. Chromatogr.*, 283, 37, 1984.
45. **Gupta, S., Peannkcoch, E., and Regnier, F. E.,** High performance cation-exchange chromatography of proteins, *Anal. Biochem.*, 128, 196, 1983.
46. **Kopaciewicz, W. and Regnier, F. E.,** Mobile phase selection for the high-performance ion-exchange chromatography of proteins, *Anal. Biochem.*, 128, 196, 1983.
47. **Kato, Y., Nakamura, K., Yamazaki, Y., and Hashimoto, T.,** Comparison of high-performance ion-exchange chromatography and gel electrophoresis in protein separations, *J. Chromatogr.*, 318, 358, 1985.
48. **Imamura, T., Sugihara, J., Yokota, E., Kagimoto, M., Naito, Y., and Yanase, T.,** Analytical peptide mapping by ion-exchange high-performance liquid chromatography: application to haemoglobin variants, *J. Chromatogr.*, 305, 456, 1984.
49. **Takahashi, N., Ishioka, N., Takahashi, Y., and Putnam, F. W.,** Automated tandem high-performance liquid chromatographic system for separation of extremely complex peptide mixtures, *J. Chromatogr.*, 326, 407, 1985.
50. **Mant, C. T. and Hodges, R. S.,** Separation of peptides by strong cation-exchange high-performance liquid chromatography, *J. Chromatogr.*, 327, 147, 1985.
51. **Offord, R. E.,** Electrophoretic mobilities of peptides on paper and their use in the determination of amide groups, *Nature*, 591, 591, 1966.
52. **James, S. and Bennett, H. P. J.,** Use of reversed-phase and ion-exchange batch extraction in the purification of bovine pituitary peptides, *J. Chromatogr.*, 326, 329, 1985.
53. **Kato, Y., Kitamura, T., and Hashimoto, T.,** High-performance hydrophobic interaction chromatography of proteins, *J. Chromatogr.*, 266, 49, 1983.
54. **Melander, W. R., Corradini, D., and Horvath, C.,** Salt-mediated retention of proteins in hydrophobic-interaction chromatography; application of solvophobic theory, *J. Chromatogr.*, 317, 67, 1984.

55. **Gooding, D. L., Schmuck, M. N., and Gooding, K. M.,** Analysis of proteins with new, mildly hydrophobic high-performance liquid chromatography packing materials, *J. Chromatogr.*, 296, 107, 1984.
56. **Fausnaugh, J. L., Pfannkoch, E., Gupta, S., and Regnier, F. E.,** High-performance hydrophobic interaction chromatography of proteins, *Anal. Biochem.*, 137, 464, 1984.
57. **Fausnaugh, J. L., Kennedy, L. A., and Regnier, F. E.,** Comparison of hydrophobic-interaction and reversed-phase chromatography of proteins, *J. Chromatogr.*, 317, 144, 1984.
58. **Kato, Y., Kitamura, T., and Hashimoto, T.,** New support for hydrophobic interaction chromatography of proteins, *J. Chromatogr.*, 292, 418, 1984.
59. **Miller, N. T., Feibush, B., and Karger, B. L.,** Wide-pore silica-based ether-bonded phases for separation of proteins by high-performance hydrophobic-interaction and size-exclusion chromatography, *J. Chromatogr.*, 316, 519, 1985.
60. **Miller, N. T. and Karger, B. L.,** High-performance hydrophobic-interaction chromatography on either-bonded phases; chromatographic characteristics and gradient optimization, *J. Chromatogr.*, 326, 45, 1985.
61. **Ingraham, R. H., Lau, S. Y. M., Taneja, A. K., and Hodges, R. S.,** Denaturation and the effects of temperature on hydrophobic-interaction and reversed-phase high-performance liquid chromatography of proteins, *J. Chromatogr.*, 327, 77, 1985.
62. **Corbett, R. J. T. and Roche, R. S.,** Use of high-speed size-exclusion chromatography for the study of protein folding and stability, *Biochemistry*, 23, 1888, 1984.
63. **Mahieu, J. P., Sebille, B., Craescu, C. T., Rhoda, M. D., and Beuzard, Y.,** Determination of the dissociation constant of oligomeric proteins by size-exclusion high-performance liquid chromatography: application to human haemoglobin, *J. Chromatogr.*, 327, 313, 1985.
64. **Hakvoort, T. B. M., Sinjorgo, K. M. C., Van Gelder, B. F., and Muijsers, A. O.,** Separation of enzymically active bovine cytochrome *c* oxidase monomers and dimers by high performance liquid chromatography, *J. Inorg. Biochem.*, 23, 381, 1985.
65. **Co-Sarno, M. E., Tapang, M. A., and Luckhurst, D. G.,** Determination of polymer and purification of albumin by high-performance liquid chromatography, *J. Chromatogr.*, 266, 105, 1983.
66. **Berchtold, M. W., Wilson, K. J., and Heizmann, C. W.,** Isolation of neuronal parvalbumin by high-performance liquid chromatography; characterization and comparison with muscle paravalbumin, *Biochemistry*, 21(25), 6552, 1982.
67. **Hunkapiller, M. W., Lujan, E., Ostander, F., and Hood, L. E.,** Isolation of microgram quantities of proteins from polyacrylamide gels for amino acid sequence analysis, *Methods Enzymol.*, 91, 227, 1983.
68. **Heukeshoven, J. and Dernick, R.,** Reverse-phase high-performance liquid chromatography of virus proteins and other large hydrophobic proteins in formic acid containing solvents, *J. Chromatogr.*, 252, 241, 1982.
69. **Heukeshover, J. and Dernick, R.,** Reverse-phase high-performance liquid chromatography of virus proteins and other hydrophobic proteins, in *Protides of the Biological Fluids*, Peters, H., Ed., Pergamon Press, Oxford, 1984, 1085.
70. **Wessel, D. and Fluegge, U. I.,** A method for the quantitative recovery of protein in dilute solution in the presence of detergents and lipids, *Anal. Biochem.*, 138, 141, 1984.
71. **Nikodem, V. and Fresco, J. R.,** Protein fingerprinting by SDS-gel electrophoresis after partial fragmentation with CNBr[1], *Anal. Biochem.*, 97, 382, 1979.
72. **Sonderegger, P., Jaussi, R., Gehring, H., Brunschweiler, K., and Christen, P.,** Peptide mapping of protein bands for polyacrylamide gel electrophoresis by chemical cleavage in gel pieces and re-electrophoresis, *Anal. Biochem.*, 122, 298, 1982.
73. **Rittenhouse, J. and Marcus, F.,** Peptide mapping on polyacrylamide gel electrophoresis after cleavage at aspartyl-prolyl peptide bonds in sodium dodecyl sulfate-containing buffers, *Anal. Biochem.*, 138, 442, 1984.
74. **Lischwe, M. A. and Ochs, D.,** A new method of partial peptide mapping using N-chlorosuccinimide/urea and peptide silver staining in sodium dodecyl sulfate-polyacrylamide gels, *Anal. Biochem.*, 127, 453, 1982.
75. **Saris, C. J. M., Eenbergen, J. V., Jenks, B. G., and Bloemers, H. P. J.,** Hydroxylamine cleavage of proteins in polyacrylamide gels, *Anal. Biochem.*, 132, 54, 1983.
76. **Jay, D. G.,** A general procedure for the end labeling of proteins and positioning of amino acids in the sequence, *J. Biol. Chem.*, 259, 15572, 1984.
77. **Jue, R. A. and Doolittle, R. F.,** Determination of the relative positions of amino acids by partial specific cleavages of end-labeled proteins, *Biochemistry*, 24, 162, 1985.
78. **Vandekerckhove, J., Bauw, G., Puype, M., Van Damme, J., and Van Montagu, M.,** Protein-blotting on polybrene-coated glass-fiber sheets. A basis for acid hydrolysis and gas-phase sequencing of picomole quantities of protein previously separated on sodium dodecyl sulfate/polyacrylamide gel, *Eur. J. Biochem.*, 152, 9, 1985.
79. **Aebersold, R. H., Teplow, D. B., Hood, L. E., and Kent, S. B. H.,** Electroblotting onto activated glass: high efficiency preparation of proteins from analytical SDS-polyacrylamide gels for direct sequence analysis, *J. Biol. Chem.*, 261, 4229, 1986.

Chapter 2

MANUAL METHODS FOR PROTEIN/PEPTIDE SEQUENCE ANALYSIS

G. E. Tarr

TABLE OF CONTENTS

I. INTRODUCTION

If protein sequencing has suffered a loss of esteem in recent years because of the great success of DNA sequencing, then manual protein sequencing must be assumed to be moribund, wounded by the same general decline and by an almost universally unfavorable comparison with automatic methods, most notably, "gas-phase" sequencing. Such comparisons are almost always unfair, being based on historical impressions rather than on current capability, and two advantages of manual methods over automatic ones remain: a low capital investment and the ability to process many samples at the same time. The establishment of service facilities for protein and nucleic acid chemistry has made the expensive machines available to everyone (for a price), but there is no current instrument that compromises the second advantage in any way. Even the most advanced machines still have cycle times that severely limit throughput, so only the largest laboratories can, for example, examine automatically all the peptides purified from a tryptic digest of a modestly sized protein. This is a simple task for the manual sequencer. However, since a multiple-sample instrument can be easily designed, one presumably will come to market and manual sequencing will remain then only for instruction and for experimental and occasional chemistries, or for research units that are truly impoverished.

The system described below has evolved from a relatively simple state, through apparatus requiring much manufacture or modification of commercial products, back to a system that is quite simple, requiring little beyond what is readily available and no complicated procedure or "artistry". Other manual methods exist,[1-3] including a spin-off of the fabled "gas-phase" approach, which is, in turn, a mechanized version of Fraenkel-Conrat's old manual method,[4] but none offer this simplicity. Also, this method is successfully employed in, perhaps, a half dozen labs in this country and in others in Europe and Asia. Nor is it ephemeral, having been used in essentially its present form for more than 5 years.[5,6]

Unlike automatic sequencing, which has relied exclusively on the standard Edman reagent phenylisothiocyanate (PITC) and direct identification of sequencing product, manual methods have often used additional or alternative chemistry. The dansyl-Edman method,[7] long popular, is still being used, and dimethylaminoazobenzene isothiocyanate (DABITC) is favored by a few researchers.[2] Sensitive direct identification of phenylthiohydantoins (PTHs) by HPLC has nullified the sole advantage of these strategies and left only their chemical deficiencies. Such improvements as may come in the next few years are likely to be minor variations in catalysts or extraction solvents, or advances in instrumentation extrinsic to protein sequencing such as the analytical system. However, true advances in chemistry are possible, although there have not been any in 30 years, and improvements in the mechanical aspects of manual sequencing, e.g., in sample immobilization, may lead to a distinctly better system than offered here.

II. APPARATUS

A. Hardware

The main glassware required includes 6 × 50 mm culture tubes (thick-walled Pyrex® from Corning) for sequencing and (thin-walled borosilicate) for extracts/PTHs and 25-mℓ vials (Pierce) for batchwise reactions and storage of 6 × 50 mm tubes. Each sequencing tube needs a tooth on its rim to aid in fluid transfer to other tubes: heat part of the rims of two tubes in a fine flame until just soft, then stick these parts together and immediately pull apart to give broad-based needles. Flame polish to 1 to 2 mm height, leaving a sharp tooth if possible. These tubes should be labeled by etching (e.g., diamond pen), but a felt marker is adequate for extract tubes. Glassware useful for reagent and solvent reservoirs includes 1-mℓ Reactivials (Pierce), 3-mℓ screw cap vials, 13 × 100 mm screwcap tubes (any general supplier), and 40-mℓ vials (Pierce). Teflon® valve/cap closures for all of these (Pierce) should have their silicone plugs removed and their bores drilled out slightly to accommodate syringe needles. Inert atmospheres (need N_2 tank, regulator at 5 to 10 psi, valve, and needle on plastic line) are produced in capped vials by flushing through the access bore and are maintained during syringe withdrawal with a blanket of gas through the plug hole. The vacuum system should contain a vacuum gauge (P8288-52 from Scientific Products) to assess dryness, leaks, and the delicate opening of valves; two glass-TFE stopcocks (Corning rotoflo, 6-mm bore) bearing, via plastic tubing, large valve/cap closures with their sliders removed, and a cold trap with two-stage vacuum pump. The vacuum manifold may be all glass or fabricated with vacuum tubing and plastic tees. Other equipment required includes a 12-hole heating block with one quadrant drilled out to a 28-mm diameter to accommodate 25-mℓ vials (or use a water bath) and kept partially full of water for thermal contact; a swinging bucket centrifuge with some of the buckets half filled with #0 rubber stoppers for easy handling of small tubes and vials; a microtiter plate used as a tube rack; a vortex mixer; a timer, and an ice bucket.

The comparatively large and repetitive volumes of wash and extraction solvents are conveniently distributed with dispensers employing disposable plastic tips (but actual disposal is usually counterproductive). Glass capillaries and Pasteur pipettes can be used instead. The best reagent dispensers are syringes with Teflon®-tipped plungers (25, 50, and 100 μℓ Dynatech C-160 from Pierce): the needle permits ready access to reagent reservoirs and to the bottom fifth or so of the 6 × 50 mm tubes. However, the steel needle is a source of heavy metal ions when used with strong acids, even with trifluoroacetic acid (TFA), and the epoxy holding the needle dissolves in 88% formic acid (HFo). Use of glass capillaries avoids these problems, but a better solution is to replace the needle and glue with pressure-fitted plastic tubing, either Teflon® or polyethylene: the needle is pulled out and the epoxy removed by drilling and scraping. A short length of narrow-bore 1/16-in. tubing is stretched carefully above a broad flame; the cooled product is cut in the middle and the narrow end fed through from the top of the syringe barrel until it can be grasped at the bottom. The tubing is pulled further until the widening tubing fits snuggly in the top, where it is cut cleanly with a blade, pulled into the barrel a little, then forced to the zero mark with the plunger. The drawn-out tubing is cut to an appropriate length. The largest syringes are the easiest to modify and provide sufficient accuracy for most applications. Color code and consistently place syringes and other dispensers to minimize the chances of mixing them up during repetitive use.

Clean glassware is, of course, essential. New 6 × 50 mm tubes and, if they are to be recycled, ones that have contained PTHs are grabbed in a closed-packed array and rinsed with HPLC-grade water and ethanol from squirt bottles, shaken out in-between. Sequencing tubes with especially desirable teeth are generally cleaned sufficiently with HFo. Glassware may be acid cleaned with nitric/sulfuric acid, but avoid chromate. A furnace is usually safe, but sometimes, and erratically, tubes with destructive properties are produced.

B. Solvents and Reagents

HPLC grade solvent is fine for most purposes: triethylamine (TEA) and dimethylformamide (DMF) from Aldrich, HFo from Baker, heptane and acetonitrile (MeCN) from Burdick and Jackson. Sequencing grade ethyl acetate (EA), TFA, and PITC are necessary for sequencing (Pierce and B & J, or repurify these reagents yourself). Good DMF and tert-amines can be made from yellowed stocks by redistillation (*in vacuo* if the bp is high) from a little succinic anhydride. Distill 100% ethanol (EtOH, Aaper Chemical) from NaOH. Good water can be purchased, but slow redistillation of house-distilled water from $NaOH/KMnO_4$ (2 g each per 4.5 ℓ water, collect between 300 and 3500 mℓ) makes a product at least as free of interfering materials as commercial water. The buffer used in the partition method (see below) is made from 15 μℓ hexafluoroacetone trihydrate (HFA) and 22 μℓ 25% aqueous trimethylamine (TMA, both from Aldrich) per 5 mℓ of solvent, but the latter is rather variable in actual concentration, so the ratio required for dilute aqueous pH 7.2 should be empirically determined. Polybrene (Aldrich), used as a carrier in the film method, is purified by Edman cycling 10 to 20 mg in a 13 × 100 mm tube and precipitation from methanol with acetone, or by dialysis against dilute aqueous TFA using Spectrapor 3500 MW cutoff. About half the polymer is lost by either process, and the oligomer-free product, now the TFA salt, should be completely soluble in methanol. Anhydrous HCl/MeOH is made by dropwise addition sans stirring of redistilled acetyl chloride to MeOH at 0°C in a vertical 13 × 100 mm tube. Use a ratio of 0.3:3.9 for 1 *N* and 0.6:3.6 for 2 *N*; store at 0°C. The conversion reagent, concentrated HCl, and EA should be protected from development of oxidizing agents by addition of about 0.02% ethanethiol. Working stocks of all reagents and solvents should be small and replenished periodically from main reserves. TEA and PITC should be kept cold and removed under N_2 barrier, and all reserves should be stored at 0°C.

III. SEQUENCING METHODS

The choice between the following two methods depends mostly on the size of the peptide: the partitioning method is excellent for small peptides up to a dozen residues or so, ones that may wash out with the film method, but poor for large peptides and proteins. Other considerations are that partitioning gives better recoveries of Ser and Thr, lower extractive yields of Arg, and is more work and more time consuming. In general, the film method is preferred, but, unfortunately, the class of materials that is best for filming is also best for automatic sequencing. Always include a sequencing standard such as insulin B chain with each batch of samples.

A. Partition Method

Samples for partitioning need little preparation provided they are salt and detergent free (clean up by reversed-phase HPLC in a volatile buffer if necessary): dissolve and transfer to a sequencing tube 50 to 5000 pmol in 5 to 50 μℓ dilute acid in water or aqueous alcohol. HFo is always an effective solvent, but do not store the peptide in it. Redry the sample from 5 to 10 μℓ EtOH/TEA/water, 2:2:1, to remove any ammonia.

Step 1. Thiocarbamylation (coupling) — Add 10 μℓ freshly prepared EtOH/TEA/water/PITC, 7:1:1:1, to the dry peptide and mix well, but do not splash anything high in the tube. Reagent can be added to tubes within the 25-mℓ incubation vial and the set mixed by cautious lateral shaking and tilting. Incubate 8 to 10 min at 50°C, under N_2 if preferred, although it seems to make no difference. The reaction is faster with DMF as the solvent (3 min, 50°C), but more by-products are usually seen. Most amine catalysts, such as TMA, are less efficient than TEA. Vacuum dry until all PITC is gone (will *look* dry, provided salt has not accumulated); this will take 5 to 10 min at room temperature, but only 2 to 4 min at 50°C, which can be arranged by locating part of the vacuum system strategically above the heating block. Longer drying is not harmful.

Step 2. Washing — Add 2.5 μℓ water low in the tube and about 200 μℓ heptane/EA, 15:1, containing 0.5% TMA (this serves to exchange residual TEA, creating a more hydrophilic, less extractable, peptide salt). Vortex well and centrifuge briefly. Decant upper phase, being careful not to tip the tube so much that the lower phase also decants. Wash two to three times in the same manner with heptane/EA, 7:1 (with no amine). This is a good place to add an internal standard for normalization of analytical results. Vacuum well until dry.

Step 3. Cleavage — Add low in tube 5 μℓ concentrated HCl (11 to 12 *M*) with glass capillary or plastic needle syringe; an alternative that reduces any chance of cross contamination is to add the HCl high in the tube and bring it down by centrifugation. Mix well, avoiding splashing, and stand 5 min at 22°C or 1.5 to 2 min at 50°C; an inert atmosphere *may* improve efficiency and/or background, but is not essential. Almost all pairs of residues cleave well at room temperature, only Pro-X requires 50°C, which composition or previous experience will indicate. Using a vacuum port different from the one used to dry the coupling medium, pull vacuum carefully to avoid bumping (gentle shaking may help) and leave on full vacuum 3 to 5 min 22°C.

Step 4. Stabilization — Add 5 to 10 μℓ MeOH and incubate 3 min at 50°C. This treatment opens the thiazolinone product of cleavage and prevents it from hydrolyzing in Step 5, which would prevent the extraction of His and Arg. It is not absolutely necessary if there are none of these residues in the peptide. Vacuum dry.

Step 5. Extraction — Add 4 μℓ water and 40 μℓ benzene/MeCN, 2:1, mix well, and centrifuge well. Heptane/EA, 1:5, can be used instead if one worries more about the toxic benzene than the peroxide-prone EA. Decant upper phase carefully into 6 × 50 mm tubes containing 50 nmol thiourea in 2 to 5 μℓ MeOH. Extract next with 40 μℓ 0.02 *M* HFA/0.016 *M* TMA in benzene/MeCN, 2:1 (see solvents), combining extracts. When processing a batch of samples, be compulsively organized to prevent errors: line up the extraction tubes in order in the microtiter plate and line up the sequencing tubes in a parallel row; always use the same hand to grasp each kind of tube, in the same order, and proceed systematically along the rows. If a peptide contains no very polar residues, a decrease in the strength of the extraction solvent may give less washout and, therefore, higher recoveries toward the end of the peptide (e.g., heptane/EA, 1:2). Dry lower phase and return n-1 peptide to Step 1.

Step 6. Conversion — The extract is dried carefully with weak vacuum or aspirator, then with full vacuum. Add 15 to 20 μℓ 1 *N* HCl/MeOH and incubate at 65°C for 10 min, or the same volume of 2 *N* HCl/MeOH and 10 min at 50°C. (If your analytical system does not accommodate Asp and Glu methyl esters, try using 25% aqueous TFA at 65°C instead.) Be careful not to wash off the labels when adding reagent. The incubation can be done individually — stopper the tube with a large glass bead — or batchwise. The 50°C conversion avoids a second heating block, but does not convert Pro as well; this is not all bad, though, for the unconverted PTC-Pro methyl ester elutes conveniently and distinctively before Met (see below). Store in freezer until ready for analysis.

B. Film Method

This method is essentially the same as that employed by automatic machines, both spinning cup and gas phase. In place of a lower aqueous phase, there is a layer of peptide plus hydrated carrier, Polybrene. Discrimination between by-products or sequencing derivatives and peptide is still by polarity, so the strength of the organic solvent is once again balanced for what will remove the comparatively hydrophobic products, but not the peptide. Samples for filming are best if salt and detergent free: final purification by RP-HPLC in a volatile buffer, exhaustive dialysis against 50% EtOH (even if the protein precipitates), or precipitation from concentrated aqueous solution with acetone or MeCN or from a small volume of HFo with EA. HFo and HFA are good general solvents for sample transfer; although HFA is more toxic and expensive, it is gentler and compatible with long-term storage.

Step 1. Precycle — Add 50 to 2000 pmol or 100 μg, whichever comes first, protein/peptide to 50 μg purified Polybrene in 6 × 50 mm tube, dry, and go to Step 4. (Save and analyze the Cycle 0 extract if you like, but it is rarely informative.)

Step 2. Thiocarbamylation — Add 10 μℓ freshly prepared DMF/TEA/water/PITC, 7:1:1:1, tilt, and shake cautiously to wet the entire film. EtOH instead of DMF is permissible, with corresponding increase in coupling time, but the higher polarity of wash solvent means less problem with by-products and DMF is a superior solvent. Incubate 3 min at 50°C. Dry until all visible PITC is gone and then a few minutes more.

Step 3. Washing — Add about 250 μℓ heptane/EA, 15:1, and extract film by tilting and rotating the tube (do not vortex). Decant, with no caution necessary, and repeat with EA, then two to three times with EA/MeCN, 1:1. The last wash can be MeCN alone, with a slightly higher risk of washout, but in either case be careful to exclude moisture from the solvent reservoir: more than a trace may lead to washout of the Polybrene and peptide. (Special precautions are necessary only in case of high humidity.) Linear arrangement of these wash solvents and color coding will help prevent errors.

Step 4. Cyclization and cleavage — Add internal standard if desired and dry briefly. Add 8 to 10 μℓ TFA, incubate 5 min at 50°C, preferably under inert atmosphere. The time required for cleavage varies greatly according to residue pair, but as this is not generally known beforehand, the routine time is chosen to give essentially complete cleavage of all but the most resistant linkages.

Step 5. Stabilization — Dry TFA 20 sec and add 5 to 10 μℓ MeOH (containing internal standard, if not added earlier) and incubate 2 to 3 min at 50°C. Dry cautiously with full vacuum for about 1 min. The MeOH in this case serves to open the thiazolinones of Ser and Thr, which are very sensitive to decomposition in the "dry" film containing residual TFA: this is why the initial drying is limited to 20 sec and should be performed consistently. The improvement in recovery of Ser and Thr is about three times or 30 to 40% of the best recovered residues.

Step 6. Extraction — Extract with 40 μℓ benzene/MeCN, 1:1, containing 0.1% acetic acid, wetting all the film by rotating the oblique tube. Decant carefully into the correct tube containing 50 nmol thiourea; repeat, combining extracts. A larger volume or three extractions will give somewhat more complete recoveries, but the slight improvement is confounded by the larger volume that must then be dried, which has some tendency to bump as it is. Return n-1 peptide to Step 2.

Step 7. Conversion — Place extracts in 25-mℓ vial and cautiously pull vacuum: initially tilt and agitate or swirl to help prevent bumping (or use vacuum centrifuge). Convert as in partition method.

IV. DATA AND JUDGMENT

A. PTH Analysis

There are many analytical systems for PTHs described in the literature, most based on reversed-phase HPLC. Many of these seem quite adequate for use with either manual or automatic sequencing methods, the major considerations being

1. Adequate resolution of all residues, including common derivatives such as alkylated Cys and the methyl esters of Asp, Glu, and any carboxyl derivatives of Cys and Lys
2. Separation of any sequencing by-products from residues, mainly of phenylthiourea (PTU) and diphenylthiourea (DPTU), perhaps, also, the phenylthiocarbamyl (PTC) derivatives of sec- and prim-amines produced by decomposition of the tert-amine catalyst (importance dependent on the efficiency of washing and on the purity of the catalyst
3. Speed, which is more critical in manual sequencing, because samples are produced so much faster than in automatic methods
4. Sensitivity, which in practice is mostly a function of background and, therefore, of cleanliness and consistency, than of the detector or peak elution volume

The two systems used in this laboratory are based on the pioneering efforts of Zimmerman and Pisano,[8] who employed a step-gradient of buffered acetonitrile with a C18 column.

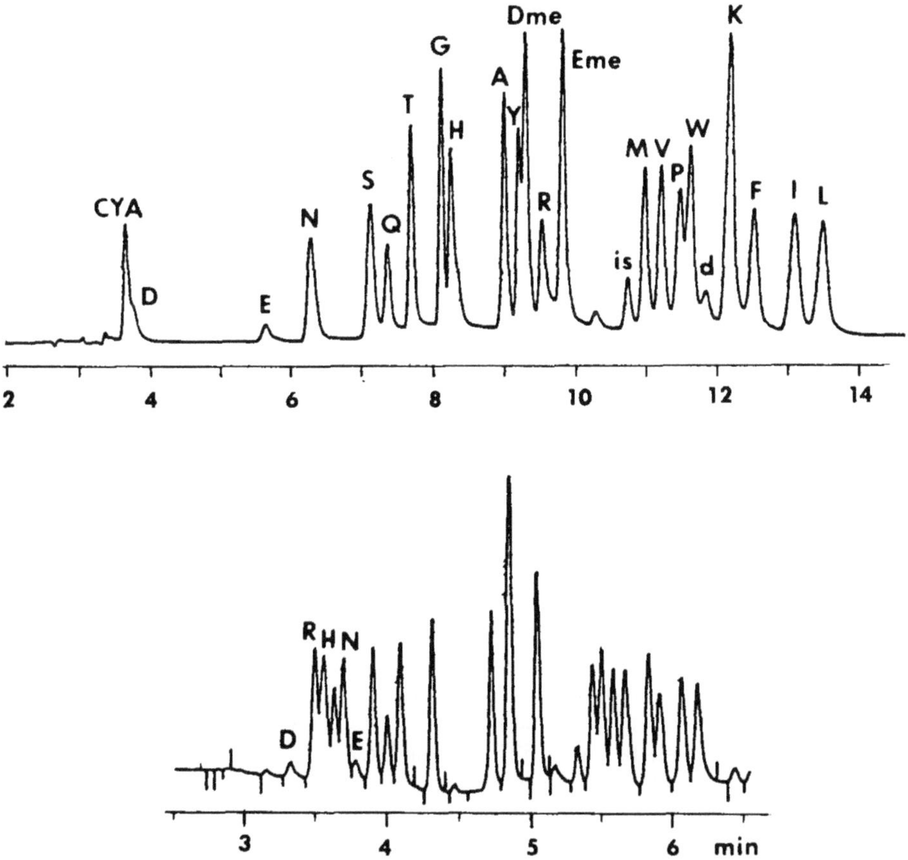

FIGURE 1. Separation of PTH amino acids by RP-HPLC. Upper trace: about 100 pmol each PTH on 4.6 × 250 mm Ultrasphere ODS, absorbance 254 nm at 0.1 AUFS, 50°C, flow 0.9 mℓ/min. Solvent a is 80 m*M* ammonium acetate, pH 4.8, containing 0.5 μM PTU to match the absorbance of solvent b (85% aqueous MeCN) and, thus, flatten the baseline: linear gradient from injection to 2 min of 30 to 60% b, begin 6 min reequilibration at 9 min (so total run time injection to injection is about 15.6 min). Waters HPLC with two 6000-A pumps, 660 solvent programmer, 440 detector, WISP 710B, Kipp & Zonen recorder; the b pump is plumbed through the a pump in the standard manner, which provides the required mixing. Standard single-letter code for most PTHs; CYA, PTH cysteic acid; Dme, PTH Asp methyl ester; Eme PTH Glu methyl ester; is, internal standard; d, DPTU. Lower trace: same standard on 4.6 × 150-mm NovaPak C18, 45°C, 1 mℓ/min; solvent a 60-m*M* TEA phosphate, pH 4.9, solvent b 84% MeCN. Gradient: initial 22% b, linear to 40% from 0 to 1.4 min after injection, 40 to 63% from 1.4 to 2 min, 63 to 65% from 2 to 4.2 min, strip with 100% for 0.8 min, reequilibrate for 4 min before next injection (total run time, 9 min). Waters gradient HPLC with 6000-A and M45 pumps configured as above, 680 controller, Beckman 160 detector, HP 3390A integrator. Only residues that elute differently are labeled; CYA elutes between H and N; Y and Dme elute essentially together, as do W and DPTU. Both aqueous buffers contain 3 to 5 m*M* KCN as a preservative. These tracings and those of the other figures are photoreproductions of actual chromatograms.

The older system comprises a 250-mm Ultrasphere column, 25 and 51% acetonitrile, with a short linear gradient between (Figure 1, upper). Salt and pH position Arg and His around the middle of the chromatogram, following the rules described in detail earlier:[9] increasing salt moves His and especially Arg toward the front of the run, while lowering the pH increases the drag of His, narrowing the gap between it and Arg. The temperature of 50°C is necessary for equal spacing in the triplet Met/Val/Pro.

The other system employs a 150-mm column of NovaPak (Figure 1, lower). The intrinsically lower absorptivity of the packing is further suppressed with a TEA buffer so that Arg and His have the position and elution order they should in a reversed-phase system. Because the column is shorter, the gradient must be run faster to achieve the same organic concentration effects (important particularly for the positioning of Tyr, Lys, and DPTU), resulting in faster runs, but lower resolution. And, unfortunately, Trp and DPTU run together, and Tyr and Asp methyl ester are inadequately separated if the column has slipped at all in plate count. This means that sometimes a sample must be rerun (60% b works for both problems), but the extra speed makes the system generally worthwhile. Ionic strength operates here as it does in the Ultrasphere system, but pH effects are reversed: lower pH brings His off earlier.

B. Data Analysis

Interpreting the chromatograms in order to read a sequence (or to conclude there is not one) is often the major effort, and one that is largely independent of the sequencing method. The examples shown in Figures 2 to 4 have been chosen to illustrate common situations and logic. It is here that the advantages of consistency become most obvious, as the best approach is to lay the chromatograms side by side and compare them in an analogous but semiquantitative, fashion. Early cycles for a reasonable quantity of a pure peptide will be obvious, as in Figure 2. Here the sequence is clearly Asn-Asp-His-Asn-Ala: the Asp on Cycle 1 comes from deamidation of Asn during conversion, while the Asn on Cycle 2 is lag from Cycle 1, out of phase or incompletely extracted or both; similarly, the Asp on Cycle 4 is partly lag from Cycle 2, but mostly deamidated in-phase Asn. Paying attention to relative quantities should avoid misassignments. Note the considerable lag by Cycle 5, probably due more to incomplete coupling of this very polar string of residues than to incomplete cleavage, which is the usual cause of being excessively out-of-phase.

Reading a sequence will be more difficult for later cycles when a high background of PTHs has appeared, or with feeble amounts read over the intrinsic sequencing background. The correct residue can be determined only by careful inspection of all peaks for the one that increases significantly over the previous cycle. One must take into account:

1. The relative yields of different residues (determined empirically by you): a small increase is more significant for Ser or Thr than for Ala or Val, which is why in Figure 3 the Thr on Cycle 1, rather than the Pro (plus unconverted Pro), goes with the major sequence
2. A residue that may be artificially inflated by temporary increases in a coeluting "garbage" peak (pay attention to sets which rise and fall together), such as the "Trp" on Cycle 3, actually only DPTU
3. Apparently legitimate peaks that are, however, present in that cycle for other peptides sequenced at the same time
4. Peaks that are "too good", that is, increased more than could be reasonably expected from the levels in preceding cycles
5. Residues that are sufficiently represented in the composition, a criterion unfortunately far more discriminating with smaller peptides (for methods attuned to sequencers, see Tarr[6])

Pay attention to out-of-phase material, using this deficiency of Edman sequencing to advantage: deep into a sequence the pattern of rise and fall should be repeated for each residue; one that shows up, however clearly, for only one cycle, while the others before and after are apparent for three, is not legitimate (go back and look for the Ser). Use numbers derived from an integrator (don't work blind: check integration marks to catch errors) or from measuring peak height (sounds primitive, but strongly recommended) to sort out possibilities like that Pro on Cycle 8, Figure 3: is it just the out-of-phase expected from the Pro on Cycle 7 (or 6), or is it another Pro (answer: it's a Pro)? How about Cycle 7? Numbers are essential

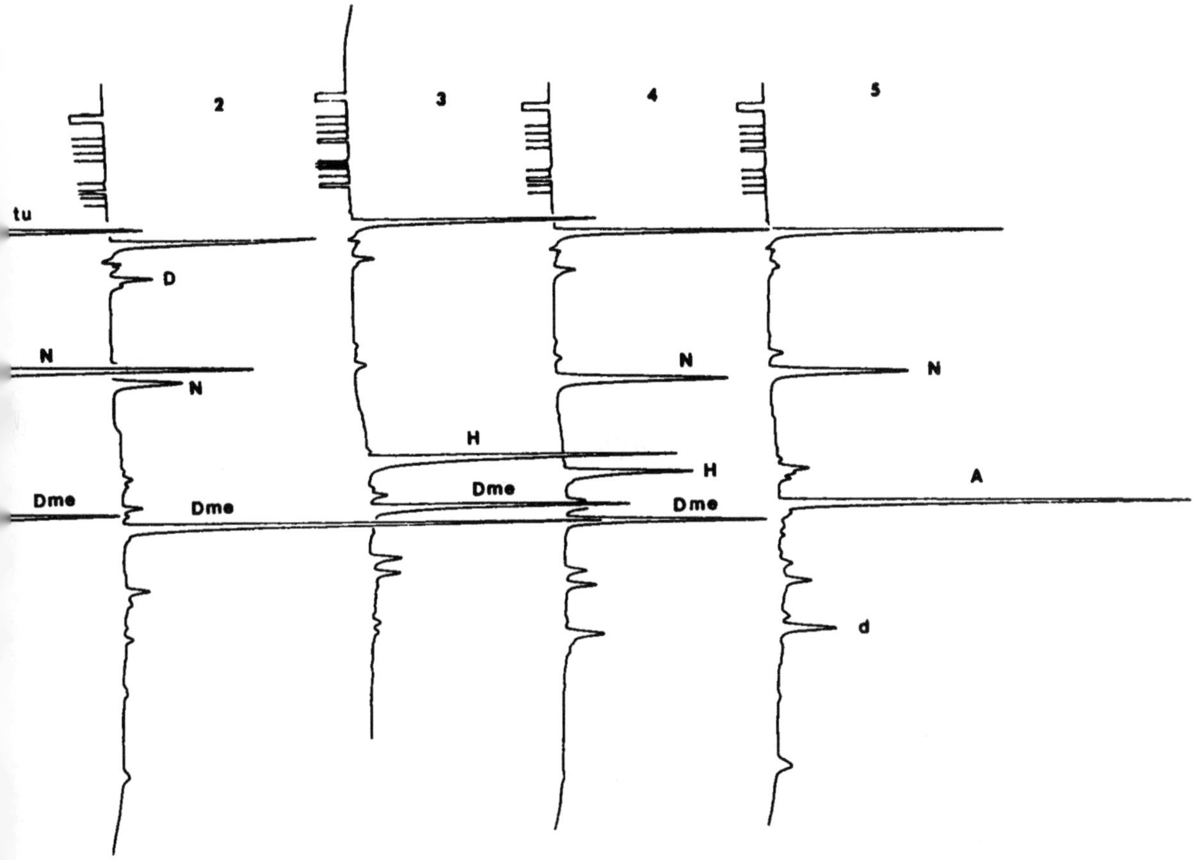

2. Degradation of a long peptide by the film method. About 4000 pmol was sequenced and 10% taken for analysis, Ultrasphere system, s as in Figure 1. Cycles 1 and 3 are displaced for clarity. tu, thiourea; other abbreviations, as in Figure 1.

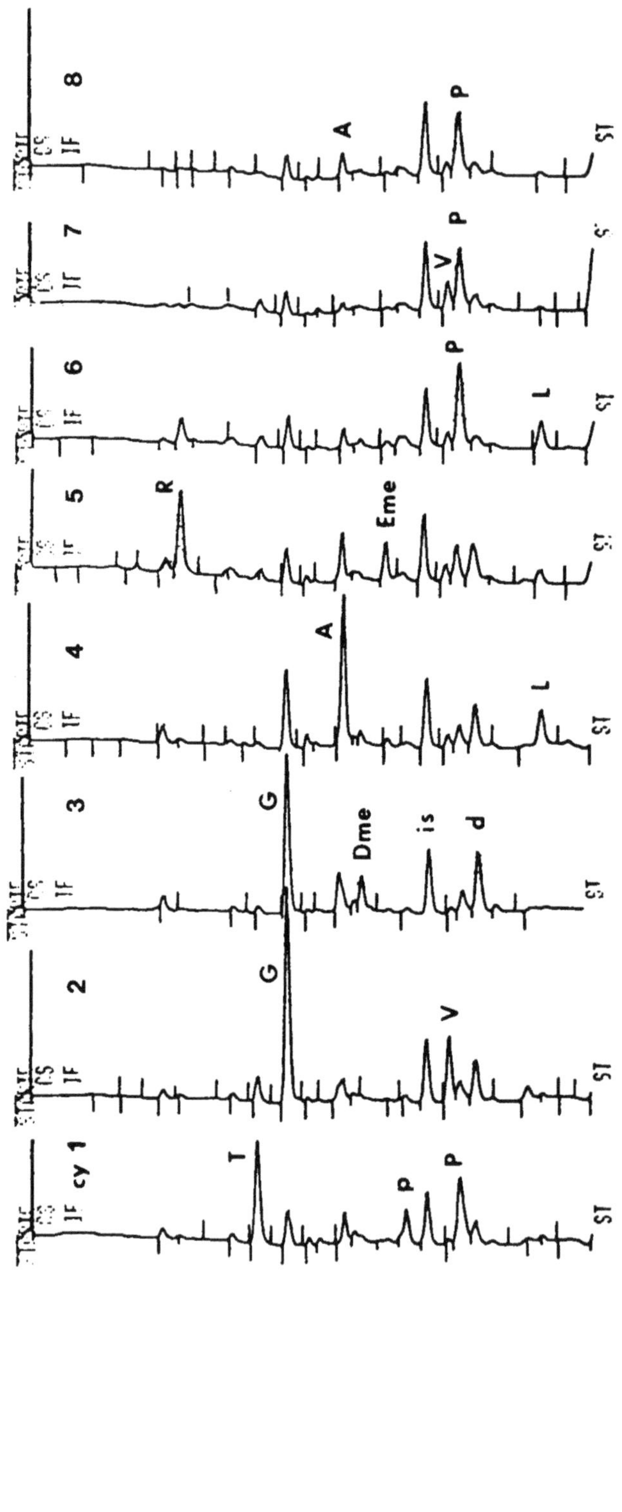

FIGURE 3. Sequencing a peptide mixture by the film method. Initial yields for the major component about 1000 pmol and for the minor about 500 pmol; 10% taken for analysis using the NovaPak system described in Figure 1. Integrator printout of elution times eliminated for clarity. p, PTC Pro methyl ester; other abbreviations, as in Figure 1.

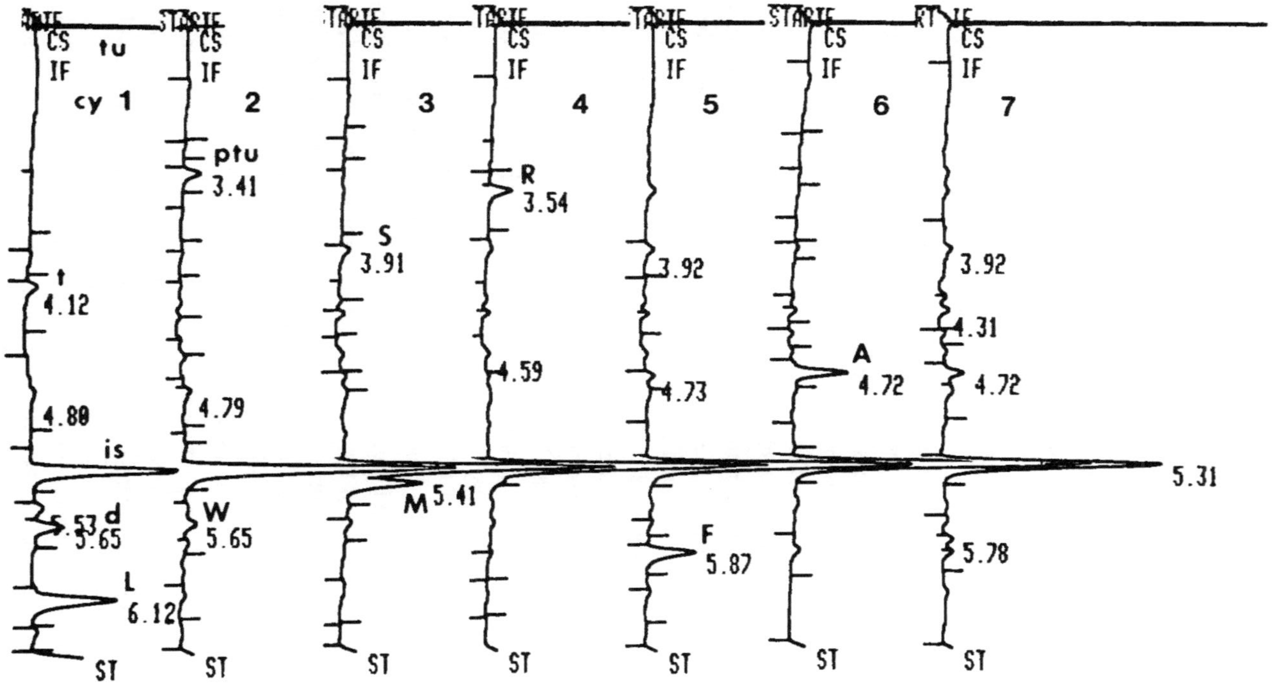

FIGURE 4. Degradation of a short peptide by the partition method. About 360 pmol of LWMRFA by Cycle 1 yield, dropping to about 150 pmol at cycle 3; 50% of sample injected, conditions as for NovaPak in Figure 1. tu, thiourea; t, by-product eluting at position of Thr; other abbreviations, as in Figure 1.

when dealing with mixtures, as in Figure 3, a situation all too frequent. Don't assume that both (or all) components will sequence with the same efficiency: the ratio of yields may change, even reverse if not too far from unity, over the course of a run.

V. PROBLEMS

Difficulties of several sorts occasionally arise. The most common is a high background of sequencing by-products that interferes with analysis. If this is confined to a particular sample or type of sample, the cause is usually contaminating materials in the samples, not the reagents, and the remedy is purification. Very large proteins can reduce the efficiency of washing if too much is taken for sequencing: the upper limit of about 100 μg of protein is probably reasonable, although more can be used if the Polybrene, surface area, reagent volumes, etc. are also increased. In a sense, too little material can cause the same problem, because there is always *some* background over which the signal must be read: a readable sequence should be routinely obtained from 50 pmol of starting material, but not from 20. Note that this level is far above the picomole or so detection limit of the HPLC, and above the 2 to 5 pmol that can be read for a sequence that is well under way. Figure 4 does not illustrate a particularly high-sensitivity run, but the problem identifying the Trp on Cycle 2 is obvious. The number or polarity of the washes can be increased for particular samples, but there is a danger of washout with smaller peptides and proteins.

A global background problem indicates something wrong in the composition of the wash solvents, a contaminated reagent, or incorrect technique. For instance, in the partitioning method, almost all of the upper phase can be removed without disturbing the interface: a too generous margin for error significantly decreases wash efficiency. TEA that has absorbed moisture or oxygen can cause a high background and a lowered sequencing efficiency. It is possible for vacuum drying and anhydrous wash solvents to be too successful in eliminating moisture under a dry atmosphere: the film must be hydrated to wash efficiently. Try varying the composition of the solvents and the number of washes to see what works.

Washout is another common difficulty and can be caused by overdoing some of the cures for high background: too much water in film-method solvents (either wash or extraction), picking up some of the interface in partitioning, or increasing the polarity of the solvents too much. Good test peptides available from Chemical Dynamics are Leu-Trp-Mer-Arg-Phe-Ala and the C-terminal octapeptide from glucagon. The former should sequence reasonably well to the end with partitioning, as in Figure 4. Note here another common problem, a sudden drop in yield due to Trp, reflected in the low yield of Met relative to Leu and, in part, by the low yield of Trp itself. This seems to result from oxidation and cyclization (without chain cleavage) that blocks a fraction of the peptide, a reaction that occurs on storage, particularly in solution, and not during sequencing. Note, too, the erratic background of PTU and low level of contaminating Ser, both are typical. The "Thr" on Cycle 1 is actually a sequencing by-product that appears erratically and is known to increase under certain solvent conditions, such as with neutralized coupling media or with DMF; if it shows up, test the coupling medium reagents. The octapeptide has a polar C terminus and should sequence to the end by both methods.

Low repetitive efficiency can be due to chemical or mechanical effects, that is, bad reagents/solvents or washout. This should not be confused with low initial yields, which are caused by chemical losses happening uniquely on Cycle 1, blocked peptide, or, most often, having less material than one thought. The most common cause of chemical inefficiency is oxidant in solvents, e.g., peroxide in EA exposed to air. This is why all such solvents should be protected from air and with scavengers such as ethanethiol. Reducing agents do not protect completely against photodecomposition products of alkyl halides, one reason chlorobutane and its relatives are not recommended, nor are they especially effective against heavy metals, the other common killer of sulfur chemistry. The thiourea added to the extracts minimizes all chemical problems from this point on, though it should not be used as a cure for impure or degraded solvents. Test all solvents and reagents (except PITC) periodically by exposing 100 pmol or so of PTH-Lys and Ala, along with an unreactive internal standard, to each of the solvents and reagents and note their relative recovery on HPLC; Lys is especially sensitive, as it possesses both a PTH and PTC functionality. Repetitive efficiency itself is classically assessed by sequencing myoglobin 12 to 15 cycles, calculating from the Val yields. In the absence of washout, efficiencies of 92 to 93% should be obtainable on 500 pmol, or only slightly lower than with the gas phase.

REFERENCES

1. **Levy, W. P.,** Manual Edman sequencing techniques for proteins and peptides at the nanomole level, *Methods Enzymol.*, 79, 27, 1981.
2. **Chang, J.-Y.,** Manual micro-sequence analysis of polypeptides using dimethylaminoazobenzene isothiocyanate, *Methods Enzymol.*, 91, 455, 1983.
3. **Mehl, E., Haniu, M., and Shively, J. E.,** Manual gas-phase sequencing on filter chips for simultaneous Edman degradation of multiple peptide samples, Abstr. 112, presented at Symp. Am. Protein Chemists, San Diego, 1985.
4. **Fraenkel-Conrat, H.,** A technique for stepwise degradation of proteins from the amino-end, *J. Am. Chem. Soc.*, 76, 3606, 1954.
5. **Tarr, G. E.,** Manual batchwise sequencing methods, in *Methods in Protein Sequence Analysis*, Elzinga, M., Ed., Humana Press, Clifton, N.J., 1982, 223.
6. **Tarr, G. E.,** Manual Edman sequencing system, in *Microcharacterization of Peptides and Proteins*, Shively, J. E., Ed., Humana Press, Clifton, N.J., 1986, 6.
7. **Gray, W. R.,** Sequence analysis with dansyl chloride, *Methods Enzymol.*, 25, 333, 1972.
8. **Zimmerman, C. L. and Pisano, J. J.,** High-performance liquid chromatography of amino acid derivatives, *Methods Enzymol.*, 47, 45, 1977.
9. **Tarr, G. E.,** Rapid separation of amino acid phenylthiohydantoins by isocratic high-performance liquid chromatography, *Anal. Biochem.*, 111, 27, 1981.

Chapter 3

LIQUID PHASE SEQUENCE ANALYSIS OF PROTEINS/PEPTIDES

A. S. Bhown and J. C. Bennett

TABLE OF CONTENTS

I. INTRODUCTION

Theoretical and practical advances in the chemistry of stepwise degradation of proteins have resulted in the development of an instrument by Edman and Begg[1] in 1967 known as a liquid phase or spinning cup sequenator, the commercial version of which is the Series 890 Beckman Automated Protein/Peptide Sequenator.

Although the original Edman degradation has changed little, the methodology has been modified and new techniques developed at an unexpected pace in order to obtain sequence information on smaller and smaller amounts of sample.[2-5] The modifications have been directed towards all aspects of the involved methodologies including peptide separation, prepurification of solvents and reagents for sequencing,[2,5,6] improvements in the automated Beckman Sequencer[2,4,5] (e.g., development of an automated conversion device[5]), and identification of phenylthiohydantoin derivatives of amino acids. These modifications have made it possible to obtain sequence information on picomole amounts[2] of proteins and peptides. Although other versions of automated sequenators, viz. solid phase (Chapter 4) and gas phase (Chapter 5), have been developed and are commercially available, this chapter will be confined to the liquid phase technique of amino acid sequence analysis.

Beckman Automated Protein/Peptide Sequenator Series 890B, and later on 890C, were two of the most widely used instruments for amino acid sequence analysis. However, in order to keep pace with the methodological advances in amino acid sequence analysis, Beckman Instruments has recently introduced a more sophisticated 890M series, which is equipped with an autoconversion device and a microprocessor which permits the operator to manipulate the degradation chemistry at each cycle as dictated by the presence or absence of a particular amino acid residue at, or adjacent to, the amino terminus in the polypeptide.

The purpose of this chapter is to acquaint the investigator with the theoretical and practical methodological aspects of stepwise degradation of proteins and peptides as performed in the Beckman 890 series, with greater emphasis on the 890M series.

II. EDMAN DEGRADATION

Following the pioneering work of Sanger[7] in determining the primary structure of insulin, the most important contribution to amino acid sequence analysis was made by Edman, who introduced[8-10] the use of phenylisothiocyanate (PITC) for sequential degradation of one amino acid at a time from the amino terminus of a polypeptide chain. Numerous other derivatives[11-17] and coupling agents[18-21] have been investigated, and some are described in detail

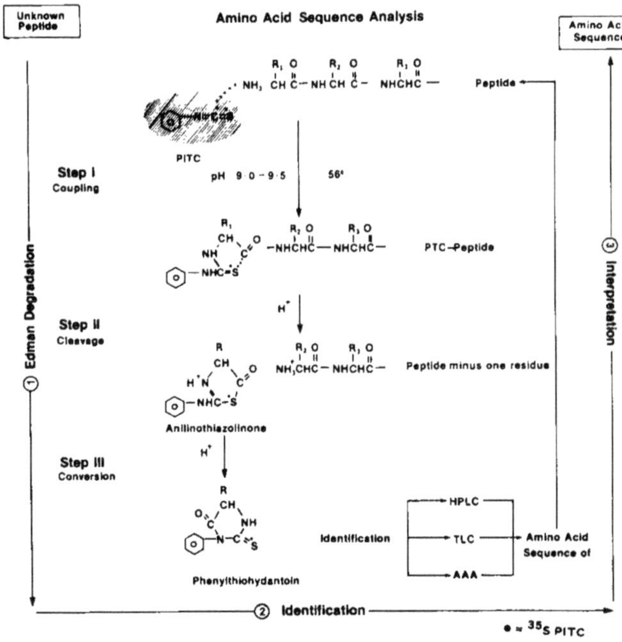

FIGURE 1. Chemistry of isothiocyanate degradation.

in Chapter 6 of this book. The isothiocyanate degradation, also referred to as sequential, stepwise, or Edman degradation, is a combination of a series of reactions (Figure 1). The first reaction is the formation of phenylthiocarbamyl derivative (PTC-peptide) of the protein/peptide by coupling the free α-amino group with PITC. This reaction is referred to as the coupling reaction. The next reaction involves the cleavage of the PTC-peptide at the peptide bond nearest to the PTC-substituent. This leads to the formation of a 2-anilino-5-thiazolinone (ATZ) derivative of the first amino acid of the peptide, and a peptide with one amino acid fewer than the original. This is the cleavage reaction. The cleaved thiazolinone (ATZ) is a derivative of the first amino acid of the polypeptide and, in principal, could be identified. However, being unstable, it is first converted to the isomeric and more stable form called 3-phenyl-2-thiohydantoin (PTH). This is known as the conversion reaction (Figure 1) and is the last of the series of reactions of Edman degradation.

A. Coupling

The first step in Edman degradation is based upon the reaction of PITC with the free α-amino group of the polypeptide chain in an inert atmosphere at pH 9.0 and a temperature of 56°C. Coupling is the limiting factor for efficient amino acid sequence analysis. Numerous factors effectively control this important initial reaction in sequential degradation. These factors are (1) sample solubility, (2) sample preparation, (3) reagents, and (4) the order in which the reagents are added.

1. Sample Solubility

Since proteins are largely polar and water soluble, and coupling reagent (PITC) is water insoluble, a mixture of an organic solvent with water is better than either alone. The one most commonly employed in the Beckman Sequencer is dilute quadrol (0.1 to 0.5 *M* in water/propanol mixture) and PITC (in heptane). Some proteins, particularly those with high molecular weights, are poorly soluble or insoluble, while others which are initially soluble become difficult to dissolve as the degradation proceeds.[22] This may be due to polymeri-

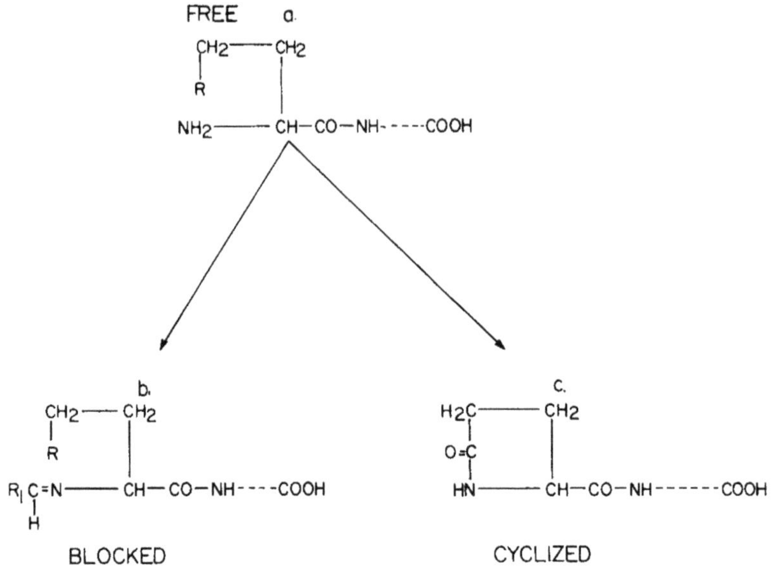

FIGURE 2. Different forms of amino termini.

zation.[23] Prior performic acid oxidation, or reduction and alkylation, of the protein may prevent this problem.[22] At times the coupling reaction seems to proceed to completion even though the protein appears to dissolve partially or not at all. Presumably, this represents a heterogeneous reaction between the protein in the solid phase and PITC in solution phase.[22] However, in the automated sequenator, the protein dissolves much more readily, primarily because it is spread as a thin film on the wall of the spinning cup. Forty percent urea[24] or 4 *M* guanidine HCl[25] has been employed to increase the coupling efficiency.

2. *Sample Preparation*

The other requirement for efficient coupling reaction is the availability of the free α-amino group at the amino-terminal end of the protein molecule. The amino termini of protein molecules have been found to exist in free, blocked, or cyclized form (Figure 2). When the amino terminus has a free α-amino group (Figure 2a) the coupling proceeds smoothly; however, with the other two forms (Figures 2b and 2c), it becomes a formidable task to couple the molecule with PITC. Polypeptides with blocked or cyclized amino group have been found to exist in native form, but also may develop during the course of purification. Certain compounds, particularly aldehydes, irreversibly block the free α-amino group by forming Schiff's base. Similarly, acidic conditions[26] or cyanocysteine cleavage[26] may cause cyclization of the exposed α-amino group. Glutamine residues are particularly susceptible to this behavior. Also, cleavage of an asparaginyl-glycyl bond as a result of rearrangement may result in cyclization of the asparagine residues.[26] Chapter 10 of this book describes how to overcome the sequencing problems under these conditions.

3. *Reagents*

In the spinning cup sequenator, the three steps of Edman degradation are performed in liquid phase; therefore, it is imperative that the reagents and solvents used should be of highest purity grade with strict quality control, especially when microsequencing is attempted. The major disadvantage of poor quality reagents is low initial and repetitive yields, which ultimately determine the sequencing efficiency. Reagents and solvents employed in the Beckman automated sequenator are commercially available in sufficiently pure form from Beckman Instruments, Pierce Chemical Company, and Burdick and Jackson. However, to achieve a high level of coupling efficiency, especially during microsequencing, further purification by redistillation has been strongly recommended by Wittmann-Liebold[27] and Hunkapiller and Hood.[28,29] Frank[30] and Bhown et al.[31] have advocated "in-line" purification of quadrol.

4. Order of Reagent Addition

One of the major drawbacks of stepwise degradation is the appearance of background during extended sequence analysis due to repeated exposure of the polypeptide to low pH during the cleavage step. This has been attributed to a low degree of nonspecific scission of peptide bonds.[32,33] Thomsen et al.[34] have proposed a program with reverse delivery of PITC and quadrol in an effort to circumvent N → O acyl shift and thus stabilize the seryl-threonyl bond in order to diminish background. Although the results are encouraging, further confirmation is pending.

B. Cleavage

Following coupling, the next reaction in stepwise degradation is the cleavage of the PTC-peptide, which is achieved by lowering the pH for a short period of time. Fluoroacids, such as heptofluorobutyric acid (HFBA) or trifluoroacetic acid (TFA), are the most commonly employed acids, because they are excellent solvents for proteins and cause few side reactions.[22] The cleavage reaction is an equilibrium reaction rather than a quantitative reaction; however, the equilibrium is close to being quantitative.[32] Cleavage is dependent upon the type of amino acid being cleaved and the amino acid adjacent to the one being cleaved.[35] When proline is the amino acid being cleaved, introduction of double cleavage followed by extraction after each cleavage and collection into the same tube improves the proline recovery.[31,35,37] Double cleavage has been proposed for aspartic acid and glutamic acid,[22] while shorter cleavage time for glutamine helps to avoid the possibility of cyclization to pyroglutamic acid.[32] The resultant ATZ derivatives of amino acids are unstable and for this reason are converted to the more stable phenylthiohydantoin (PTH) derivatives.

C. Conversion

This is the last of the series of reactions of isothiocyanate degradation. The conventional techniques for conversion involve heating the ATZ derivative under nitrogen in: (1) 1.0 *N* hydrochloric acid;[32] (2) 25% TFA;[36] (3) 1.0 *N* HCl in methanol;[37] or (4) 25% TFA in methanol.[38] Recently, we have introduced[39] the use of 10% TFA in ethyl acetate as the conversion medium.

III. AUTOMATION

Since the introduction of the automated protein/peptide sequenator by Edman and Begg,[1] a number of institutions and private enterprises have undertaken the design and construction of similar instruments,[40,41] one of which is the series 890 introduced by Beckman Instruments. All the versions of protein/peptide sequenators were capable of performing extended amino acid sequence analysis on rather large (micromole) quantities of samples; however, extended sequential degradation on these instruments of those molecules available in small (nanomole-picomole) quantities was a formidable task. This resulted in a number of instrument modifications[4,5,28,36-38,42] which led to the decreased requirement for quantities of sample. There is probably no one single reason for this decreased requirement, but rather a combination of improvements in instrumentation, chemicals, sample modification, carriers, and selective use of extrinsic and intrinsic radiolabeling techniques, as well as a vastly improved detection system for the PTH derivatives (Chapter 12). Although no fine line can be drawn

FIGURE 3. Beckman 890M Liquid Phase Sequencer.

between macrosequencing and microsequencing, the latter techniques use relatively small amounts (<10 n*M*) of protein in an automated sequencer. These modifications and improvements in the chemistry of isothiocyanate degradation and in instrumentation have resulted in four different types of sequenators which are commercially available and are most commonly employed. They are

1. Beckman 890M liquid phase sequencer (Figure 3)
2. LKB solid phase sequencer
3. Applied Biosystem gas phase sequencer
4. Porton protein sequencer

Procedural details of solid phase and gas phase sequencing are described in Chapters 4 and 5, respectively.

IV. SERIES 890M BECKMAN AUTOMATED SEQUENCER

The Beckman Automated Sequencer Series 890M essentially comprises the following modules as shown in Figure 4:

1. Reaction assembly
2. Delivery system
3. Vacuum system

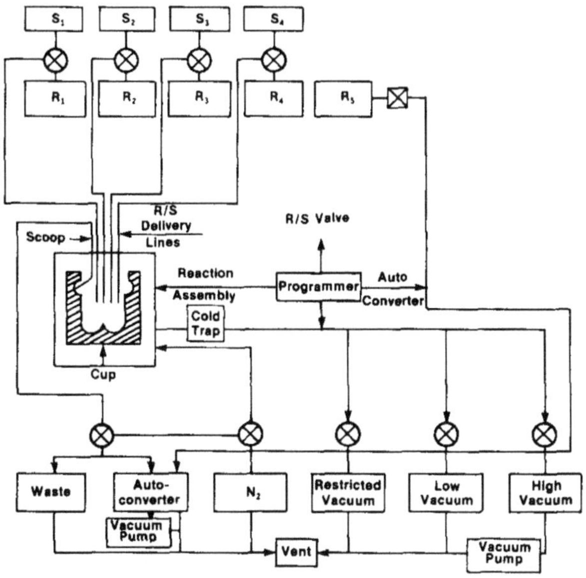

FIGURE 4. Schematics of the Beckman Automated Sequencer Model 890M.

4. Autoconverter system
5. Vent system
6. Programmer/microprocessor

A. Reaction Assembly

The reaction assembly is shown schematically in Figure 4. It consists of a specially designed glass cup, the drive, and a thermostatically controlled heater. The cup is cylindrical and constructed of precision-ground silicate glass. The inside of the cup has some important features. First, it has an undercut to retain the protein from washing. Second, the bottom corner of the cup is rounded to prevent accumulation of protein, to aid in cleaning, and to provide uniform protein film. Below the undercut, the glass is roughened to help retain the protein film. At the top of the cup is a circumferential groove from which the solvents are scooped out of the cup and, with the help of a valve, directed to waste or to a fraction collector. The cup is mounted on a shaft which is continuously spun at 1200 or 1800 rpm by a magnetic drive. In order to have an efficient coupling, the cup is kept at a constant temperature. This is accomplished by circulating hot air within the reaction assembly, which is surrounded by an electrically heated bell jar and a Perspex® cover which surrounds the reaction chamber. The cover of the cup has a cylindrical Teflon® plug which reduces the total cup volume. Inserted in this plug are the reagent and solvent delivery lines (Figure 4), a heat sensor probe, and the "scoop" which removes the solvents from the cup to waste or fraction collector.

B. Delivery System

One essential requirement for efficient degradation is the controlled delivery and removal of solvents and reagents into and out of the spinning cup. The solvents and reagents are stored in glass containers under nitrogen pressure slightly higher than the pressure inside the cup. The precise delivery is controlled by this positive nitrogen pressure and a set of solenoid valves. Accuracy of delivery is critical, since any variability in reagents will cause displacement of protein film in the cup, while variability in solvents will cause inefficient removal through the scoop. This results in turbulence inside the cup, causing protein film breakdown. The protein will then be washed out of the cup. The delivery is by liquid displacement through solenoid valves with positive pressure. The removal of reagents, solvents, and the products of degradation is accomplished by (1) washing and (2) drying under vacuum, with a continuous slow nitrogen flush inside the cup, which causes liquids to be scooped efficiently.

C. Vacuum System

In the original Edman and Begg sequenator, the aim of employing the vacuum was to dry off the bulk of the volatile components left inside the cup after coupling, cleavage, and washing. This is necessary to facilitate precipitation of the protein.[33] However, this concept is slightly modified in the new versions of sequenators. The efficient vacuum in the spinning cup is essential, not only to dry off the volatile reagents and/or solvents, but when micro-sequencing is the object, to remove the reaction products as well, which would otherwise appear as background peaks during identification by high pressure liquid chromatography. In order to attain good vacuum, Wittmann-Liebold,[5] Hunkapiller and Hood,[28] and Bhown et al.[4] have introduced modifications in the vacuum system.

Vacuum in the cup is applied with heavy duty vacuum pumps, which are well sealed, sturdy, capable of running for prolonged periods of time, and resistant to corrosion by the solvents, particularly HFBA. In order to improve the vacuum pump efficiency, Beckman initially introduced an electrically cooled cold trap placed in between the spinning cup and the vacuum pump (Figure 4) which has since been taken out. The vacuum on the cup is applied in three stages: restricted, low, and high (Figure 4). Each stage has its own opening activated by its own set of valves. It is necessary that the restricted vacuum line be kept open when low vacuum is applied, and similarly, when high vacuum is employed, both restricted and low vacuum lines must be kept open. It is also important to have a nitrogen bleed inside the cup during the high vacuum stage. Although the Beckman 890M and other 890 models have two vacuum pumps, those equipped with cold traps employ only one pump to create restricted, low, and high vacuum on the cup, while the other pump is used to dry the fractions in the fraction collector and also to activate different valves of the sequencer. However, the Beckman sequencers not equipped with a cold trap utilize one pump for high vacuum only and the other (1) to apply restricted and low vacuum, (2) to activate valves, and (3) to dry the fractions in the fraction collector. This causes a tremendous load on the pumps, resulting in poor vacuum in the cup. The introduction of the cold trap has dramatically improved the vacuum system and increased the life of the pump, which does not require frequent oil changes. Nonetheless, sequencers other than the 890M series are still equipped with an air cylinder valve (Bimba® valve) to apply high vacuum to the cup. This valve is controlled by nitrogen pressure and low vacuum. This configuration causes significant load on the low-vacuum pump. Furthermore, the air cylinder valve has two major drawbacks. It is prone to leakage because of its design, and accumulation of the solvents and reagents causes it to stick. In addition, its operation relies on activation by another solenoid valve on the nitrogen manifold, which we have found to be troublesome. The resulting decrease in the high vacuum causes incomplete drying of the sample in the cup and thus can affect the yields drastically. Bhown et al.[4] have proposed a simple modification for 890C to improve the vacuum system, adaptable to the existing one-or two-pump system with cold trap. The change involves replacement of the Bimba® valve by a stainless steel valve which is operated electrically (Figure 5A), rather than by low vacuum and nitrogen pressure as in the conventional system (Figure 5B). Model 890M is equipped with an electrically operated high vacuum valve which is more efficient than the earlier version.

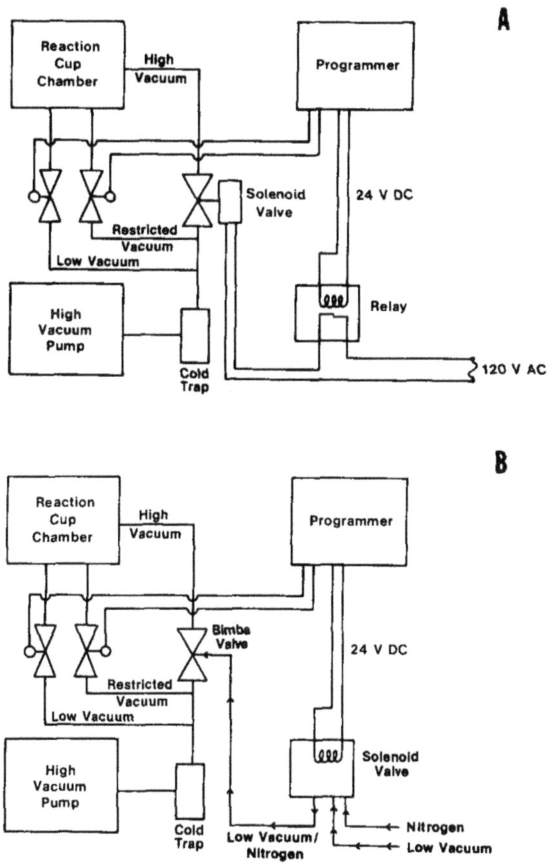

FIGURE 5. Schematic representation of the arrangement of high vacuum valve in the Beckman 890C Sequencer with cold trap: (A) after modification; (B) before modification.

D. Autoconverter

After the automation of isothiocyanate degradation, the next significant step to automate sequence analysis was the introduction of a device for automated conversion of ATZ to PTH derivative by Wittmann-Liebold et al.[36] This can be coupled to the Beckman instrument to achieve unattended conversion after each degradative cycle. The device (Figure 6) consists of a 100-mℓ conical, jacketed glass vessel positioned to the right of the spinning cup. It is maintained at 55°C by water circulated through the outer jacket and has four lines with inlet/outlet openings controlled by a series of valves. Chlorobutane extracts from the cup containing anilinothiazolinone derivatives are flushed through the sides of the flask by an inlet tube controlled by Valve A. The drying step is accomplished by flushing nitrogen over the liquid through the inlet tube controlled by Valve B and bubbling nitrogen into the liquid through Valves E and D. After drying, 20% aqueous TFA containing 0.01% dithioerythritol, stored in R4, is introduced through Inlet B, during which time Valve B is open to the nitrogen inlet and C to waste. This nitrogen is also used to purge the line containing the conversion medium. The conversion is accomplished by heating at 55°C for 20 to 30 min. In order to save time and prevent destruction of PTH derivatives by a longer exposure to acidic conditions, evaporation is initiated during the conversion by passing nitrogen through Valves B and E. Final drying of conversion medium is accomplished by applying vacuum for about 200 sec through Valve C or, alternatively, by extending the evaporation time. The dried

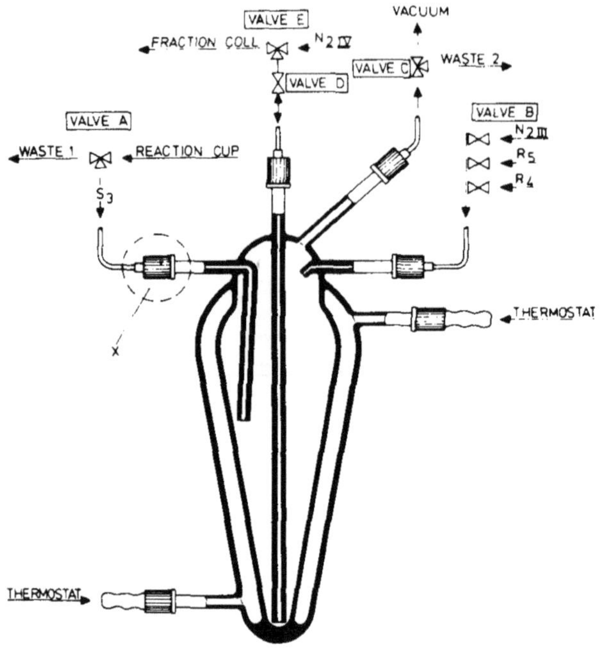

FIGURE 6. Schematic of conversion flask for the Beckman Sequencer. (From Wittmann-Liebold, B., Graffunder, H., and Kohls, H., *Anal. Biochem.*, 75, 621, 1976. With permission.)

PTH derivatives are extracted in a 1,2-dichloroethane-methanol (7:3) mixture contained in R5 and delivered in the form of a spray through Inlet B together with nitrogen. The extraction is completed in two aliquots of 2.5 mℓ or three of 1.5 mℓ each. The dissolved PTH derivative is transferred from the conversion vessel to the fraction collector by opening Valves D and E and employing nitrogen pressure inside the vessel through Valve B.

Sequemat, Inc. (Watertown, Mass.) markets a commercial version, the P-6 autoconverter, that can be attached to both liquid and solid phase sequenators. It uses a "reaction cell" as a conversion vessel and an independent programmer. Figure 7 shows the design of the P-6 autoconverter.

Beckman Instruments, in their modified 890M Sequenator, have introduced an autoconverter which is simpler than the two described above. In this version, the tube, containing the ATZ derivative in chlorobutane scooped from the spinning cup, is raised upward with the help of a small heating block. This seals the test tube at the top and simultaneously starts heating it with slow nitrogen flush. After the tube is dry, conversion fluid from R5 is delivered, and after conversion at 50°C for 20 min, the tube is dried again with nitrogen flush.

E. Ventilation System

Since the Beckman Sequenator Series 890 uses highly flammable and corrosive chemicals, a reliable system to vent all the vapors and other reaction products is warranted. The reaction assembly, all the reagent and solvent reservoirs, and the fraction collector are ventilated to the laboratory exhaust system through solenoid valves. However, vacuum pumps and waste

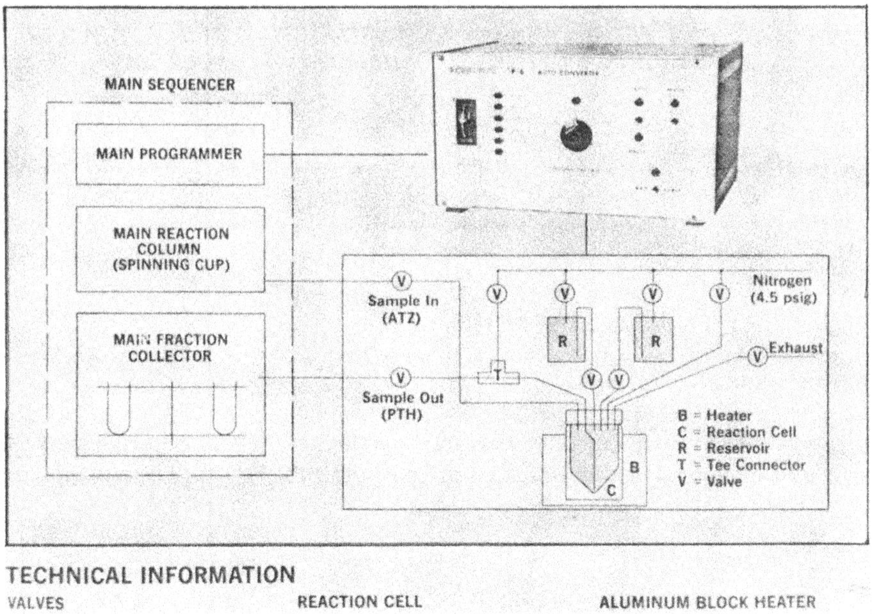

FIGURE 7. Schematic of the Sequemat P-6 Auto Converter. (Courtesy of MilliGen.)

container are always open to vent. The sections housing chemicals, mechanicals, and electronics are ventilated directly from and to the atmosphere through upper and lower fans. The upper fan, which ventilates the electronics, blows the outer air in and, thus, maintains a slight positive pressure in this chamber. On the other hand, the lower fan, which ventilates the chemicals and mechanicals housing, acts as an exhaust and thus maintains a slightly negative pressure. This arrangement prevents the escape of flammable and other vapors and gasses into the chamber which houses all the electrical and electronic components of the sequencer.

F. Controller

All operations of the Beckman Sequenator Model 890 are controlled by a single programmer unit housed in the upper right corner. It consists of a 42-channel programmer that utilizes punched mylar-coated tape on a drum tape reader. Once it is in position, the sequencing proceeds automatically cycle after cycle. The programmer controls all the functions of the sequencer without operator attention. It is also possible to override the programmer by placing any specific function control in the manual mode and by turning it off. This allows instantaneous program changes. A lighted program function panel indicates the function being performed by the sequencer at any particular moment. After sequencing conditions are established, the various steps for each cycle are punched into the tape and the two ends sealed. It is strongly recommended that abrasive paper (Beckman part #339334) be used in order to keep the contacts clean. Before using this, it is desirable that the tape be checked.

In the new version of the Beckman Sequencer 890M, the programmer and drum-type tape reader have been replaced with a microprocessor controller which controls more functions than the tape reader. Since the 890M has been designed to handle micro- and macrosequencing with an autoconversion device, the manufacturer provides built-in programs which can be used for amino acid sequence analysis. The microprocessor has manual controls and a mini-cassette tape reader, which enables the operator to manipulate the degradation cycle at any step for any type of change, such as introduction of double coupling/cleavage at any

Table 1
ARRANGEMENT OF SOLVENTS AND REAGENTS IN THE BECKMAN SEQUENATOR 890C AND 890M SERIES

Model	R_1[a]	S_1	R_2[b]	S_2	R_3	S_3	R_4	S_4	R_5	S_5
890M	1% PITC	Nitrogen	0.1 *M* Quadrol	Ethyl acetate	HFBA	N_2	Wash solvent[c]	Butyl chloride	Conversion fluid	—
890C	1% PITC	Benzene + ethyl acetate (1:1)	0.1—0.5 *M* Quadrol	Ethyl acetate	HFBA	N_2	Empty	Butyl chloride		—

[a] PITC solution is made in heptane.
[b] Quadrol solution is made in propanol/water (3:4) or can be purchased in desired concentration.
[c] Water/isopropyl alcohol (50:50; v/v).

predefined cycle(s), or extend/reduce the coupling/cleavage times. New programs suitable to the needs of the operator may be keyed in and stored on the mini-cassette tape or in the memory of the microprocessor.

V. OPERATION

In the earlier sections we have discussed isothiocyanate degradation and theoretical aspects of its automation with detailed description of the Beckman 890M Sequenator. This section, which will describe the operation of the Beckman 890M Protein/Peptide Sequenator, will be divided into two parts, preparation of the instrument and actual start-up of the sequencer.

A. Preparation

To prepare the Beckman Sequenator for protein/peptide sequence analysis, it is advisable to begin in a certain order. Although there are no set guidelines, we recommend the following sequence:

1. Load the reagent and solvent bottles.
2. Clean and prepare the reaction cup.
3. Load the fraction collector.
4. Prepare and load the sample.
5. Check vacuum.
6. Check cold trap.
7. Check waste bottle.
8. Check oil filter.
9. Check upper and lower fans.
10. Check nitrogen.

1. Loading the Reagent and Solvent Reservoirs

All the reagent and solvent bottles must be washed with Chromerge® and rinsed thoroughly with deionized water, then with acetone, and finally dried in an oven. Cleaning with any kind of detergent is strongly discouraged.

Table 1 gives the contents of the reagent and solvent bottles. Most of the reagents and all solvents are supplied by Beckman Instruments in ready-to-use state. All reagents supplied by Beckman Instruments are sealed under nitrogen in glass ampules. It is very important that one be extra careful when breaking open these ampules, because very small pieces of broken glass can be transferred into the reagent bottles and finally find their way into the solenoid valves resulting in their malfunction. It is therefore recommended that one carefully observe the bottom of the ampule before pouring the liquid. One way of avoiding the broken glass pieces is to aspirate the reagents from the ampules rather than pouring from them. As mentioned above, Beckman Instruments supplies these chemicals in a ready-to-use state; however, if purchased from other sources, reagents of the proper concentration (Table 1) can be easily prepared.

Once the reagents and solvent bottles are properly filled, the next step is to pressurize the bottles, which is accomplished as follows. Locate and close all the vent valves for reagents and solvents on the backside of the bottle cabinet. They are numbered. Open the vent valve and activate it by turning on the corresponding reagent/solvent transfer switch on the main control panel. This allows the excess pressure inside the bottle to be released. On the main control panel, turn the "bottle" switch "nitrogen" to the "on" position. The nitrogen will begin to bubble through the liquid. Continue for 30 sec. Excess time will cause evaporation and change the concentration. Close the vent valve (which was open). Allow 3 to 5 min for equilibration of pressure within the bottle. Check for bottle leak by observing the inlet line. If no bubbles seep out, the vent valve is closed and there is no leak. If the nitrogen bubbles continue to seep, it indicates a leak which may occur for a number of different reasons. The most common is a loose stopper or dirty valve. Correct the problem by tightening the stopper or cleaning the valve. After all leaks are stopped, the nitrogen flow in the reagent/solvent bottles is adjusted by slowly opening the valve so that the vent rate in the flowmeter is at 70 to 75 cc. This procedure should be repeated for all bottles in which the chemicals have been replenished.

Once all the reagent/solvent bottles to which fresh chemicals have been added are purged with nitrogen, the next step is to purge their respective delivery lines. The delivery lines should be purged as follows, in the following order: R_1/S_1; R_2/S_2; R_3/S_3; and S_4.

On the main control panel, turn the cell vacuum (restricted, low, and high) switches and cell drive to "off" and cell nitrogen to "on". After 5 to 10 sec, when the vacuum inside the cup is released, turn the nitrogen cell switch off again. Rotate the scoop counterclockwise until it stops, and then remove the locking screw (red). Raise the delivery head out of the cup. Place a 50-mℓ beaker under the delivery lines to collect the fluid delivered during purging of the lines. To purge the lines turn R1 on, and after about 15 sec turn the nitrogen bottle switch to "on" and the nitrogen cell switch to "on". This will start the delivery of R1; hold until about 1 mℓ has been delivered, then turn off the switches: first nitrogen cell, followed by nitrogen bottle, and then R1. Repeat the procedure for the remaining solvents and reagents in the order mentioned above. Replace the cup cover and lock it by pushing down the red screw until hand tight. Turn the cell drive switch to high and restricted vacuum switch to "on" for 2 min, and then turn them off. This will pull out the solvent left between the delivery valve and the tube end, depositing it in the cup. Turn on the low vacuum for 5 min, followed by high for 5 min, then off. When high vacuum is applied, it is advisable to have restricted and low vacuum on, also. The next step is to clean the cup and prepare the reaction assembly for sequence analysis. A preset program performs all these functions automatically in 890M sequencer.

2. Cleaning and Preparing the Cup

The cup is stopped and cooled by turning off the heater. The cup enclosure is removed after first retracting the scoop and then unlocking the red screw. While cleaning, place paper towels around the cup. Spin the cup at low speed and add 0.75 to 1.0 mℓ of 0.2 *N* NaOH and clean the inside of the cup with a cotton swab. Stop the cup. Aspirate the liquid and repeat. Subsequently, clean the cup in the following order: water, acetic acid, and water. The final wash may be done with acetone and drying with Kimwipes®. An alternative procedure to clean the cup is to wash it twice with a 50% solution of isopropyl alcohol/water at low speed. Clean the bell jar and "O" rings with Kimwipes® sprayed with acetone. To reassemble, coat the "O" rings with a very thin film of silicone vacuum grease and put them in place. Wash the cup cover plug by dipping it into a beaker containing acetone, and wipe it with Kimwipes®. Cover the cup, tighten it with the red screw, and start the chamber heater, set the cell drive to high speed, and then apply low vacuum for 3 min followed by high vacuum for 5 to 10 min. The cup is now ready to start the sequencing run.

Under most circumstances, it is advantageous to use polybrene[42] as a carrier in the cup. This helps in retaining the sample in the cup during wash cycles. In order to make sure that polybrene does not interfere in the sequencer run or the identification procedure, it is recommended that the polybrene be washed by subjecting it to one complete sequencer run, which is accomplished as follows. A solution of polybrene containing 1 mg/100 μℓ of water is prepared and stored with refrigeration. After the temperature of the reaction assembly has reached 56°C (±1°C), the cup is opened (make sure the scoop has been retracted), and 1 to 3 mg of polybrene is added to the contents with a pipette while the cup is spinning. The cup is immediately closed and the Sample Application Subroutine (SAS) program on the 890C is started. After the polybrene is dried, one cycle of the sequencing program is started. However, the 890M Sequenator is programmed to run Programs #1 and #2, one time each. These programs dry and wash the polybrene inside the cup. The cup is then ready to load the sample for sequencing.

3. Loading the Fraction Collector

Once the cup has been cleaned, the next step is to put the tubes in the fraction collector. Each hole in the fraction collector is numbered. It is always good to start the fraction collector from #1. Each tube is numbered and must be inserted into the hole with the corresponding number. This is advantageous if the tube numbers get erased. It is strongly recommended that while taking the tubes out of the fraction collector, the numbers be carefully checked and the tubes arranged in serial order.

4. Sample Preparation

If the sequencer has not been used for some time (a week or so), it is necessary to run a standard protein, e.g., sperm whale myoglobin or insulin, to check the performance and efficiency of the sequencer. The standard or unknown protein sample for automated degradation must be free of nonvolatile salts as well as denaturing agents. Large amounts tend to precipitate the protein and form an impermeable covering on the protein film inside the cup. Oxidizing agents and heavy metals interfere with the chemical reaction of the degradation and, therefore, cannot be tolerated even in trace amounts. Ammonium salts are undesirable, because ammonia reacts with PITC to form phenylthiourea.

The protein to be used as a standard for the calibration run, and the unknown sample (<10 n*M* for microsequencing or >10 n*M* for macrosequencing) may be dissolved or suspended in approximately 500 μℓ or less of a suitable volatile solvent. We prefer to use water or 50% acetic acid. Use of formic acid should be avoided, because the aldehydes present in it may block the free α-amino groups if certain pH and temperature conditions exist. It is not necessary that the protein be completely soluble; a suspension may be transferred into the sequenator cup. However, this should be avoided as much as possible. Samples with low solubility may be dissolved in large volume and added in several portions over a period of time, with application of vacuum between each addition. Very insoluble proteins may be dissolved in HFBA. A list of solvents which could be used to dissolve the samples is given below:

1. Water
2. Dilute acetic acid (20 to 50%)
3. 1 to 2% triethylamine
4. HFBA

5. Dilute pyridine
6. Dilute TFA

Insoluble samples could be added dry into the cup and then dissolved into the coupling buffer (such as quadrol). If this needs to be done, stop the cup before adding the sample and quadrol, and then start the regular sequencing, but omit the quadrol delivery.

5. Vacuum Check

In the original design of the Edman and Begg sequenator, the vacuum steps during the sequencer run were primarily to dry the solvents from the cup. The need for better vacuum was realized only with microsequencing. In order to improve sequencing efficiency at levels of less than 10 n*M*, it is imperative that vacuum should be very good; otherwise reaction products accumulate and appear as background peaks during the identification procedure. It is, therefore, important that the vacuum on the cup should be checked before loading the sample, even the polybrene. A vacuum gauge located on the top right corner of the main panel indicates the vacuum on the cup in millitorr. A reading of 50 millitorr ($\pm$ 15 millitorr) indicates a good vacuum. Values above that suggest (1) a leak into the vacuum system and/or (2) malfunction of vacuum pumps. The first place to look for vacuum leaks is the "O" rings of the drive assembly, followed by the different sets of valves which control the vacuum. Invariably, dirty oil in the vacuum pumps is the main culprit for their malfunction and changing the oil takes care of the problem. Changing oil once in 2 months avoids this problem.

6. Cold Trap Check

A cold trap located between the vacuum pump and the reaction assembly used to be standard on the 890M and can be added onto the 890C without much plumbing. The cold trap acts as a barrier to prevent corrosive vapors from entering the vacuum pump, thus significantly increasing the efficiency and life of the pumps. It is, therefore, necessary that the cold trap be properly maintained. If the sequencer is used constantly, it is advisable to defrost the trap once a week and change the alcohol inside the cold trap once every 2 months. These precautions help maintain the cold trap in good working condition.

7. Waste Bottle Check

Before starting the sequencer run, make sure that the waste bottle is not full.

8. Oil Filter Check

The oil filter located in the back of the sequencer should be regularly checked and replaced with a new one when needed.

9. Upper and Lower Fans

As mentioned earlier, there are two fans located in the upper and lower sections of the sequencer. A failure of the lower fan will shut off the sequencer. This is for safety. The upper fan has its own filter, clogging of which will cause the fan to stop. It is, therefore, suggested that the filter, which can be removed, should be cleaned frequently.

10. Nitrogen Check

In the Beckman Sequencer, the entire Edman degradation is accomplished under nitrogen, and all the valves controlling the delivery of solvents and reagents are activated by nitrogen. It is, therefore, essential that proper supply of nitrogen be maintained. For this purpose, two cylinders containing highly purified nitrogen are attached to the sequencer. One acts as a reserve, while the second (main) one is used to operate the sequencer, which has a safety feature enabling the sequencer to draw nitrogen from the reserve tank when the nitrogen pressure in the main tank drops below 40 psi. It is highly desirable to check the available supply of nitrogen in both tanks before starting the run. If the main nitrogen tank has just enough pressure, it should be replaced with a new cylinder. The leftover nitrogen can be used elsewhere in the laboratory. We also strongly recommend the use of "oxy traps" in-between the nitrogen tanks and sequencer. These "oxy traps" dry the nitrogen and absorb residual oxygen.

B. Start-Up

Once the instrument preparation has been completed, the next step is to perform an actual run of a standard protein or unknown sample. It is advisable that the operator start the run as early in the day as possible to observe the first couple of cycles in their entirety. This will ensure that there are no mechanical problems. The actual start-up is as follows.

Clean the cup as described earlier and add 1 to 3 mg of polybrene dissolved in water. Start the SAS program followed by a complete degradation cycle. The sample is then loaded in the cup and the SAS program started. As mentioned earlier, the SAS program is used in the 890C series, while the 890M has a built-in SAS program and polybrene wash cycles (Programs #1 and #2). If the sample has been dissolved in an acidic solution, it is advantageous to dry the sample completely in the cup before starting the sequencer run, since even trace amounts of acid cause a drop in pH of the coupling buffer. Niall[22] has suggested washing with ethyl acetate to remove the last traces of acid. The instrument is then set to run a definite number of cycles, the program tape inserted, and the run started. Although different programs could be employed for different samples and the sequencing strategy altered for certain amino acids, the salient features in an automated cycle are as follows. The cycle is started by introducing a solution of 1 to 5% PITC in heptane sufficient to rise to the undercut of the cup. Heptane is removed by application of low vacuum and nitrogen flush. This leaves PITC droplets all along the inside of the cup. A mixture of 0.1 to 0.5 *M* quadrol in propanol and water mixture is then introduced at high speed, in sufficient quantity to cover the protein film. The sample will dissolve in the buffer, followed by the PITC, which goes into solution more slowly. The coupling reaction is carried out at 56°C for 15 to 30 min. It is helpful to change the cup speed during coupling. This helps mix the PITC and the protein sample in the buffer. At the end of coupling, the cup is set at low speed so that the protein film settles at a lower level than at the start. After coupling, a brief flush of ethyl acetate (S2) is carried out to clean the quadrol delivery line. The volatile components, such as propanol, heptane, and water, are removed by vacuum. The nonvolatile reagents (quadrol and reaction products such as diphenylthiourea and phenylthiourea) are removed by extraction with benzene (S1) followed by ethyl acetate (S2), or a mixture of ethyl acetate/benzene (50:50) (S1) in the conventional system. In the new Beckman Sequencer 890M series, heptane (S1) is used in place of benzene, while S2 contains ethyl acetate.

Most of the proteins are insoluble in benzene and heptane, and washing with these two solvents does not require delay time for protein precipitation. However, ethyl acetate, being a polar solvent, has a tendency to wash the protein out of the cup. Therefore precipitation delay time of 5 to 10 sec shall be introduced into the program to reduce sample losses. During these wash cycles, the scoop is open to waste. After the wash, the leftover solvents are removed by applying restricted, restricted and low, and all three vacuums together, with nitrogen flush. At this point the cup should look completely dry. However, occasionally, presence of polybrene makes it appear a little wet. This should not be any cause of concern. On the other hand, a glossy or transparent appearance indicates residual quadrol in the cup. Causes of and suggestions to overcome this problem are discussed later in this chapter (see Section X). Anhydrous HFBA is then introduced into the cup at low speed to a level slightly higher than the protein film. It is usually difficult to see the upper limit of the acid layer, since acid vapors rise from the liquid and may appear to dissolve protein. HFBA delivery should always be followed by nitrogen flush of the delivery line to clear the line. The delivery time of HFBA should be adjusted so that the acid level reaches the desired height after nitrogen flush. It is important that the solvent bottle corresponding to HFBA bottle should be empty or contain a small volume of ethyl acetate. The nitrogen inlet tube should be below the level of ethyl acetate, and the delivery line, above the liquid level. This configuration will allow the nitrogen flush through the HFBA delivery line. Addition of HFBA starts the cleavage reaction and is allowed to proceed for 2 to 3 min. The excess acid is removed by applying vacuum starting with restricted, followed by low and high together. Drying time after cleavage is very critical. If completely dried, the cleaved anilinothiazolinone (ATZ) becomes insoluble in chlorobutane. On the other hand, if not dried enough, the protein becomes soluble and is washed out of the cup with chlorobutane. A vacuum time of 1 to 2 min is usually enough. The ATZ derivative is extracted with 3- to 4-min delivery of chlorobutane and collected into the tube in the fraction collector. The cup is dried under vacuum and the sample is ready for the next cycle. In the 890C series, the tube in the fraction collector is dried by blowing nitrogen and simultaneously applying vacuum. The fraction collector is kept at 4°C in these models. The dried ATZ derivative is converted to PTH form as follows: 100 μℓ of 10% TFA in ethyl acetate is added to each tube, flushed with nitrogen, capped, vortexed, and incubated for 10 min at 80°C. After this the reagent may be dried under a stream of nitrogen or carefully freeze dried. The dried PTH derivative can then be subjected to identification (Chapter 12). However, in the 890M series, the fraction collector is at room temperature. It is equipped with a heating block on which the tube containing the chlorobutane extract rests. After the delivery of chlorobutane, the tube is raised by raising the heating block, which seals the tube at the top. The heat, nitrogen flush, and vacuum dries the tube. Ten percent TFA in ethyl acetate (R_5) is delivered into this tube and the ATZ derivative allowed to convert to PTH derivative at 55°C for 20 min. After conversion, the reagents are dried and the fraction collector moves to collect the next cycle. Results of amino acid sequence analysis of peptides with 85 to 90% repetitive yield and proteins with 95 to 98% repetitive yield are shown in Figures 8 and 9, respectively.

VI. ALTERNATIVE OPERATIONAL PROCEDURES

One of the major drawbacks of liquid phase sequencing is repeated washing with polar and nonpolar solvents, which significantly contributes to the sample loss from the cup. It has been reported that most of the losses occur during the butylchloride extraction, less with ethyl acetate, and least with benzene. The losses in the latter solvents can be minimized by reducing the quadrol concentration or changing to a volatile buffer system. Dimethylallylamine (DMAA) and dimethylbutylamine (DMBA)[46,47] are the two commonly used volatile buffers, particularly for peptides. However, since the introduction of polybrene by Tarr et al.,[43] use of dilute (0.1 *M*) quadrol as buffer, and wash with benzene/ethyl acetate (1:1) mixture by Brauer et al.,[44] peptides as small as six to eight residues can be sequenced with 88 to 90% repetitive yield.

Accumulation of background residues due to nonspecific cleavage is the other drawback of isothiocyanate degradation. Thomsen et al.[34] have proposed an additional mechanism involving seryl/threonyl residues in an N → O acyl shift, resulting in a cleavage of the peptide bonds on the amino side of these amino acids. In an effort to avoid such a shift, they[34] have developed a program in which quadrol is delivered first, followed by PITC, which reverses the N → O acyl shift.

In an alternative effort to eliminate the background problem, Bhown et al. have described the use of fluorescamine[45] and *o*-pthaldialdehyde[52] to block the newly generated amino termini at proline residues in the sample. The procedure is as follows.

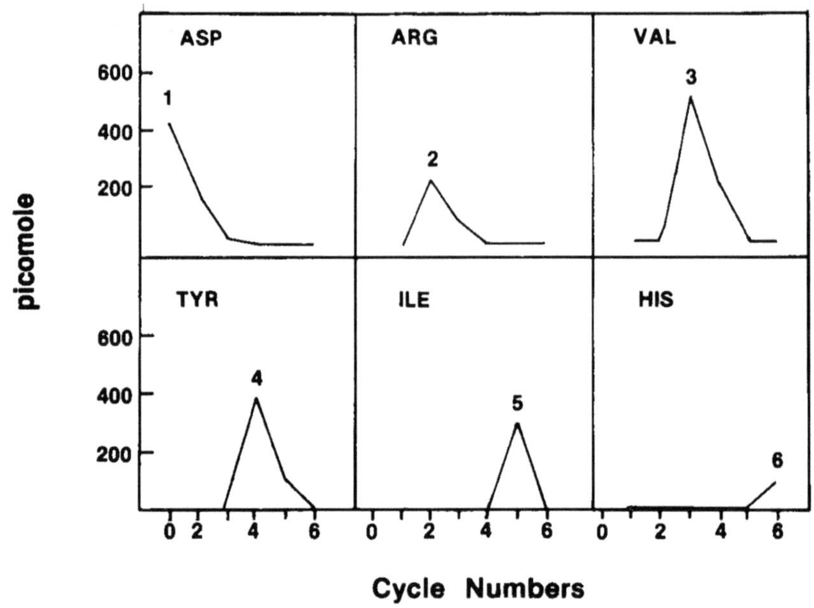

FIGURE 8. Amino acid sequence analysis of angiotensin II (1 n*M* put into the cup).

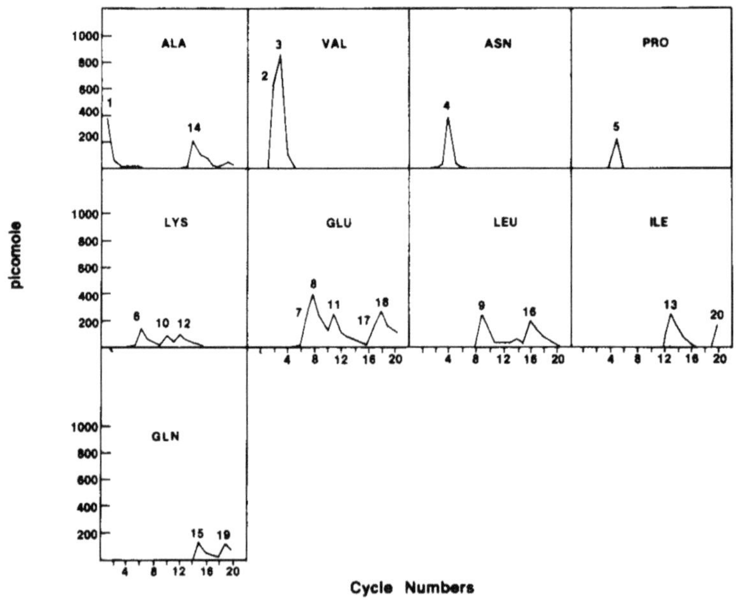

FIGURE 9. Amino acid sequence analysis of an unknown viral protein (1 n*M* put in the cup).

Initially, the sample is subjected to Edman degradation and as many positions of proline ascertained as possible. The sample is sequenced a second time, but the analysis is interrupted at proline residues. At each proline position, the cup is cooled to room temperature and PITC delivery switched off. The regular sequencing program is then started and stopped after the delivery of quadrol. The cup, while spinning, is opened and 100 μℓ of acetone containing fluram (3 to 4 mg/mℓ) added. The cup temperature is raised to the normal coupling temperature (56°C) before the start of the wash cycle, the cycle completed, and the chlorobutane extract discarded. The PITC switch is turned on and the regular cycle started with a double cleavage on the first cycle. In place of fluram, *o*-pthaldialdehyde[52] can be employed.

VII. SEQUENCE ANALYSIS OF SUBSTITUTED PROTEINS

Often protein/peptide molecules have nonproteineous groups such as phosphates, carbohydrates, etc. linked to certain specific amino acid residues. These residues, depending upon the degree of their substitution, cause serious problems during amino acid sequence analysis. Although no definite methods are available to sequence such molecules, there are reports that phosphoprotein has been dephosphorylated and glycoproteins deglycosylated, and amino acid sequence obtained. Butler et al.[49] have reported dephosphorylation of rat dentin phosphoproteins in order to obtain amino acid sequence information. The following procedure was adopted for dephosphorylation.

Forty milligrams of protein is dissolved in 50 mℓ of 0.1 *M* TRIS-HCl buffer, pH 8.0, and incubated at 37°C in a shaking water bath with 50 units of bovine intestinal alkaline phosphotase conjugated to agarose beads (P-0762; Sigma Chemical Co., St. Louis, Mo.). The rate of dephosphorylation is determined by assaying the release of inorganic phosphate as described by Linde et al.[50] The dephosphorylated protein is recovered by removing the conjugated enzyme by centrifugation, dialyzing the supernatant against water for long periods of time, and lyophilizing.

Glassman et al.[51] have suggested deglycosylation by treatment with hydrofluoric acid (HF) or methanolic hydrogen chloride. The procedure reported by these authors is described below.

A. HF Deglycosylation

The apparatus used for HF treatment is purchased from Peninsula Laboratory, Inc. (San Carlos, Calif.). Dried protein (82 mg) and 3 mℓ of anisole scavenger are added to the Kel-F reaction vessel and treated with 10 mℓ of anhydrous liquid HF (previously dried over cobalt trifluoride).

Solvolysis was allowed to proceed for 2 hr at room temperature. HF is completely removed over a calcium oxide trap. It is then solubilized in 44% formic acid containing 2% SDS, dialyzed against 44% formic acid to remove SDS and anisole, and centrifuged to remove insoluble material.

B. Methanolic Hydrogen Chloride Deglycosylation

Forty-four milligrams of protein is dissolved in 1 mℓ of water and aliquoted in ten 50-mℓ ampoules and slowly mixed with constant vortexing at room temperature with 40 mℓ of pretreated methanolic hydrogen chloride (see Glassman et al.[51] to prepare this). The ampules are sealed and heated at 80°C for 17 hr with intermittent vortexing. The sample is dried under nitrogen, dissolved in water, and lyophilized.

Using this procedure, Glassman et al.[51] could obtain amino acid sequence of carcinoembryonic antigen (CEA).

VIII. QUANTITATION

Efficiency of the sequencer may be obtained by calculating the yield of a given PTH residue at two places in the run and then estimating the repetitive yield between these two residues, this being the absolute repetitive yield. Alternatively, the number of PTH residues at each cycle is plotted against the cycle number and a regression line (the best fit line for the points) is drawn, and the repetitive yield calculated between any two points on this.

This is the average repetitive yield, which is lower than the absolute repetitive yield. The formula for calculating the repetitive yield is

$$\%\text{R. Y.} = \left(\frac{B}{A}\right)^{\frac{1}{NB-NA}} \times 100$$

where R.Y. = repetitive yield, B = yield at higher cycle, A = yield at lower cycle, N_B = higher cycle number, and N_A = lower cycle number.

IX. OVERLAP

It has been observed that occasionally PTH amino acid residue in one cycle also appears in the next cycle along with the residue of that cycle. This carryover from one cycle to the next is referred to as *overlap*. Overlap can be caused by any factor which influences Edman degradation. It is cumulative and it should not be greater than 0.5 to 1.0% per cycle at the most. The most frequent cause for overlap is incomplete coupling and/or cleavage due to incorrect adjustment of the relative heights of the coupling buffer and cleavage acid. It is necessary that all the protein should be covered by the buffer and acid during coupling and cleavage. Occasionally, presence of a particular residue may cause serious asynchrony. This results in a significant amount of overlap. One way of overcoming such asynchrony is to perform double cleavage at proline residues. This significantly reduces the overlap. A double cleavage program has been suggested for aspartic and glutamic acid residues.[22]

X. QUADROL RETENTION

Quadrol is a polar base which has a tendency to bind to acidic groups on the protein. When the HFBA is added, the bound quadrol is displaced and extracted with 1-chlorobutane. Hence, a small amount of quadrol always finds its way into the tube in the fraction collector. The amount normally present is no cause of concern, because it does not interfere with the identification procedure. However, if microsequencing is performed when detection is carried out at a high sensitivity, such small amounts of quadrol may cause some concern. On the other hand, failure to remove quadrol adequately by the solvent extractions results in excess amounts of it being accumulated in the test tube. This affects, sometimes severely, the identification of the residues. The quadrol peak elutes at different times depending on the system of identification employed.

Quadrol retention should be suspected if at the end of the wash cycle after coupling there is a visible clear glassy film inside the cup instead of a dry white residue. If not noticeable at this point, then presence of a thick viscous liquid in the fraction collector tube left after drying indicates quadrol retention.

There are several reasons which cause quadrol retention. Most common is inadequate wash after coupling or excess delivery of quadrol. This problem is invariably caused by malfunction of the delivery valves or nitrogen pressure inside the solvent bottle. Infrequently, quadrol retention may appear because of the quadrol interaction with protein. Under these circumstances, an alternate coupling buffer such as DMAA or DMBA should be tried.

Brauer et al.[44] have suggested use of dilute quadrol (0.1 *M*) and combined wash of ethyl acetate to benzene, 50:50, to overcome this problem. Guyer et al.[48] have used boric acid to extract quadrol.

ACKNOWLEDGMENTS

The authors wish to express their thanks to Robert W. Herrick for science writing, to

James Wayland and Lisa Kallman for their expert technical assistance, as well as to Sandra Reid for typing this manuscript. The work was supported by Grants AM-03555 and AM-20614.

REFERENCES

1. **Edman, P. and Begg, G.,** A protein sequenator, *Eur. J. Biochem.*, 1, 80, 1967.
2. **Hunkapiller, M. W. and Hood, L. E.,** New protein sequenator with increased sensitivity, *Science*, 207, 523, 1980.
3. **Bhown, A. S., Mole, J. E., Hunter, F., and Bennett, J. C.,** High sensitivity sequence determination of proteins quantitatively recovered from sodium dodecyl sulfate gels using an improved electrodialysis procedure, *Anal. Biochem.*, 103, 184, 1980.
4. **Bhown, A. S., Cornelius, T. W., Mole, J. E., Lynn, J. D., Tidwell, W. A., and Bennett, J. C.,** A simple modification on the vacuum system of the Beckman Automated Sequencer to improve the efficiency of Edman degradation, *Anal. Biochem.*, 102, 35, 1980.
5. **Weittmann-Liebold, B.,** Amino acid sequence studies on ten ribosomal proteins of *E. coli* with an improved sequenator equipped with an automatic conversion device, *Hoppe-Seyler's Z. Physiol. Chem.*, 354, 1415, 1973.
6. **Frank, G.,** A cheap and simple method to achieve and maintain the necessary purity of reagents and solvents for automated amino acid sequence determination with the sequenator, *Hoppe-Seyler's Z. Physiol. Chem.*, 360, 997, 1979.
7. **Sanger, F.,** The free amino groups of insulin, *Biochem. J.*, 39, 507, 1945.
8. **Edman, P.,** A method for the determination of the amino acid sequence in peptides, *Arch. Biochem. Biophys.*, 22, 475, 1949.
9. **Edman, P.,** Method for determination of the amino acid sequence in peptides, *Acta Chem. Scand.*, 4, 283, 1950.
10. **Edman, P.,** Sequence determination, in *Protein Sequence Determination*, Needleman, J. B., Ed., Springer-Verlag, New York, 1970, 211.
11. **Leguin, R. M. and Niall, H. D.,** The application of a fluorinated isothiocyanate as coupling agent in the Edman degradation, *Biochim. Biophys. Acta*, 257, 76, 1972.
12. **Robertson, B. H., Wolcot, R. M., and Bennett, J. C.,** A preliminary study using electron capture detection of pentafluorophenylthiohydantoin amino acid derivatives, *Proc. Soc. Exp. Biol. Med.*, 155, 287, 1977.
13. **Preriero, A. and Pechere, J. F.,** Cyanomethyldithiobenzoate as a new reagent for the sequential degradation of peptides, *Biochem. Biophys. Res. Commun.*, 40, 549, 1970.
14. **Chang, J. Y.,** High sensitivity sequence analysis of peptides and proteins by 4-NN dimethylaminoazobenzene 4′-isothiocyanate, *Biochem. J.*, 163, 517, 1977.
15. **Doolittle, L. R., Mross, G. A., Fothergill, L. A., and Doolittle, R. F.,** A simple solid phase amino acid sequencer employing a thioacetylation stepwise degradation procedure, *Anal. Biochem.*, 78, 491, 1977.
16. **Datta, S., Datta, S. C., and Sen Gupta, R.,** Para-phenylazophenyl isothiocyanate as a possible reagent in the N-terminal amino acid sequence determination of proteins, *Biochem. Biophys. Res. Commun.*, 72, 1296, 1976.
17. **Chang, J. Y.,** Microsequence analysis of polypeptide using dimethylaminoazobenzene isothiocyanate, in *Methods in Protein and Peptide Sequence Analysis*, Birr, C., Ed., Elsevier/North-Holland Biochemical Press, Amsterdam, 1980, 115.
18. **Bentley, K. W. and Creaser, E. H.,** Qualitative determination of N-terminal amino acids of peptides and proteins with cobalt-(III) chelates, *Biochem. J.*, 135, 507, 1973.
19. **Ivanov, C. P. and Mancheva, I. N.,** A study of the interaction of 2-p-isothiocyanophenyl-3-Phenylindone with peptides and proteins, *Anal. Biochem.*, 53, 420, 1973.
20. **Prager, E., Arnheim, N., Mross, G., and Wilson, A.,** Amino acid sequence studies on bobwhite quail egg white lysozyme, *J. Biol. Chem.*, 247, 2905, 1972.
21. **Niall, H. D.,** Automated methods for sequence analysis, in *The Proteins*, Vol. 3, Neurath, H. and Hill, R. L., Ed., Academic Press, New York, 1977, 180.
22. **Niall, H. D.,** Automated sequence analysis of proteins and peptides, *Agric. Food Chem.*, 19, 638, 1971.
23. **Edman, P.,** Phenylthiohydantoin in protein analysis, *Ann. N.Y. Acad. Sci.*, 88, 602, 1960.
24. **Blomback, B. and Yamashina, I.,** Amino acid sequence analysis of proteins and peptides, *Ark. Kemi*, 12, 299, 1958.

25. **Harris, H. W., Jr. and Lux, S. E.,** Structural characterization of the phosphorylation sites of human erythrocyte spectrin, *J. Biol. Chem.*, 255, 11512, 1980.
26. **Konigsberg, W. H. and Steinman, H. M.,** Strategy and methods of sequence analysis, in *The Proteins*, Vol. 3rd ed., Neurath, H. and Hill, R. L., Eds., Academic Press, New York, 1977, 1.
27. **Wittmann-Liebold, B.,** Amino acid sequence studies on 10 ribosomal proteins of *E. Coli* with an automatic conversion device, *Hoppe-Seyler's Z. Physiol. Chem.*, 354, 1415, 1973.
28. **Hunkapiller, M. W. and Hood, L. E.,** Direct microsequence analysis of polypeptides using an improved sequenator, a nonprotein carrier (polybrene), and high pressure liquid chromatography, *Biochemistry*, 17, 2124, 1978.
29. **Hunkapiller, M. W. and Hood, L. E.,** New protein sequenator with increased sensitivity, *Science*, 207, 523, 1980.
30. **Frank, G.,** A cheap and simple method to achieve and maintain the necessary purity of reagents and solvents for automated amino acid sequence determination with the sequenator, *Hoppe-Seyler's Z. Physiol. Chem.*, 360, 997, 1979.
31. **Bhown, A. S., Mole, J. E., and Bennett, J. C.,** An improved procedure for high sensitivity microsequencing: use of aminoethyl aminopropyl glass beads in the Beckman Sequencer and the Ultrasphere ODS column for PTH amino acid identification, *Anal. Biochem.*, 110, 355, 1981.
32. **Edman, P.,** Sequence determination, in *Protein Sequence Determination*, 2nd ed., Needleman, S. B., Ed., Springer-Verlag, Berlin, 1975, 232.
33. **Niall, H. D.,** Automated Edman degradation: the protein sequenator, in *Methods in Enzymology*, Vol. 27, Hirs, C. H. W., Timeasheff, S. N., Eds., Academic Press, 1973, 942.
34. **Thomsen, J., Bucher, D., Brunfeldt, K., Nexo, E., and Olesen, H.,** An improved procedure for automated Edman Degradation used for determination of the N-terminal amino acid sequence of human transcobalamin I and human intrinsic factor, *Eur. J. Biochem.*, 69, 87, 1976.
35. **Wittmann-Liebold, B.,** Current advances in sequencing as applied to the structure determination of ribosomal proteins, in *Polypeptide Hormones*, Beers, R. F., Jr. and Bessett, E. G., Eds., Raven Press, New York, 1980, 87.
36. **Wittmann-Liebold, B., Graffunder, H., and Kohls, H.,** A device coupled to a modified sequenator for the automated conversion of anilinothiazolinones into PTH amino acids, *Anal. Biochem.*, 75, 621, 1976.
37. **Margolies, M. N., Brauer, A., Oman, C., Klapper, D. G., and Horn, M. J.,** Improved automatic conversion for use with a liquid phase sequenator, in *Methods in Protein Sequence Analysis*, Elzinga, M., Ed., Humana Press, Clifton, N.J., 1982, 189.
38. **Bhown, A. S. and Bennett, J. C.,** unpublished observation.
39. **Bhown, A. S. and Bennett, J. C.,** A modified system for thiazolinone conversion to thiohydantoin derivatives and their separation by high pressure liquid chromatography, *Anal. Biochem.*, 150, 457, 1985.
40. **Lynn, J. D. and Bennett, J. C.,** A noncommercial protein sequencing instrument, *Anal. Biochem.*, 45, 498, 1972.
41. **Waterfield, M. D., Corbett, C., and Haber, E.,** A protein sequencing device of simple construction, *Anal. Biochem.*, 38, 475, 1970.
42. **Knight, E., Jr., Hunkapiller, M. W., Korant, B. D., Hardy, R. W. F., and Hood, L. E.,** Human fibroblast interferon: amino acid analysis and amino terminal amino acid sequence, *Science*, 207, 525, 1980.
43. **Tarr, G. E., Beecher, J. F., Bell, M., and McKean, D. J.,** Polyquarternary amines prevent peptide loss from sequenators, *Anal. Biochem.*, 84, 662, 1978.
44. **Brauer, A. W., Margolies, M. N., and Haber, E.,** The application of 0.1 *M* quadrol to the microsequence of proteins and the sequence of tryptic peptides, *Biochemistry*, 14, 3029, 1975.
45. **Bhown, A. S., Bennett, J. C., Morgan, P. H., and Mole, J. E.,** Use of fluoroescamine as an effective blocking reagent to reduce the background in protein sequence analyses by the Beckman Automated Sequencer, *Anal. Biochem.*, 112, 158, 1981.
46. **Capra, J. D., Wilde, C. E., and Klapper, D. G.,** Comparison of solid phase and spinning cup sequencing methodologies II: the use of polybrene for the automated sequence analysis of small peptides with a forty minute DMAA program in the spinning cup sequencer, in *Solid Phase Methods in Protein Sequence Analysis*, Previero, A., and Coletti Previero, M. A., Eds., North-Holland, Amsterdam, 1977, 69.
47. **Hermodson, M. A., Ericsson, L. H., Titani, K., Neurath, H., and Walsh, K. A.,** Application of sequenator analyses to the study of proteins, *Biochemistry*, 11, 4493, 1972.
48. **Guyer, R. L., Schnute, W. C., Jr., Todd, C. W., and Lovius, R. E.,** The removal of contaminating quadrol by boric acid extraction prior to mass spectral analysis of amino acid thiazolinones, *Anal. Biochem.*, 66, 623, 1975.
49. **Butler, W. T., Bhown, M., DiMuzio, M. T., Cothran, W. C., and Linde, A.,** Multiple forms of rat dentin phosphoproteins, *Arch. Biochem. Biophys.*, 225, 178, 1983.
50. **Linde, A., Bhown, M., and Butler, W. T.,** Noncollagenous proteins of dentin: a re-examination of proteins from rat incisor dentin utilizing techniques to avoid artifacts, *J. Biol. Chem.*, 255, 5931, 1980.

51. **Glassman, J. N. S., Todd, C. W., and Shively, J. E.,** Chemical deglycosylation of carcinoembryonic antigen for amino acid sequence studies, *Biochem. Biophys. Res. Commun.*, 85, 209, 1978.
52. **Bhown, A. S., Cornelius, T. W., Volanakis, J. E., and Bennett, J. C.,** A comparison of fluorescamine and o-phthaldialdehyde as effective blocking reagents in protein sequence analyses by the Beckman Sequencer, *Anal. Biochem.*, 131, 337, 1983.

Chapter 4

SOLID-PHASE METHOD APPROACHES TO PROTEIN MICROSEQUENCE ANALYSIS

J. J. L'Italien

TABLE OF CONTENTS

I. INTRODUCTION

During the past 20 years the direct sequence analysis of polypeptides has undergone many improvements in three principle areas which make up the sequencing process: methods of polypeptide purification, instrumentation for the Edman degradation, and methods of PTH analysis. The advent of automated liquid-phase sequencers, introduced by Edman in 1967, permitted then unparalleled efficiency for the extended sequence analysis of intact proteins and large fragments.[5] The analysis of peptides, however, was somewhat less reliable due to the loss of sequenceable material in the extraction steps of the Edman process. In 1971, an automated solid-phase sequencer was introduced by Laursen to provide a solution to the problem of sample extraction loss experienced with liquid-phase instruments at the time.[13] Solid-phase methods address the problem of sample washout by covalent immobilization of the polypeptide to a solid support, thereby eliminating the problem of removal of excess reagents or recovery of the PTH-amino acid from the remainder of the polypeptide. Once peptides or proteins were immobilized, solid-phase approaches were a successful method for automated sequence analysis of peptides. The advantages of solid-phase Edman degradation are only realized, however, when efficient covalent immobilization of the peptide of interest is achieved. The initial polypeptide immobilization procedures, while efficient, had problems with side-chain carboxyl group interactions with the support which led to termination of the sequence.[14] This problem was circumvented through the development of alternative immobilization methods which took advantage of the chemical properties of the specific amino acid side chains of lysine and homoserine. These methods, the phenylene diisothiocyanate (DITC) coupling procedure for lysine[15] and the lactone coupling procedure for homoserine,[7] were found to be very efficient for peptides, but limited to situations where one of these residues was at or near the carboxyl terminus of the peptide. These limitations precluded the solid-phase approach from becoming a more universal sequencing method, because of the lack of a generalized immobilization method. It was still quite useful as an adjunct to manual and automated liquid-phase methods and by 1978 it had been used in the sequence determination of over 76 polypeptides as referenced in 85 publications.[16] It should be pointed out that at this time, automated liquid phase methods were not efficient for peptide sequencing and usually required the aid of an alternative approach (either manual or solid-phase) to determine peptide sequences. In 1978, with the introduction of polybrene by Tarr et al., a method was available which permitted more efficient sequence analysis of smaller peptides in liquid-phase instruments by reducing extraction losses.[33]

By the late 1970s, HPLC was having a profound impact on sequence analysis in two major areas. The first was in the area of PTH analysis, where HPLC permitted the rapid, quantitative analysis of all PTHs.[2,39] The ability to quantitate the recovery of all PTHs without derivatization, as in the case of GC[30b,30c] or back hydrolysis to the parent amino acid and amino acid analysis,[29,31] generally permitted an increase in the length of readable sequence and a decrease in the amount of material needed for identification. The second area where HPLC had a profound impact was in the area of polypeptide purification. The net effect of HPLC peptide purification on sequencing strategy was to favor approaches where smaller peptides would be generated to take advantage of the efficiency of HPLC.[8,22] This change in strategy emphasized the need for rapid peptide sequencing methods which were HPLC compatible. In 1981, a reinvestigation of generalized C-terminal carboxyl group immobilization resulted in the optimization of a generalized solid-phase immobilization procedure which was efficient and was without side-chain coupling problems.[22] The method was demonstrated to permit the direct, covalent immobilization of HPLC purified peptides less than 30 residues without intermediate handling steps such as removal of nonvolatile buffer salts.[22,25] This method, which permits a general strategy for the immobilization of peptides, and the development of milder conditions for the immobilization of larger fragments and proteins[19,35] have effectively eliminated immobilization as the principle stumbling block for the general application of solid-phase methods to protein microsequence analysis.

With the resolution of the problems of polypeptide immobilization, it brings to the forefront many of the inherent advantages of carrying out chemical reactions on immobilized polypeptides. Several of these advantages are derived from the ability to use a "column-like" reaction vessel to perform the Edman degradation on immobilized polypeptides. This reaction vessel arrangement facilitates the removal of excess coupling reagents from the polypeptide by promoting efficient solvent washes between steps and circumvents the problem of concentrating trace buffer or solvent impurities in the reaction vessel which can impede the sequencing and/or detection of subnanomole samples. It should be noted that the selection of wash solvent is not limited to those solvents in which the peptide/protein is insoluble, as is the case with other automated sequencers, but may include the use of polar solvents which can result in cleaner samples. The use of immobilized samples also enables the cleaved ATZ from each cycle to be removed from the remainder of the polypeptide chain in liquid trifluoroacetic acid. This eliminates the recovery problems associated with the solvent extraction of positively charged or posttranslationally modified amino acid residues that are often encountered with other automated sequencers.[30]

The third modern automated approach for protein sequence analysis, the gas-phase sequencer, was introduced in 1981.[6] This instrument incorporates a flow-through reaction vessel, one of the major advantages of solid-phase instruments, with the miniaturization of all plumbing and value blocks to create the state-of-the-art, commercially available instrument. In current gas-phase instruments, the polypeptide to be sequenced is retained on a polybrene-coated glass membrane in the reaction chamber, thus circumventing the need for covalent attachment of the polypeptide, but limiting the liquids which pass through the reaction chamber to solvents in which the polypeptide sample is insoluble. Thus, the acid and base reagents which are required in the Edman sequencing process must be supplied in a gas-phase so that the polypeptide sample will not be extracted from the reaction vessel.

At present, commercially available gas-phase sequenators using noncovalent methods of retaining the sample in the reaction chamber, and, pending the skill of the operator and quality of the sample, permit routine high-sensitivity sequencing in the less than 10-pmol range.[32] It is interesting to note that when covalently immobilized polypeptides have been sequenced in an unmodified, gas-phase instrument using standard programs, they resulted in cleaner samples with slightly higher repetitive yields than identical peptides sequenced using polybrene (noncovalent) sample retention.[32] Thus, it is clear that the miniaturization and design of the gas-phase instrument are the major contributing factors to routine low picomole sequencing capability. The advantages of higher repetitive yield, cleaner samples, and the ability to use liquid acid and base reagents for decreasing cycle time and improving the recovery of unusual amino acids will likely establish a trend towards using gas-phase instruments for solid-phase sequencing.

There are several recent reviews of solid-phase sequence analysis emphasizing microsequencing[19] and extended sequence runs.[26] Earlier reviews cover much of the background, chemistry, and earlier approaches to immobilization strategy.[12,16] This review will cover selected topics in each of the major areas which comprise sequence analysis: sample preparation, solid-phase Edman degradation, and PTH analysis.

II. MICROPURIFICATION METHODS FOR SOLID-PHASE MICROSEQUENCE ANALYSIS

Polypeptide purification is now considered to be an integral part of protein microstructural analysis. The methods for approaching complete protein purification have been completely revolutionized in recent years with the development of such modern microanalytical purification techniques as bioaffinity chromatography, electrophoresis, and HPLC. The relevance of some aspects of these methods to solid-phase microsequence analysis has been described in a recent review.[19]

III. MICROSEQUENCING BY SOLID-PHASE METHODS

A. Microscale Methods of Polypeptide Immobilization

At the present time, there are three preferred methods of polypeptide immobilization for solid-phase microsequence analysis. The important criteria in the selection of which these methods to employ for a particular application depend on the size of the polypeptide and its method of purification. Peptides less than 5000 daltons (generally all peptides resulting from complete proteolytic digestion of a protein or large polypeptide fragment) purified by HPLC are immobilized to polystyrene or glass supports through activation of their C-terminal carboxyl group. Polypeptides greater than 5000 daltons, including intact proteins and/or their large fragments resulting from limited proteolysis or chemical cleavage, which have been purified by HPLC, are immobilized in limited side-chain attachment through lysine or the side-chain carboxyl groups of aspartic and glutamic acid. For polypeptides whose preferred mode of purification is gel electrophoresis, the preferred method of immobilization is direct electrotransfer to DITC-glass. The selection of polypeptide immobilization procedures based on these criteria has proven to be very successful.

1. Immobilization of Peptides Less Than 5000 Daltons

The preferred method for all peptides less than 5000 daltons is activation of the C-terminal carboxyl group by water-soluble carbodiimide for covalent attachment to aminopolystyrene or arylamine glass. The major advantage of this attachment method which is applicable to all peptides (except those without a free C-terminal carboxy group) is that it effectively permits selective immobilization through the C-terminal carboxyl group without complications with the other amino acid side chains. This method also permits the direct immobilization of HPLC-purified peptides in both volatile and salt-containing buffer systems without the need for desalting or buffer exchange. The applicability of the water-soluble carbodiimide method for direct immobilization of HPLC-purified peptides was demonstrated in 1981.[22] In a subsequent study,[25] quantitative coupling data for 38 peptides resulting from HPLC purification of peptides resulting from tryptic or *Staphylococcus aureus* proteolytic digests were presented. From this report, the average immobilization yields for 25 peptides which varied in length from 4 to 14 residues were 57% of starting material (range was 30 to 87%), while 13 peptides with 15 or more residues were coupled in an average of 41% yield (range 23 to 70%). As was the case in the 1981 paper, the 1982 paper reported that in every instance in which this attachment method was used, the peptide was immobilized and the attachment was highly selective for the C-terminal carboxyl group, as evidenced by the fact that both aspartic and glutamic acid residues were readily detectable and there was no detectable loss of peptide in the sequenator cycle following either of these residues, as would be expected if they were a primary attachment site.

The applicability of this combined approach of HPLC peptide purification and solid-phase microsequence analysis following direct immobilization of the peptides to aminopolystyrene via carbodiimide activation was also demonstrated at the subnanomole level.[25] This study was performed on the tryptic peptides which resulted from a 1-nmol complete tryptic digestion of a 12,000-molecular weight cyanogen bromide fragment of Elongation Factor Tu. The five peptides from 11 to 19 residues long which resulted were collected following the separation of a 1-nmol tryptic digest by reverse-phase HPLC. These peptides were recovered, as seen in Table 1, in yields of 90, 90, 90, 50, and 25%, respectively. The low yields of the last two peptides in the table resulted, in large part, because the peaks were shaved to

Table 1
MICROSEQUENCING OF HPLC-PURIFIED TRYPTIC PEPTIDES

Peptide	Recovered from HPLC[a] (pmol)	Coupling yield[b] (% pmol)	Sequenceable yield[c] (% pmol)	No. of residue identified	Yield (%)
ALEGDAEWEAK	900	55(500)	60(300)	11/11	91 ± 1
VGEEVEIVGIK	900	50(450)	50(270)	11/11	91 ± 1
ELSVYDFPGDDTPIVR	900	61(550)	64(350)	16/16	91 ± 1
ILELAGFLDSYIPEPER	500	60(350)	57(200)	16/17	91 ± 1
AIDKPFLLPIEDVFSISGR	250	60[d]	63(95)	14/19	91 ± 1

[a] Average of two sample injections of 1 nmol each onto reverse-phase column. Recovery determined by amino acid analysis.
[b] Average of two samples determined by acid hydrolysis of acid-washed resins followed by amino acid analysis. The number in parentheses is the actual picomole amount.
[c] Sequenceable yield is based on area measurements of PTH-amino acid residues following automated Edman degradation and is a measure of the amount of immobilized peptide capable of being sequenced.
[d] Determined from a duplicate experiment. In this experiment the amount of peptide coupled was below the limit of detection of the amino acid analyzer.

From L'Italien, J. J. and Strickler, J. E., *Anal. Biochem.*, 127, 198, 1982.

minimize cross contamination of these peaks which eluted as a doublet. Each of these five peptides was coupled to amino polystyrene in 50 to 60% yield by the carbodiimide method. When sequenced the average recovery of PTH in Cycle 1 for each of these peptides was 60% of the amount of immobilized peptide. Thus, sequence analysis was carried out on approximately 30 to 36% of the amount of each peptide which was recovered from the HPLC. These immobilization/sequencing yields were sufficient to generate complete sequence on the first three peptides in the table, which had first-cycle PTH recoveries of 270 to 350 pmol. The fourth peptide was sequenced 16 of 17 residues starting with 200 pmol of PTH in Cycle 1, while the fifth peptide was sequenced through 14 of 19 residues with recovery of 95 pmol in Cycle 1. All of these peptides were sequenced on an unmodified, commercially available solid-phase sequencer (Mini-15, Genetic Design). The instrument, as used here, did not have the advantages of a miniaturization of plumbing, which permits more efficient delivery of reagents and promotes more efficient recovery of anilinothiazolinones, thereby increasing the instrument sensitivity.

More recently, aryl glass supports have been studied and solid-phase sequence analysis has been performed in both a more efficiently plumbed Mini-15 sequencer and in a commercially available gas-phase sequencer. As seen in Table 2, the results obtained with the aryl glass support parallel those described in Table 1. On average we are able to recover 30 to 40% of the amount of starting peptide (before coupling) based upon PTH analysis of the first sequencer cycle. These recovery values were quite similar to our previous observations using polystyrene supports and working in the 250- to 900-pmol range. The advantages of aryl glass manifest themselves in the ability to incorporate several solvent and instrument changes as described below.

Contrary to previous reports,[16] arylamine supports may be the optimal support for the covalent immobilization of peptides by carboxyl activation. The rationale for this statement is that at pH 5, where these coupling reactions are performed, the arylamine becomes a preferred nucleophile because of its pKa ≅ 5 relative to other amines (pKa > 8) in the reaction mixture. The deprotonated arylamine preferentially reacts with the activated carboxyl groups to form a stable amide bond. Undesired condensation reactions are unfavored without the need for covalent protection of α and ε amino groups due to the pH of the reaction

Table 2
WATER-SOLUBLE CARBODIIMIDE IMMOBILIZATION OF PEPTIDES TO ARYLAMINE GLASS

	Initial amount of peptide (pmol)	PTH recovery[a] at cycle 1 (pmol)	No. of residues identified
ALEGDAEWEAK	500	200	11/11
ALEGDAEWEAK	100	32	11/11
VGEEVEIVGIK	100	30	11/11
AFPAMSLSGLFANAVLR	100	27	14/17
DLEEGILALMR	100	35	10/11

[a] PTH recovery is based on area measurements of PTH-amino acid residues following automated Edman degradation. These values represent a cumulative value for the coupling yield (percentage of initial amount of peptide immobilized) and the sequenceable yield (percentage of immobilized peptide which is capable of being sequenced).

mixture and the pK of the peptide amino groups. The apparent selectivity of attachment for the C-terminal carboxyl group is probably dependent upon both the peptide conformation, which is determined by the reaction buffer and conditions, and by the state of deprotonation of the peptide carboxyl groups. The coupling buffer used, 1 *M* pyridine chloride (pH 5.0)-dimethylformamide (1:1 by volume) has demonstrated excellent solubility characteristics for peptides. In addition, when aminopolystyrene is used, this buffer permits maximal exposure of resin amino groups by swelling the support in dimethylformamide.

2. Immobilization of Proteins and Large Peptides

DITC glass is both the preferred support and attachment method for the immobilization of all polypeptides greater than 5000 daltons. One advantage of this method is that polypeptide immobilization may take place under a wide range of solvent conditions, from aqueous to completely organic. Solvent selection is determined by the ability to maintain the solubility of the polypeptide in the presence of base (such as triethylamine) to facilitate the coupling process. Aqueous coupling in the presence of SDS or GnCl, to maintain polypeptide solubility, has been the most commonly used immobilization approach.[19,26] A wide range of organic solvent conditions have been successfully used for proteins and fragments which are not soluble in detergent containing aqueous solutions. The list of successful combinations includes chloroform-methanol (3:1) for protolipid fragments[17] and trifluoroethanol for hydrophobic membrane proteins and their fragments.[28] As stated in the reports of these non-aqueous coupling methods, the hydrophobic nature of these proteins or fragments precluded their being sequenced by liquid phase approaches where the polypeptides were routinely extracted from the reaction chamber. Because the problem with obtaining sequence on these samples occurred at the PTH extraction step, it presumably would adversely affect gas-phase sequencing as well.

For proteins and large fragments, the selection of SDS or GnCl in the coupling buffer has a significant effect on the degree of immobilization at each potential attachment site. As seen in Table 3, the observation has been made that the addition of GnCl in the coupling buffer promotes complete immobilization at each lysine residue,[25] while immobilization in the presence of 5% SDS promotes immobilization in comparable yields to GnCl, but exhibits a "random" use at the potential immobilization sites (the α and ϵ amino groups of the polypeptides).[19,35] Thus, the use of the SDS method for immobilization of polypeptides permits efficient coupling yields while permitting the identification of the amino terminal

Table 3
IMMOBILIZATION OF POLYPEPTIDES TO DITC GLASS

Polypeptide	Method[a]	Initial amount of sample (pmol)	PTH recovery[b] (pmol)	Identification of N-terminal amino acid
Recombinant bovine growth hormone	GnCl to DITC	100	80 (2)	No
	SDS to DITC	100	82 (1)	Yes
Lysozyme	GnCl to DITC	100	60 (2)	No
	SDS to DITC	100	58 (2)	No
Sperm whale apomyoglobin	GnCl to DITC	100	75 (2)	No
	SDS to DITC	100	75 (1)	Yes
β-Galactosidase	GnCl to DITC	100	60 (2)	No
	SDS to DITC	100	65 (1)	Yes

[a] Refers to the immobilization procedure used as described in the text.
[b] PTH recovery is based on area measurements of PTH-amino acid residues following Edman degradation: (1) indicates recovery at first cycle; (2) indicates that the first cycle did not yield an identifiable PTH residue and the recovery for these samples was based on the second cycle.

residue as well as internal lysine residues.[19] This coupling procedure is not adversely affected by the presence of residual SDS or Coomassie stain from gel-eluted samples,[9] but ion-pair extraction[11] generally improves the interpretation of PTHs in the first few sequencer cycles. As will be described below, however, the immobilization of polypeptides following electrophoretic purification can be facilitated by direct electrotransfer to a DITC glass fiber filter, thereby eliminating the need for electroelution.

An alternative coupling method for the immobilization of proteins and larger fragments is carbodiimide activation of side-chain carboxyl groups. This can be accomplished by using water-soluble carbodiimide and slightly harsher conditions than those used to get a random side-chain immobilization. The use of carbodiimide activation to aminoaryl glass in the presence of SDS generally permits the identification of every residue in the sequence with good overall coupling yields. If gaps appear in the sequence, with either the DITC glass or carbodiimide-to-glass procedure, these two methods may each be used in separate experiments to generate overlapping information on the attachment sites.

3. *Electrotransfer to DITC-Activated Glass Fiber*

At present, the method of choice for the recovery of polypeptides from electrophoresis for sequence analysis is direct electrotransfer from the polyacrylamide matrix used in electrophoresis to a glass fiber filter.[1] This published method for direct electrotransfer describes both low and high pH techniques for noncovalent attachment of electrophoretically purified polypeptides to glass fiber filter paper. An alternative to these noncovalent attachment methods is the use of DITC glass fiber filter for direct covalent electrotransfer. A method for covalent electrotransfer has been recently developed[21] which promises to become a powerful addition to the covalent immobilization methods available for solid-phase sequence analysis. The described system makes use of the phosphate gel system[36] on gradient pore gels for separation of polypeptides which have been lightly prestained by reaction with FITC prior to electrophoresis. The phosphate system circumvents the presence of primary amines in the gel/gel buffers which would compete with the primary amino groups on the protein for the reactive DITC sites on the glass filter. The use of gradient gels permits sharper resolution of the polypeptides in this gel system than normally available, because there is no isotachaphoretic stacking with this gel system. Staining with FITC enables direct visualization of the proteins during electrophoresis, as well as a visual check on the completeness

Table 4
ELECTROBLOTTING OF PROTEINS TO DITC GLASS

Protein	Mol wt (kdalton)	PTH recovery Cycle 1 (pmol)	No. of residues determined
Candida krusei cytochrome *c*	12.6	17	19/20
Bovin growth hormone	21	19	30/33
Bovine rhinotracheitis virus gP-16	54	22	12/12
Bovine rhinotracheitis virus gP-11	70	19	19/20
Bovine rhinotracheitis virus gP-11A	72	30	26/30
E. coli β-galactosidase	116	15	20/20
Bovine rhinotracheitis virus gP-6	124	25	15/15

of electrotransfer and location of the desired areas of the filter for sequence analysis. Transfer by this method has proven to be fast and efficient. Some of the proteins purified, immobilized, and sequenced by this method are given in Table 4.

4. Immobilization of Peptides Using Carbodiimide Activation

All peptides less than 30 residues were coupled to aminopolystrene or aminoaryl glass (Genetic Design) by carbodiimide activation of C-terminal carboxyl groups. For immobilization of 100 pmol to 2 nmol of peptide to aminopolystyrene, 500 μℓ of 1.0 *M* pyridine chloride (pH 5.0) (pyridine redistilled from phthalic anhydride) was added to 3 mg of aminopolystyrene (Sequemat) and the resin was stirred at 30°C for 10 min or until a color transition of the resin (from brown to violet) was observed. The resin was then washed two times with 1 mℓ of water and two times with 200 μℓ of dimethylformamide (redistilled from P_2O_5). The lyophilized peptide was dissolved in 80 μℓ of 1.0 *M* pyridine chloride (pH 5.0) and transferred to the swollen resin. The original sample tube was rinsed with 20 μℓ of 1.0 *M* pyridine chloride followed by 50 μℓ of dimethylformamide (DMF) and both rinses were added to the peptide resin mixture. A fresh solution of 1-ethyl-3,3′-dimethylamino-propylcarbodiimide · HCl (Sigma) was prepared such that 1 mg of carbodiimide was in 25 μℓ of water: DMF (1:4) was added to each tube. The reaction was allowed to proceed at 30°C for 1 hr with stirring, after which the sample was centrifuged and the supernatant removed. The pH of the resin was raised to approximately 9 by adding 50 μℓ of 1.0 *M N*-methylmorpholine (Pierce) and 50 μℓ of DMF, and allowing the resin to stir at 30°C for 10 min. The sample was then centrifuged and the supernatant was removed. Excess reagents were removed by washing the support with 1 mℓ of methanol (Baker, reagent) three or four times. The last traces of methanol were removed by drying under vacuum (water aspirator). We usually perform six coupling reactions at a time and store the coupled peptide at −20°C until sequence analysis. The blocking of excess amino groups is performed on the sequencer by doing a double PITC coupling on the first sequencer cycle.

5. Immobilization of Large Polypeptides to DITC Glass

All polypeptides (proteins or their fragments) greater than 50 residues were immobilized to DITC glass which was either prepared as described by Machleidt and Wachter[28] or purchased from Genetic Design. Following lyophilization (HPLC samples are not dried completely), the samples were immobilized to DITC (diphenylisothiocyanate) glass by dissolving the protein in 50 μℓ of 5% SDS, 25% ethanol, and 5% triethylamine, prior to adding 10 mg of DITC glass and allowing the reaction to proceed at 50°C for 1 hr. The support was then washed with Buffer A (0.1% SDS, 25% ethanol, 5% triethylamine) prior to blocking the excess reactive sites. The blocking reaction was performed by suspending the glass in 40 μℓ of Buffer A, adding 10 μℓ of 100 m*M* β-alanine (Sigma), and placing the mixture at 50°C for 45 min. The support was washed three times with Buffer A, then with water and methanol, and dried. If more complete immobilization is desired at each attachment site, the polypeptide was solubilized in 20 μℓ of 6 to 8 *M* GnCl (Pierce) before being diluted to 3 to 4 *M* GnCl by the addition of 20 μℓ of water. The DITC glass (10 mg) was then added to the sample followed by 20 μℓ of 4 *M* GnCl which was 5% in triethylamine (redistilled). The reaction was allowed to proceed at 50°C with occasional shaking. After 1 hr, the sample was centrifuged, the supernatant removed, and the support washed once with 50 μℓ of 4 *M* GnCl · triethylamine. Excess DITC groups were then blocked by adding 40 μℓ of 4 *M* GnCl · triethylamine and 10 μℓ of 10 m*M* β-alanine (Sigma). The blocking reaction was carried out at 50°C for 45 min. The support was then washed two times with 4 *M* GnCl, three times with water, four times with methanol, and dried under vacuum as above.

6. Immobilization by Direct Electrotransfer to DITC Glass Fiber Filter

The electroblot procedure will be described in detail in a manuscript that is in preparation.[21]

B. Reagents

The selection of solvents and reagents used in solid-phase sequencers depends upon several factors, including instrument design and choice of solid support used to immobilize the polypeptide. In general, most of the chemicals used in solid-phase instruments are reagent grade. This stems from the ability to use scavengers for potential amine-reactive impurities in sequencer solvents and the ability to scavenge potential oxidizing impurities by the large excess of phenylthiocarbamyl groups on the support relative to the amount of phenylthiocarbamyl peptide. In addition, the inherent design advantages of performing the Edman degradation on an immobilized polypeptide eliminates the need for all solvent extraction steps (to remove excess reagents and the ATZ). The design also eliminates the drying steps necessary to remove excess solvent from the reaction vessel which also concentrates less volatile solvent impurities in the reaction vessel of liquid-phase sequencing instruments. Using reagent-grade chemicals and appropriate scavengers, several investigators have reported stepwise yields[27] of greater than 96% and extended sequences of greater than 70 amino acid residues.[10]

In virtually all solid-phase instruments, the cleavage reagent used is trifluoroacetic acid. The advantages of trifluoroacetic acid are that it is more volatile, less noxious, and less expensive than other available perfluorinated acids. An undesirable side effect of trifluoroacetic acid is *O*-trifluoroacetylation of serine. This problem can be effectively eliminated by including a methanol wash following trifluoroacetic acid cleavage[16] or by the use of methanol as a coupling step solvent.[26]

The current Edman reagent of choice for solid-phase sequence analysis is phenylisothiocyanate, which can be delivered either neat to a mixing tee prior to the reaction column[26] or by dilution into acetonitrile to a final concentration of 5 to 10%.[16] The use of alternative sequencing reagents, such as 4-N,*N*-dimethylaminoazo-benzene-4′isothiocyanate (DABITC), presents no real advantage over PITC, because of slower reactivity and instability of the compound which necessitates that each DABTC coupling step be chased with PITC.[30a] The instability and reactivity problems negate any advantage in detection which might otherwise be obtained by the fact that the extinction coefficient of DABITC is approximately twice that of PITC. Furthermore, the observed detection limits[3] for DABTHs are about the same as those of PTHs, using HPLC for residue identification.[25] It is clear, however, that the most direct approach to improving the sensitivity of automated protein microsequencing with currently available instrumentation is to enhance the sensitivity of PTH identification for direct analysis by HPLC. Towards this end we have introduced a new class of isothiocyanates which overcome many of the problems previously encountered with DABITC.

These new phenylisothiocyanates are substituted in the 4-position of the phenyl ring with a protected amine and have the general structure tert-Boc-N H-$(CH_2)_n$-PITC.[24] The advantages of these novel reagents are that they eliminate many of the problems such as compound instability, poor solubility in solvents compatible with automated sequencers, and decreased reaction rates, by not introducing the additional moieties necessary to increase detectability until the postsequence identification step. The stability and reactivity of these compounds is similar to PITC and efforts are underway to optimize the reagent and conditions in automated sequencers.

Unlike the cleavage acid and Edman reagent, the selection of wash solvent and coupling buffer can depend upon the type of solid support used for covalent immobilization. When aminopolystyrene is used the wash solvents and coupling buffer of choice are methanol, dichloroethane, and 1 *M* *N*-methylmorpholinium trifluoroacetate (pH 8.1)-pyridine (2:3, by volume), respectively.[13] The use of two wash solvents, which alternatively swell (dichloroethane) and shrink (methanol) the resin, promotes efficient washing of the polystyrene support. The aqueous coupling buffer, 1 *M* *N*-methylmorpholinium trifluoroacetate (pH 8.1)-pyridine (2:3, by volume), used for polystyrene supports contains pyridine, which also caused the support to swell, permitting maximum exposure and reactivity of the immobilized peptide. The use of glass supports for covalent immobilization eliminates the solvent restrictions imposed by aminopolystyrene, which requires swelling solvents for efficient coupling and washing. Thus, solvents like pyridine, which are strongly UV absorbing and may interfere with PTH analysis, can be eliminated by using rigid glass supports. Methanol remains as a recommended solvent for any type of solid support, because of its favorable solvent properties and its ability to reduce problems with *O*-trifluoroacetylation of serine. The use of a second wash solvent such as dichloroethane or benzene, in addition to methanol, has been observed to reduce non-PTH background.[34] The need for a second, inert solvent may be circumvented, however, by the use of an inert gas.[26] The importance of the inert solvent or gas is to remove all traces of methanol from the support prior to the acid cleavage step to reduce methanolysis (hydrolysis of internal peptide bonds during the acid cleavage step of the Edman degradation[16]). Both aqueous and nonaqueous coupling buffers can be used with rigid glass supports.

The preferred chemicals and solvents are as follows. The wash solvents for all solid supports are methanol (Baker reagent grade stored over aminopropyl glass) and dichloroethane (Baker, reagent grade). The dichloroethane is passed over alumina and checked for acidity by extracting an aliquot with an equal volume of water and measuring the pH. If the pH of the sample is more acidic than a water blank, the material is not used. The coupling and cleavage reagents are also the same for all solid supports. Phenylisothiocyanate (Pierce, 5-mℓ ampules sequence grade) is used as a 5% solution in acetonitrile (Baker, HPLC grade) and trifluoroacetic acid is purchased from Genetic Design or redistilled. The aqueous coupling buffer, 1 *M* *N*-methyl morpholinium trifluoroacetate (pH 8.1)-pyridine (2:3, by volume) is used for sequencing aminopolystyrene supports and 2% triethylamine, 0.5% *N*-propylamine in methanol is used as a coupling medium for glass-immobilized polypeptides. For automated conversion of thiazolinones to PTH amino acids, the conversion reagent used is 1 *M* methanolic HCl prepared from acetylchloride (Genetic Design) and the wash solvent is methanol (Baker, reagent grade) which contains 2 mg of dithiothreitol (Calbiochem) per milliliter. Recoveries of PTH-Ser are higher when the conversion reagent in the instrument is changed every other day (keeping only 20 mℓ in the reservoir) while storing the remainder at −20°C.

C. Instrumentation

1. Hardware

Much of the solid-phase microsequencing work which has been published was performed on solid-phase sequencers which were not designed for subnanomole sequence analysis. Thus, the fact that routine sequence analysis has been performed in the 200- to 900-pmol range of immobilized polypeptide[25] highlights the potential of solid-phase methods for microsequence analysis. The sequenceable yield for immobilized polypeptides has generally been observed as 60% of the immobilized starting material. Therefore, on average, one would recover and identify 120 pmol of the first amino acid in the first sequencer cycle and proceed until the signal fell below background or below the limit of detection. The sequence analysis of <100 pmol of immobilized peptide on unmodified, commercially available solid-phase instruments at the present time is not routine, but may be performed if extra care is taken in the collection and analysis of the PTH-AA. As illustrated in Table 1, satisfactory results have been obtained with samples as small as 45 pmol by direct UV detection without carrier proteins. Other reports have claimed results at the 1- to 2-pmol level using intrinsically labeled proteins with carrier proteins at the nanomole level with results based on recovery of radioactivity.[7] Several changes to facilitate microsequencing can be made in commercially available solid-phase instruments. These changes include replumbing of the instrument to eliminate dead volume wherever possible, placing all Teflon® tubing inside of oxygen-impermeable tubing, and changing the fraction collector. In addition, older instruments can be retrofitted with a new microprocessor (Sequetronic M215, Genetic Design) which gives programming flexibility within a sequencer run at the discretion of the operator. Mini sequencing columns which permit the use of less reagent reducing time and background can also be used.[7a] The sum of these changes has improved the ability to sequence immobilized samples in the <100-pmol range, as seen in Table 2. These changes have also permitted an extension of the lower limit of sequence analysis to the 30-pmol range prior to covalent immobilization (recovery in Cycle 1 of 20 pmol).

The sequencer programs used in the experiments are similar to those previously published,[23] but have been modified for the first sequencer cycle to permit a double PITC coupling. This has been found to be more convenient than prereaction of the polypeptide with PITC prior to putting it in the sequencer to insure that all excess attachment sites on the solid support were blocked. In addition, it may permit higher sequenceable yields of peptide compared to the previous method. Most automated sequencing instruments are currently using programs with cycle times of approximately 1 hr. Using solid-phase approaches, however, it is possible to shorten the cycle time dramatically. Walker et al.[35] have reported a 34-min per cycle program which they have successfully used for sequencing proteins attached to DITC glass. The shorter cycle time was accomplished by shortening both the TFA cleavage time and decreasing the number and length of wash steps between the coupling and cleavage steps. This accelerated program is made possible because of the advantages of sequencing immobilized polypeptides which permits the efficient removal of excess coupling reagents and recovery of the ATZ in TFA, eliminating the need for solvent extraction. The utility of immobilized polypeptides for low-level sequence analysis has also been demonstrated in commercially available gas-phase sequencers.[32] These authors had some difficulty with the efficiency of immobilization at low levels (range from 10 to 100 pmol). They concluded that "In general, sequencer runs on solid-phase samples gave 'cleaner' analysis and slightly higher repetitive yields (1—2%) . . . With improved coupling efficiencies, solid-phase techniques would provide an alternative to immobilization of peptides in Polybrene films for low-picomole level gas-phase sequencing". These authors were able to sequence several immobilized peptides to completion at the 10-pmol level (Recovery Step 1). More recently, samples which have been electroblotted to DITC glass fiber have been sequenced at the 10- to 30-pmol level (Recovery Step 1), Table 4. These proteins and large fragments have been sequenced in the gas-phase instrument with standard programs (02NVAC) or solvent modifications. These results illustrate that routine sequence analysis is possible in the 10-pmol range with immobilized polypeptides and indicate that more mechanically efficient solid-phase sequencers should be able to duplicate these results.

2. PTH Analysis

Each cycle from the solid-phase sequenator (after conversion) was dried twice under N_2 or dried in a speed-vac (Savant) under vacuum with heat to 50°C. The resulting PTH-amino acids were identified by either gradient or isocratic reverse-phase HPLC. For the gradient system, each sample was dissolved in 50 μℓ of methanol and applied to an Altex Ultrasphere® RP-18 column (5-μm particle size) on a Waters liquid chromatography system consisting of two Model 6000A pumps, a Model 7101B WISP, a Model 720 system controller, a Model 730 data module, a Model 440 detector, and a Gilson Holochrome® detector. The PTHs were eluted at room temperature with a series of linear gradients of acetonitrile (Solvent B) into 100 m*M* sodium acetate (pH 4.1) (Solvent A): 0 to 10 min (27 to 45% B), 10 to 15 min (45 to 50% B), 15 to 20 min (isocratic at 50% B), 20 to 25 min (50 to 60% B), 25 to 27 min (isocratic at 60% B), 27 to 33 min (60 to 27% B). Both PTH-Asp and PTH-Glu were detected as methyl esters resulting from conversion to PTHs in the methanolic · HCl. The PTHs for all of the amino acids were detected at 254 nm. The isocratic system consisted of a single M-600 pump (Waters Assoc.), a WISP (Waters model 710B), and a Waters Model 440 detector. The separation of all common PTH-amino acids was achieved in 30 min using 30 m*M* sodium acetate (pH 4.7), which was 33% in acetonitrile and 0.4% in dichloroethane at a flow rate of 1.2 to 1.4 mℓ/min over 2 × 25-cm supelcosil C-18 columns (Supelco 5-8298) connected in series at 50°C. More recently, we have also been using the gradient PTH system described by Hunkapiller.[9a] We find that it is advantageous to maintain two different PTH analysis systems in the laboratory to confirm the identity of uncertain residues.

IV. SUMMARY

This report has outlined the major microanalytical methods for microscale protein purification and has illustrated the direct compatibility of solid-phase microsequence analysis with these methods for the investigation of specific protein structural problems. This integrated approach has proven successful in the analysis of such problems as the location of specific genes through DNA probes based upon protein sequence and the confirmation of DNA-generated protein sequence. These methods have also been used in structure-function studies using temperature-sensitive mutants to identify single amino acid substitutions which alter the thermal stability of several DNA-binding proteins.[4,37] The identification of post-translational modifications in mature polypeptides by comparative peptide mapping points out the utility of HPLC for this type of study.[25] Several of the important advantages of solid-phase microsequence analysis manifest themselves when this method is used in conjunction with the previously described micropurification methods. The compatibility of solid-phase immobilization procedures with volatile as well as nonvolatile HPLC buffer systems gives the investigator a wide range of options for optimization of chromatographic separation. This presents an advantage over other sequencing approaches which require salt-free polypeptides for sequence analysis, by eliminating the need for desalting samples and, thus, cutting down on handling loss and streamlining the overall purification process. The immobilization of peptides by carbodiimide activation has been demonstrated to be efficient and routine in the subnanomole range.[25] The immobilization of larger polypeptides in SDS has been shown to randomly couple Lys residues and, thus, allow positive identification of the attachment sites while having little effect on the overall efficiency of coupling. Solid-phase sequence analysis is the most preferred chemical method for the direct identification of highly charged and/or covalently modified residues, because the anilinothiazolinone which results from each Edman cycle is removed from the remainder of the polypeptide chain in trifluoroacetic acid, thus, eliminating the solvent extraction problems which are experienced with other types of automated sequencer.

REFERENCES

1. **Abersold, R. H., Teplow, D. B., Hood, L. E., and Kent, S. B. H.,** Electroblotting onto activated glass, *J. Biol. Chem.*, 261, 4229, 1986.
2. **Bhown, A. S., Mole, J. E., Weissinger, A., and Bennett, J. C.,** Methanol solvent system for rapid analysis of phenylthiohydantoin amino acids by high-pressure liquid chromatography, *J. Chromatogr.*, 148, 532, 1979.
3. **Chang, J.-Y.,** N-Terminal sequence analysis of polypeptide at the picomole level, *Biochem. J.*, 199, 557, 1981.
4. **Chase, J. W., L'Italien, J. J., Murphy, J. B., Spicer, E. K., and Williams, K. R.,** Characterization of the *Escherichia coli* SSB-113 mutant single-stranded DNA-binding protein, *J. Biol. Chem.*, 259, 805, 1984.
5. **Edman, P. and Begg, G.,** A protein sequenator, *Eur. J. Biochem.*, 1, 80, 1967.
6. **Hewick, R. M., Hunkapiller, M. W., Hood, L. E., and Dreyer, W. J.,** A gas — liquid solid phase peptide and protein sequenator, *J. Biol. Chem.*, 256, 7990, 1981.
7. **Horn, M. J. and Laursen, R. A.,** Solid-phase Edman degradation: attachment of carboxyl-terminal homoserine peptides to an insoluble resin, *FEBS Lett.*, 36, 285, 1973.

7a. **Horn, M. J. and Bonner, A. G.,** Sensitivity enhancement in solid phase protein sequence analysis — an examination of methodology, *Methods in Protein Sequence Analysis,* Elzinga, M., Ed., Humana Press, Clifton, N.J., 1982, 159.

8. **Hughes, G. J. and Wilson, K. J.,** High-performance liquid chromatography: analytic and preparative applications in protein structure determination, *Methods Biochem. Anal.*, 29, 59, 1983.
9. **Hunkapiller, M. W., Lujan, E., Ostrander, F., and Hood, L. E.,** Isolation of microgram quantities of proteins from polyacrylamide gels for amino acid sequence analysis, *Methods Enzymol.*, 91, 227, 1983.

9a. **Hunkapiller, M. W., Granlund-Moyer, K., and Whiteley, N. W.,** Analysis of phenylthiohydantoin amino acids by HPLC, *Micromethods of Polypeptide Characterization,* Shively, J. J., Ed., Humana Press, Clifton, N.J., 1986.

10. **Karr, S. R. and Foster, J. A.,** Primary structure of the signal peptide of tropoelastin b, *J. Biol. Chem.*, 256, 5946, 1981.
11. **Konigsberg, W. H. and Henderson, L.,** Removal of sodium dodecyl sulfate from proteins by ion-pair extraction, *Methods Enzymol.*, 91, 254, 1983.
12. **Laursen, R. A.,** Solid-phase sequencing of peptides and proteins, *The Peptides,* Gross, E. and Meienhofer, J., Eds., Academic Press, New York, 1981, 261.
13. **Laursen, R. A.,** Solid-phase Edman degradation — an automatic peptide sequencer, *Eur. J. Biochem.*, 20, 89, 1971.
14. **Laursen, R. A.,** *J. Am. Chem. Soc.*, 88, 5344, 1966.
15. **Laursen, R. A., Horn, M. J., and Bonner, A. G.,** Solid-phase Edman degradation. The use of *p*-phenyldiisothiocyanate to attach lysine- and arginine-containing peptides to insoluble resins, *FEBS Lett.*, 21, 67, 1972.
16. **Laursen, R. A. and Machleidt, W.,** Solid-phase methods in protein sequence analysis, *Methods Biochem. Anal.*, 26, 201, 1980.
17. **Lees, M. B., Chao, B., Laursen, R. A., and L'Italien, J. J.,** A hydrophobic tryptic peptide from bovine white matter proteolipid, *Biochem. Biophys. Acta,* 702, 117, 1982.
18. **L'Italien, J. J.,** Microscale elucidation of an N-linked glycosylation site by comparative HPLC peptide mapping, *J. Chromatogr.*, 359, 213, 1986.
19. **L'Italien, J. J.,** Solid-phase methods in protein microsequence analysis, *Micromethods of Polypeptide Characterization,* Shively, J., Ed., Humana Press, Clifton, N.J., 279, 1986.
20. **L'Italien, J. J.,** Applications of immunoaffinity chromatography to the purification of polypeptides for microsequence analysis, *Proteins: Structure and Function,* L'Italien, J. J., Ed., Plenum Press, New York, in press.
21. **L'Italien, J. J.,** manuscript in preparation.
22. **L'Italien, J. J. and Laursen, R. A.,** The amino acid sequence of elongation factor Tu of *Escherichia coli*: the large cyanogen bromide peptides, *J. Biol. Chem.*, 256, 8092, 1981.
23. **L'Italien, J. J. and Laursen, R. A.,** Protein microsequencing by HPLC peptide purification and solid-phase Edman degradation, *Methods in Protein Sequence Analysis,* Elzinga, M., Ed., Humana Press, Clifton, N.J., 1982, 383.
24. **L'Italien, J. J. and Kent, S. B. H.,** Protein microsequencing with post-column fluorescent phenylisothiocyanate analogues, *J. Chromatogr.*, 283, 149, 1984.
25. **L'Italien, J. J. and Strickler, J. E.,** Application of high performance chromatographic peptide purification to protein microsequencing by solid-phase Edman degradation, *Anal. Biochem.*, 127, 198, 1982.
26. **Machleidt, W.,** The current status of automated solid-phase sequencing, *Modern Methods in Protein Chemistry,* Tschesche, H., Ed., Water deGruyter, Berlin, 1983, 267.

27. **Machleidt, W. and Hofner, H.,** Progress in solid-phase sequencing: extended degradations of peptides and proteins, *Methods in Protein Sequence Analysis,* Elzinga, M., Ed., Humana Press, Clifton, N.J., 1982, 173.
28. **Machleidt, W. and Wachter, E.,** New supports in solid-phase sequencing, *Methods Enzymol.,* 47, 263, 1977.
29. **Mendez, E. and Lai, C. Y.,** Regeneration of amino acids from thiazolinones formed in the Edman degradation, *Anal. Biochem.,* 68, 47, 1975.
30. **Merrill, B. M., Williams, K. R., Chase, J. W., and Konigsberg, W. H.,** Photochemical cross-linking of *Escherichia coli* single-stranded DNA-binding protein to oligodeoxynucleotides, *J. Biol. Chem.,* 259, 10850, 1984.
30a. **Salnikow, J., Lehman, A., and Wittmann-Liebold, B.,** Improved automated solid-phase microsequencing of peptides using DABITC, *Anal. Biochem.,* 117, 433, 1981.
30b. **Pisano, J. J.,** Gas-liquid chromatography (GLC) of amino acid derivatives, *Methods Enzymol.,* 25, 27, 1972.
30c. **Pisano, J. J. and Bronzert, T. J.,** Analysis of amino acid phenylthiohydantoins by gas chromatography, *J. Biol. Chem.,* 244, 5597, 1969.
31. **Smithies, O., Gibson, D., Fanning, E. M., Goodfliesch, R. M., Gillman, J. G., and Ballantyne, D. L.,** Quantitative procedures for use with Edman-Begg Sequenator. Partial sequences of two unusual immunoglobulin light chains, Rzf and Sac, *Biochemistry,* 10, 4912, 1979.
32. **Strickler, J. E., Hunkapiller, M. W., and Wilson, K. J.,** Utility of gas-phase sequencer for both liquid- and solid-phase degradation of proteins and peptides at low picomole levels, *Anal. Biochem.,* 140, 553, 1984.
33. **Tarr, G. E., Beecher, J. F., Bell, M., and McKean, D.,** Polyquarternary amines prevent peptide loss from sequenators, *Anal. Biochem.,* 84, 622, 1978.
34. **Wachter, E. and Werhahn, R.,** Attachment of tryptophanyl peptides to 3-aminopropyl-glass suited for subsequent solid-phase Edman degradation, *Anal. Biochem.,* 97, 56, 1979.
35. **Walker, J. E., Auffret, A. D., Carne, A., Gurnett, A., Hanisch, P., Hill, D., and Saraste, M.,** Solid-phase sequence analysis of polypeptides eluted from polyacrylamide gel, *Eur. J. Biochem.,* 123, 253, 1982.
36. **Weber, K. and Osborn, M.,** The reliability of molecular weight determinations by dodecyl sulfate-polyacrylamide gel electrophoresis, *J. Biol. Chem.,* 224, 4406, 1969.
37. **Williams, K. R., L'Italien, J. J., Guggenheimer, R. A., Sillirud, L., Spicer, E., Chase, J., and Konigsberg, W. H.,** Comparative peptide mapping by HPLC: identification of single amino acid substitutions in temperature sensitive mutants, *Methods in Protein Sequence Analysis,* Elzinga, M., Ed., Humana Press, Clifton, N.J., 1982, 499.
38. **Wittmann-Liebold, B. and Lehmann, A.,** *Solid-Phase Methods in Protein Sequence Analysis,* Laursen, R. A., Ed., Pierce Chemical Co., Rockford, Ill., 1975, 81.
39. **Zimmerman, C. L., Apella, E., and Pisano, J. J.,** Rapid analysis of amino acid phenylthiohydantoin by high-performance liquid chromatography, *Anal. Biochem.,* 77, 569, 1977.

Chapter 5

GAS PHASE SEQUENCE ANALYSIS OF PROTEINS/PEPTIDES

M. W. Hunkapiller

TABLE OF CONTENTS

I. INTRODUCTION

The chemical process employed by automated protein/peptide sequencers is derived from the technique originated by Edman in the 1950s for the sequential degradation of peptide chains.[1,2] The first step in this degradation is selective coupling of a peptide's amino-terminal amino acid with the Edman reagent, phenylisothiocyanate (PITC), a reaction catalyzed by an organic base delivered with the coupling reagent. The second step is cleavage of this derivatized amino acid from the remainder of the peptide, a reaction effected by treating the peptide with a strong organic acid. Each repeated coupling/cleavage cycle occurs at the newly formed amino-terminal amino acid left by the previous cycle. Thus, repetitive cycles provide sequential separation of the amino acids which form the primary structure of the peptide.

The sequencing process is not completed by the Edman degradation alone. Once the amino acids are removed from the sample, they must be analyzed to determine their identity. Since the cleaved amino acid derivative, the anilinothiazolinone (ATZ), is not generally suitable for analysis, it is converted to the more stable phenylthiohydantoin (PTH) form before analysis is attempted. In modern sequencers,[3,4] this conversion is accomplished automatically in a reaction vessel separate from that in which the Edman degradation occurs. The ATZ produced at each degradation cycle is extracted from the peptide with an organic solvent, transferred to the reaction vessel, and treated with an aqueous solution of a strong organic acid to effect conversion to the PTH. The PTHs produced from each degradation cycle may be transferred to fraction collector vials until several are manually collected and prepared for analysis. Alternatively, the PTHs may be transferred directly and automatically from the sequencer conversion vessel to an on-line analysis system.[5-7]

Although a variety of analytical procedures have been used to identify the amino acids released during the Edman degradation, only high performance liquid chromatography (HPLC) is currently in widespread use. In fact, HPLC on reverse-phase, silica-based packings has revolutionized peptide sequencing. It provides rapid, sensitive, and quantitative analysis of

PTH amino acids and is the only technique used for PTH analysis that can reliably resolve all of the PTH amino acids in a single chromatographic run. Moreover, because it provides quantitative data at the picomole level, HPLC is the only analytical method suitable for microsequencing by the latest generation of automated Edman sequencers.

The reliability, accuracy, and sensitivity of any Edman degradation scheme will ultimately be limited by the weakest of its components. The weakest component may be sample preparation, the Edman chemistry itself, or PTH analysis. This chapter describes a combination of automated gas phase Edman chemistry and on-line PTH analysis that is suitable for sequencing of both proteins and peptides down to the low picomole level.

II. AUTOMATED EDMAN DEGRADATION

A. Instrumentation

The instrumentation used for automated Edman degradation in the studies described in this chapter is the Applied Biosystems Model 470A Gas Phase Protein/Peptide Sequencer. This sequencer is similar in design to the prototype instrument constructed at the California Institute of Technology.[4] The use of gas phase coupling and cleavage reagents in this sequencer provides retention of the sample on a solid support (a small disk of glass-fiber filter paper) without necessity of covalent chemical attachment to the support. The filter is housed in a simple glass reaction cartridge set in a thermostatted environment. At the end of each degradation cycle, the ATZ of the terminal amino acid is removed from the sample and automatically transferred from the cartridge to a second, separately heated chamber (a small glass flask) for conversion to the more stable PTH amino acid. The PTH is then removed from the conversion flask and transferred either to the sequencer fraction collector for subsequent analysis, generally by conventional HPLC, or to the Applied Biosystems Model 120A PTH Analyzer for immediate, on-line analysis.

The major hardware components of the Model 470A sequencer are depicted schematically in Figure 1. Chemicals are moved through the sequencer by a microprocessor-controlled, pneumatic pressure-driven chemical delivery system. The components necessary for chemical delivery include a regulated pressure source, reagent and solvent reservoirs, delivery valves, the reaction chambers (filter cartridge and conversion flask), a vented waste bottle, and the interconnecting delivery lines. All inner surfaces of the chemical delivery system that can affect the integrity of the reagents and solvents are made of inert materials (Pyrex® glass, Teflon®, Kel-F®, or Kalrez®). All delivery lines are made of narrow-bore Teflon® tubing with inert end fittings. The length of this tubing is kept short to minimize "dips" where liquids or condensation products can collect and interfere with gas flow.

1. Filter Cartridge

The filter cartridge (Figure 2) is composed of two cylindrical glass blocks that are mounted in a steel cartridge holder in a temperature-controlled, forced-air oven chamber. A pathway is created for reagent and solvent flow by a narrow bore that extends through both cartridge blocks. The reaction chamber is created by the shallow conical recess on the inside face of each cartridge block. When the two blocks are pressed together, a 150-$\mu\ell$ reaction chamber is formed between the two recesses. The recess in the upper block is slightly indented and holds the filter disk and the sample adsorbed on it. A narrow circular lip extends above the surface of the lower block and presses against the upper cartridge block.

A porous Teflon® membrane is placed between the two blocks. In the cartridge holder, a leak-tight seal is formed between the upper cartridge block and the lip of the lower cartridge block, with the crushed Teflon® membrane acting as a gasket. The Teflon® membrane also supports the filter while allowing chemicals to flow through the reaction chamber. With the cartridge blocks in the cartridge holder, a keyed metal washer is placed above the cartridge

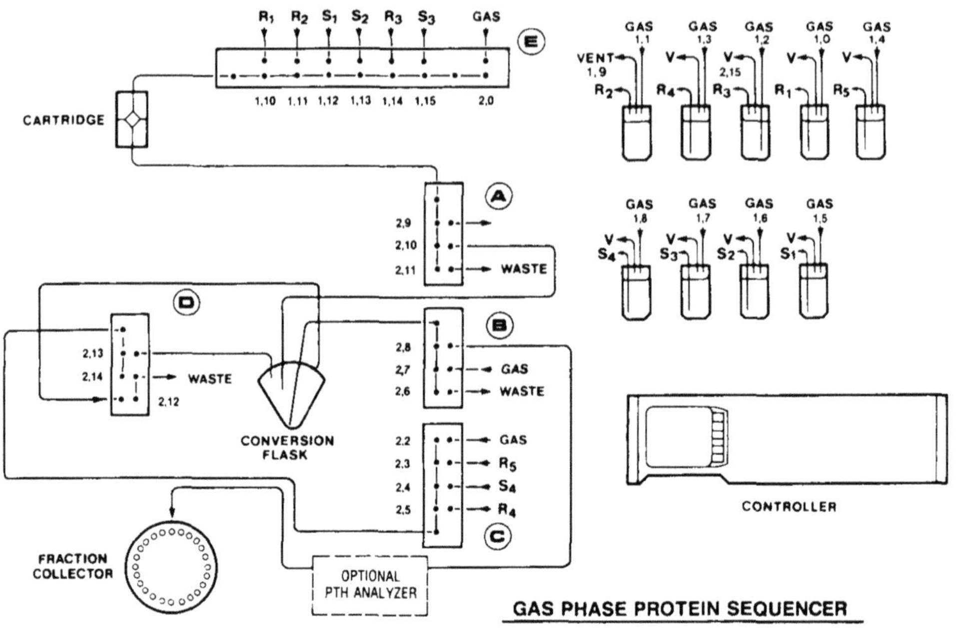

SEQUENCER FUNCTIONS

FILTER CARTRIDGE

1. Prep R_1	6. Deliver R_3	11. Prep S_3	16. ATZ Purge
2. Deliver R_1	7. Prep S_1	12. Deliver S_3	17. ATZ Collect
3. Prep R_2	8. Deliver S_1	13. Prep transfer	18. Pressurize
4. Deliver R_2	9. Prep S_2	14. Transfer w/S_3	19. Argon dry
5. Prep R_3	10. Deliver S_2	15. Transfer w/Ar	20. Pause
			21. Cartridge Vent

CONVERSION FLASK

1. Prep R_4	6. Deliver R_5	11. Pressurize	16. LC Start
2. Load R_4	7. Prep S_4	12. Collect	17. Pause
3. Deliver R_4	8. Load S_4	13. Empty	
4. Prep R_5	9. Deliver S_4	14. FC Step	
5. Load R_5	10. Argon Dry	15. Line Purge	

FIGURE 1. Schematic diagram of Model 470A Gas Phase Protein/Peptide Sequencer.

to prevent the cartridge blocks from cracking or rotating when the upper locking ring is tightened. A Belleville spring washer is placed between the keyed washer and upper locking ring to help maintain the seal provided by the crushed Teflon® membrane.

The ends of the cartridge inlet and outlet lines are fitted into cylindrical Kel-F® ferrules that are pressed firmly against the cartridge blocks by finger-tightening the inlet and outlet line connectors. Deliveries of argon, coupling reagents R1 and R2, cleavage reagent R3, and solvents S1, S2, and S3 through the cartridge inlet line and the cartridge assembly are controlled by Valve Block E. Exiting of chemicals from the cartridge is controlled by Valve Block A. From the cartridge, flow can be directed to the waste system, to the conversion flask, or to an external fraction collector. The cartridge is lighted from behind for visual examination of reagent and solvent flows.

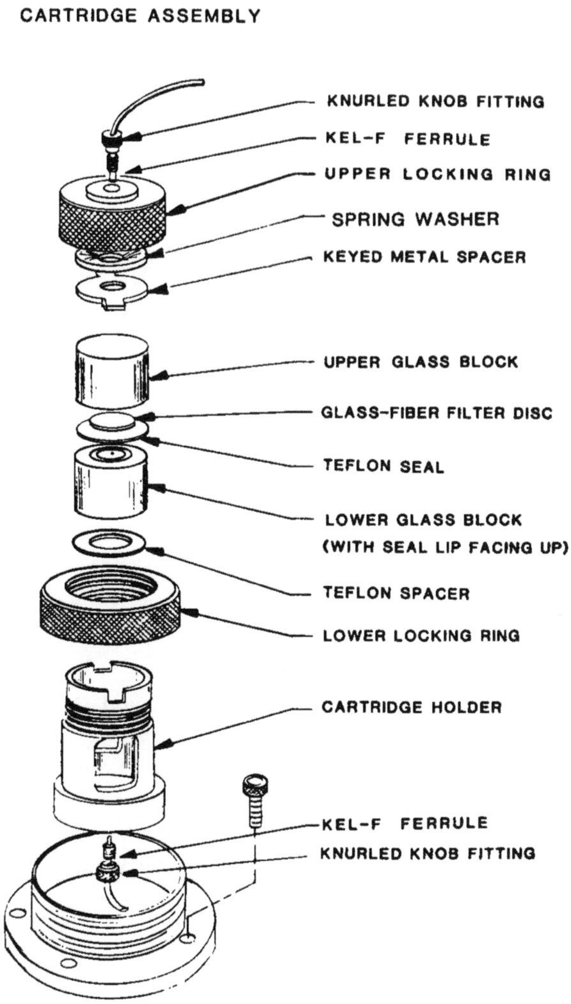

FIGURE 2. Gas phase sequencer filter cartridge.

2. *Conversion Flask*

The conversion flask (Figure 3) is a conical glass vial with an internal volume of approximately 1 mℓ. Chemicals and argon enter or exit the flask via four tubes which pass through the flask's Teflon® cap. Each tube serves a different function:

- The *pickup tube* extends to the bottom of the flask and is used as both an inlet and outlet to the flask. Bubble-argon flows through Valve Block B to the flask pickup tube to mix the flask contents during the deliveries and to speed evaporation during the drying steps. After conversion, PTH amino acids dissolved in S4 flow out of the flask through the pickup tube to the fraction collector. Alternatively, the contents of the flask can be sent to waste.
- The *cartridge transfer tube* extends about 1 mm into the flask. The ATZ amino acid, dissolved in S3, transfers from the cartridge to the flask via this tube and Valve Block A.
- The tip of the *reagent delivery tube* is placed above the cartridge transfer tube. Entry of conversion reagent R4, PTH Standard R5, solvent S4, and flask argon entering

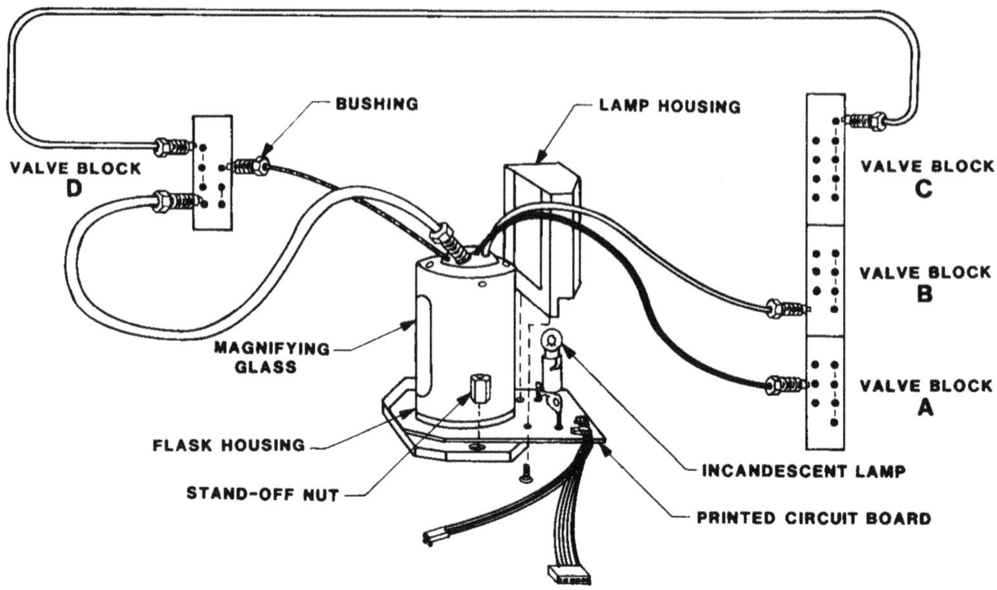

FIGURE 3. Gas phase sequencer PTH conversion flask.

through this tube is controlled by Valve Blocks C and D. The position of the tube aids washing any residue down to the tip of the flask.

- The *vent tube* acts as an exhaust vent. Valve Block D controls this outlet.

The conversion flask is placed in a temperature-controlled steel housing. Lighted from behind, the flask can be monitored through a magnifying window for timing reagent and solvent deliveries, drying steps, etc.

B. Edman Chemistry

During sequencing, the protein is held by a 12-mm-diameter disk of porous glass-fiber filter in a heated, argon-purged reaction chamber. First, the filter is pretreated with BioBrene Plus® (Applied Biosystems Part Number 400385) and then cycled through a few repetitions of the Edman chemistry to reduce contaminants and improve the efficiency of subsequent sample sequencing. At this point, a solution of the sample protein or peptide (typically, 20 pmol to 20 nmol) is loaded onto the filter and dried to leave the sample embedded in the film of pretreated BioBrene® on the glass-fiber disk. Covalent attachment of the sample to the filter is generally unnecessary, since the Model 470A standard chemistry uses relatively apolar solvents in which proteins and peptides (even most short, hydrophobic peptides) are poorly soluble.

The large reaction surface of the filter, coupled with the low reaction chamber volume and simple flow-through design of all valve and reaction chambers, facilitates addition and removal of both gaseous and liquid reagents and solvents during the chemical reactions. This results in low rates of chemical consumption and reduced risk of sample loss or contamination during sequencing.

All reagents and solvents should be specially purified for microsequencing applications. Those used in the studies described in this chapter were supplied by Applied Biosystems (Part Number 400340). A list of reagents and solvents used with the sequencer degradation cycles listed in the next section is given in Table 1.

The steps of the Edman degradation are listed schematically in Figure 4. Their implementation in the sequencer programs is described in detail below.

Table 1
LIST OF SEQUENCER REAGENTS AND SOLVENTS

R1	5% Phenylisothiocyanate (PITC) in *n*-heptane
R2	12.5% Trimethylamine (TMA) in water
R3	Trifluoroacetic acid (TFA), neat, with 0.002% dithiothreitol (DTT)
R4A	25% TFA in water, with 0.01% DTT
R5	Acetonitrile, with 0.001% DTT (for use with the Model 120A PTH Analyzer)
S1	*n*-Heptane
S2	Ethyl acetate
S3	1-Chlorobutane
S4A	Acetonitrile, with 0.001% DTT (for use when the Model 120A is not used)
S4B	20% Acetonitrile in water (for use with the Model 120A)

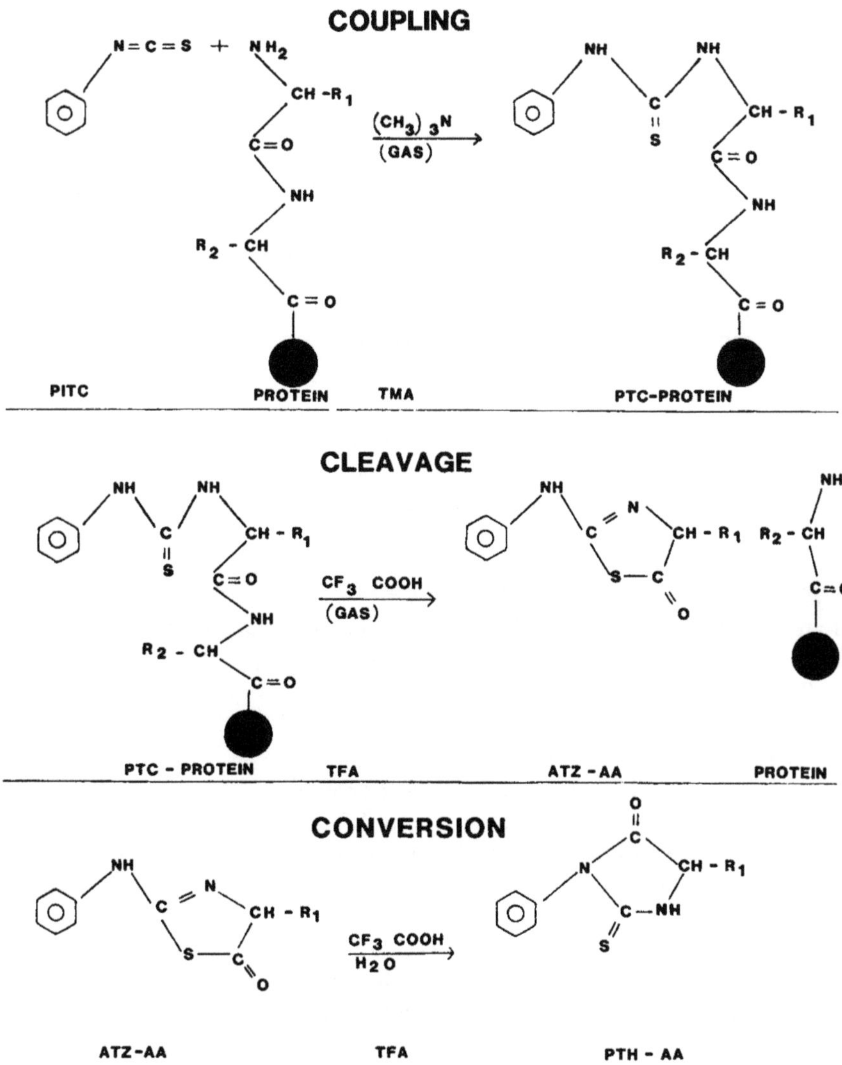

FIGURE 4. Edman chemistry.

1. Coupling

***a. R1 (PITC in** n*-Heptane)

R1 is a sweet smelling, volatile liquid. In the presence of base, the PITC in it reacts quantitatively with the free amino-terminal amino group of a peptide (as well as amino side chain groups) to form the phenylthiocarbamyl-protein (PTC-protein). In the standard degradation cycles for the sequencer, R1 is delivered to the cartridge to saturate the filter (usually a 3-sec delivery is sufficient). Three separate deliveries are made in-between deliveries of the basic R2. These repeated R1 deliveries increase the reagent-protein contact and improve the reaction yield. After each R1 delivery, argon removes the nonreactive heptane.

b. R2 (TMA in Water)

The TMA in R2 is a strongly basic amine, identifiable by its strong "ammonia" odor. It is highly volatile at room temperature and is delivered as gas vapors carried by a stream of argon swept through the R2 reservoir. It provides the basic environment necessary for PITC to react with the amino terminus of a peptide. In the standard degradation cycles, R2 is delivered to the filter cartridge for about 22 min. The R2 delivery is interrupted by three R1 deliveries/argon dry steps.

c. Precoupling Base Treatment

Most protein samples sequence more effectively when a short R2 delivery precedes the initial R1 delivery for the coupling reaction. This precoupling base treatment helps to reverse acid-catalyzed N $\rightarrow$ O acyl shifts which occur at trifluoroacetylated seryl residues and ultimately lead to extraneous cleavage along the peptide chain. This cleavage results in increased PTH background, and the acylated side-chain hydroxyl can participate in a side reaction that results in loss of free amino groups (and, hence, a drop in repetitive yield) when the serine is the amino-terminal amino acid. If the protein being sequenced is very rich in serine, then the precoupling R2 delivery should be increased and the cartridge temperature reduced (Cycle 03CSER, Run 03RSER).

d. Oxidation

Once coupling has occurred, the PTC-protein is susceptible to desulfurization. Replacement of sulfur with oxygen inhibits cyclization and cleavage, reducing yields. For this reason, all solvents and reagents are kept under argon and should be argon purged whenever refilled.

e. Solvent Extraction

Solvent are delivered through the filter in order to extract unwanted materials that could hinder later reactions or form by-products that could interfere with the chromatographic analysis of PTH amino acids. The solvents help remove:

- PITC by-products, generally thioureas and ureas
- Residual PITC to avoid additional thiourea formation
- Residual TMA to avoid salt formation during R3 delivery
- Residual water (for which BioBrene® has an affinity) to minimize acid-catalyzed hydrolysis during cleavage

S1 (*n*-heptane) removes most of the excess PITC. S2 (ethyl acetate) removes the remainder of the unwanted materials. The amount of S2 delivered can be increased relative to that called for in the standard sequencer cycles in order to minimize unwanted residue levels, but excess S2 delivery can cause extractive loss of some small peptides.

2. Cleavage

a. R3 (TFA)

R3 is a strong, highly volatile organic acid. The TFA is used to cleave the PITC-coupled amino acid residue from the amino terminus of the protein, thus producing the ATZ of the amino acid. The remaining protein is left with a new amino terminus, ready for the next degradation cycle. R3 is delivered as gas vapors carried by a stream of argon swept through the R3 reservoir. The R3 delivery is followed by argon drying which removes most residual R3, minimizing acidificiation of S3 and subsequent extractive sample loss during the initial phase of the ATZ transfer. Excessive argon drying, however, can result in poor transfer of the ATZs of histidine and arginine.

The cleavage conditions in the primary sequencer cycle, 03CPTH, may not be optimum for all samples. Several special cycles and runs are included on the standard sequencer tape to accommodate these samples. Synthetic, resin-bound peptides generally require more vigorous cleavage conditions, both longer R3 delivery time and higher temperature. These and other changes are incorporated in Cycle 03CREZ, Run 03RREZ. Some prolyl and glycyl residues may also require longer cleavage times. If the positions of these residues in the sequence are known, their cleavage times can be extended on an individual basis by modifying Run 03RPTH to include 03CPRO cycles at the appropriate places. If several prolyl and glycyl residues are present in unknown positions, then the cycle temperature for the entire run (03RPRO) can be increased to minimize lag. The use of higher temperature is preferable to extending the cleavage for every cycle in the run.

b. ATZ Transfer

S3 (1-chlorobutane) is used to extract the cleaved ATZ amino acid from the protein and transfer it from the filter cartridge to the conversion flask. In the standard sequencer cycles, the extraction is begun by delivering just enough S3 to soak the filter. The solvent is allowed to soak the filter for 20 sec. Additional S3 delivery then transfers the ATZ amino acid to the conversion flask. The last portion of this solution is transferred to the flask with argon.

Precision in the initial delivery of S3 is important. The S3 should reach and soak the filter, but should not begin to flow into the capillary of the bottom glass cartridge block. The 20-sec pause is crucial to prevent excessive loss of sample, particularly small peptides, that might otherwise occur if the S3 flow proceeded uninterrupted. The pause step gives time for dilution of residual R3 left on the filter, R3 that might be sufficiently concentrated in the initial wave of S3 to dislodge the sample from the filter.

A small amount of R4A (aqueous TFA) is delivered to the flask just before the extraction solution reaches it. This R4A commences conversion and minimizes exposure of the labile ATZ amino acids to the hot walls of the flask. The presence of dithiothreitol in the R4A also minimizes oxidation of the ATZ amino acids.

Additional S3 is delivered through the cartridge to the waste bottle. This delivery is essential to rinse R3 from the cartridge waste line and prevent salt formation that can block this line.

3. PTH Conversion

a. R4A (25% TFA in Water)

The transferred extraction solution (S3 + ATZ amino acid), along with most of the initial delivery of R4A, is evaporated from the conversion flask. An additional aliquot of fresh R4A is then delivered to the flask to effect the conversion of the ATZ amino acids to the more stable PTH amino acids. In the standard sequencer cycles, the conversion takes place during about 20 min at 55°C. These conditions are sufficient to give virtually complete conversion for all amino acids except glycine (about 15% of which remains in the partially converted PTC-glycine form).

b. Reconstitution and Collection

After conversion is complete, R4A is evaporated from the flask, leaving behind a dried residue of the PTH amino acids. The residue is brought back into solution for transfer from the flask to the sequencer fraction collector for short-term storage (S4A, acetonitrile) or to the Model 120 PTH Analyzer for immediate analysis (S4B, 20% aqueous acetonitrile).

In the standard sequencer cycles, S4A or S4B is delivered to the flask, mixed, and allowed to "pause" for dissolution to occur. The solution is then transferred to the appropriate fraction collector vial and/or to the Model 120A. A second solvent delivery is made to pick up any residual PTH amino acid and to rinse the lines between the conversion flask and the fraction collector. This solution is added to the previous collection in the collector vial. In preparation for the next conversion, S4A or S4B is then delivered to fill the flask. The flask contents are mixed with bubble argon and then emptied to waste.

c. PTH Standard

R5 (acetonitrile) can be used to introduce an aliquot of internal standard or PTH standard to the flask for subsequent HPLC.

C. Sequencer Programs

The system controller manages and monitors all sequencer activity. Most of the software is stored in 32 kilobyte (32 K) words of permanent read-only memory (PROM) where it is unaffected by power failures. Additional software is stored on a removable 195-K magnetic tape cassette, as are all standard and user-generated sequence files. The software is "menu driven". Once the software is loaded into the microprocessor, the CRT will display a series of "menus" which provide both current sequencer status information and a list of available options for the operator. When the selection key for an option is pressed, any appropriate information is transferred to the microprocessor, and then the action called for by the selection is carried out. The sequencer can be programmed for unattended operation in understandable chemical and functional terms, i.e., "deliver R1". These terms are used to define the steps of sequencer cycles and runs. Cycles and runs are easily modified to tailor sequencer operation for particular types of protein or peptide samples. The individual sequencer valves may also be controlled for testing and troubleshooting, and an electronic "self test" is available for verification of the sequencer's electronic and electromechanical components.

The Edman chemistry is automatically performed through commands stored on the sequencer cassette tape. These commands are called steps, cycles, and runs.

- *Step*: a single step is made of one cartridge and one flask function, activated together for a selected time.
- *Cycle*: a cycle is a series of steps designed to deliver a set of reagents and solvents in order to accomplish a collection of defined chemical reactions, such as coupling, cleavage, and conversion, that typically result in one complete degradation reaction.
- *Run:* a run is a number of cycles carried out in sequence to perform many repetitions of the degradation reaction.

A set of standard sequencer cycles and runs have been created and stored on the standard sequencer cassette tape (Applied Biosystems Part Number 400305). These provide optimal sequencing for most samples. The two most commonly used runs and their corresponding cycles are listed in Tables 2 to 6. Additional cycles and runs that you might need for your particular applications can be created on a blank sequencer tape using the FILE EDITOR menus.

Accompanying each run listing are notes that describe the particular application for which the run has been optimized. Accompanying each cycle listing are notes that describe particular

Table 2
LISTING FOR SEQUENCER RUN *03RPTH*

Cycles	Cycle name	Cartridge temp (°C)	Flask temp (°C)
1—1	03CBGN	45	55
2—100	03CPTH	45	55

Note: This is the primary run program for general sequencing experiments. 03CBGN cycle provides R3 treatment and S3 extraction of the sample to help remove contaminants, as well as a short R1/R2 coupling and S1/S2 extraction prior to commencing the repetitions of the primary run cycle, 03CPTH. The first fraction removed from the conversion flask should be blank. The second will contain a PTH standard if one is present in the R5 reservoir, or will be a blank. The first sample residue will be in fraction three. When the run is stopped after an 03CPTH cycle, a mixture of R4A/S3 with the ATZs from that cycle remains in the flask. If the Model 120A PTH Analyzer is being used, start the LC program *before* cycle 03CBGN starts.

Table 3
LISTING FOR SEQUENCER RUN *03RPRE*

Cycles	Cycle name	Cartridge temp (°C)	Flask temp (°C)
1—3	03CFIL	45	55

Note: This run is the primary filter conditioning cycle, designed to precycle the BioBrene, preparatory to loading and running the protein or peptide sample. All three fractions removed from the conversion flask will contain a PTH standard if one is present in the R5 reservoir, or will be a blank. If the Model 120A PTH Analyzer is being used, start the LC program *after* the first 03CFIL cycle has started.

steps where times or solvent flows must be individually set by the user for optimum performance.

D. Sample Loading

The first step in sequencing a sample is to treat a clean glass-fiber filter with BioBrene® so that the sample, when added to the filter, will be immobilized during the Edman chemistry. Since the aqueous BioBrene® solution may contain small amounts of compounds that could interfere with sequencing, several short cycles of the Edman chemistry are run to wash and condition the BioBrene®-coated filter. The protein sample is then loaded onto the conditioned filter and the actual sequencing begun.

1. Filter Conditioning

A stepwise procedure for loading the filter with BioBrene® is given below. Refer to Figure 4 for a diagram of the cartridge assembly.

Table 4
LISTING FOR SEQUENCER CYCLE
03CPTH

Step	Cartridge function	Flask function	Time
1	Deliver R2	Argon dry	120
2	Prep R1	Argon dry	6
3	Deliver R1	Argon dry	3
4	Argon dry	Argon dry	50
5[a]	Deliver R2	Argon dry	60
6	Deliver R2	Load R4	6
7	Deliver R2	Argon dry	4
8	Deliver R2	Pause	330
9	Deliver R1	Pause	3
10	Argon dry	Pause	50
11	Deliver R2	Pause	400
12	Deliver R1	Pause	3
13	Argon dry	Pause	50
14	Deliver R2	Pause	400
15	Argon dry	Line purge	60
16[b]	Deliver S1	Argon dry	100
17[c]	Deliver S2	Argon dry	240
18	Argon dry	Argon dry	120
19[d]	Deliver R3	Argon dry	40
20	Deliver R3	LC start	10
21	Deliver R3	Collect	60
22	Deliver R3	Load S4	6
23	Deliver R3	Argon dry	4
24	Deliver R3	Load S4	6
25	Deliver R3	Argon dry	4
26	Deliver R3	Pause	540
27[e]	Deliver R3	Collect	14
28	Deliver R3	Load S4	6
29	Deliver R3	Argon dry	4
30	Deliver R3	Load S4	6
31	Deliver R3	Argon dry	4
32	Deliver R3	Collect	30

Table 5
LISTING FOR SEQUENCER CYCLE
03CBGN

Step	Cartridge function	Flask function	Time
1	Argon dry	Empty	30
2	Argon dry	Deliver S4	25
3	Argon dry	Argon dry	10
4	Argon dry	Empty	40
5	Argon dry	LC start	10
6	Argon dry	Collect	30
7	Argon dry	Load S4	6
8	Argon dry	Argon dry	4
9	Argon dry	Load S4	6
10	Argon dry	Argon dry	4
11	Deliver R3	Pause	240
12	Argon dry	Pause	80
13	Prep S3	Pause	6
14[a]	Deliver S3	Pause	12
15	Pause	Pause	20
16	Deliver S3	Pause	100
17	Argon dry	Pause	120
18[b]	Deliver R2	Collect	14
19	Deliver R2	Load S4	6
20	Deliver R2	Argon dry	4
21	Deliver R2	Load S4	6
22	Deliver R2	Argon dry	4
23	Deliver R2	Pause	60
24	Deliver R2	Collect	40
25	Deliver R2	Deliver S4	25
26	Deliver R2	Argon dry	5
27	Deliver R2	Empty	40
28	Deliver R2	FC step	2
29	Deliver R2	Argon dry	45
30	Prep R1	Pause	6
31	Deliver R1	Pause	3
32	Argon dry	Pause	50

Table 6
LISTING FOR SEQUENCER CYCLE
03CFIL

Step	Cartridge function	Flask function	Time
1	Prep R2	Empty	6
2	Deliver R2	Empty	24
3	Deliver R2	Prep S4	6
4	Deliver R2	Deliver S4	24
5	Deliver R2	Argon dry	10
6	Deliver R2	Empty	30
7	Deliver R2	Prep R5	6
8	Prep R1	Load R5	6
9	Deliver R1	Argon dry	3
10	Argon dry	Argon dry	50
11	Deliver R2	Argon dry	90
12	Deliver R2	Prep R4	6
13	Deliver R2	Load R4	6
14	Deliver R2	Argon dry	4
15	Deliver R2	Pause	400
16	Argon dry	Line purge	60
17	Prep S1	Pause	6
18	Prep S2	Pause	6
19	Deliver S1	Pause	100
20	Deliver S2	Argon dry	300
21	Argon dry	Argon dry	120
22	Deliver R3	Argon dry	300
23	Deliver R3	LC start	10
24	Deliver R3	Collect	60
25	Deliver R3	Load S4	6
26	Deliver R3	Argon dry	4
27	Deliver R3	Load S4	6
28	Deliver R3	Argon dry	4
29	Argon dry	Pause	80
30	Prep S3	Pause	6
31[a]	Deliver S3	Pause	12
32	Pause	Pause	20

33	Deliver R3	Deliver S4	20
34	Deliver R3	Argon dry	5
35	Argon dry	FC step	2
36	Argon dry	Empty	40
37	Transfer w/Ar	Pause	30
38	Prep transfer	Pause	6
39[f]	Transfer w/S3	Pause	15
40	Pause	Load R4	6
41	Pause	Argon dry	4
42	Pause	Line purge	10
43[g]	Transfer w/S3	Pause	50
44	Pause	Argon dry	20
45	Transfer w/Ar	Pause	30
46	Deliver S3	Argon dry	30
47	Argon dry	Argon dry	120

Total time for cycle: 52 min 7 sec

[a] Extend time of Step 1, if necessary, so that all visible bubbling of liquid in flask has stopped by end of Step 5.

[b] Adjust S1 regulator pressure so S1 flow during step is 0.6—0.8 mℓ.

[c] Adjust S2 regulator pressure so S2 flow during step is 1.5—1.8 mℓ.

[d] Extend time of step, if necessary, so that all liquid in flask is gone by end of step.

[e] If using on-line PTH analysis, adjust time to 2 sec less than that required for liquid to just emerge from end of injector backpressure line.

[f] Adjust time to midpoint between S3 reaching and leaving filter cavity.

[g] Adjust S3 regulator pressure so S3 begins to drip into flask 18—22 sec after start of step.

33	Deliver R2	Pause	400
34	Argon dry	Pause	60
35	Deliver S1	Pause	120
36	Deliver S2	Pause	300
37	Argon dry	Pause	90
38	Argon dry	Load R5	6
39	Argon dry	Argon dry	4
40	Argon dry	Line purge	20

Total time for cycle: 34 min 13 sec

[a] Adjust time to midpoint between S3 reaching and leaving filter cavity.

[b] If using on-line PTH analysis, adjust time to 2 sec less than that required for liquid to just emerge from end of injector backpressure line.

33	Deliver S3	Pause	120
34	Argon dry	Pause	120
35	Deliver R2	Pause	60
36	Deliver R1	Pause	3
37	Argon dry	Pause	50
38	Deliver R2	Pause	60
39[b]	Deliver R2	Collect	14
40	Deliver R2	Load S4	6
41	Deliver R2	Argon dry	4
42	Deliver R2	Load S4	6
43	Deliver R2	Argon dry	4
44	Deliver R2	Pause	60
45	Deliver R2	Collect	40
46	Deliver R2	Deliver S4	25
47	Deliver R2	Argon dry	10
48	Deliver R2	FC step	2
49	Deliver R2	Pause	280
50	Argon dry	Pause	60
51	Deliver S1	Pause	100
52	Deliver S2	Pause	300
53	Argon dry	Pause	120
54	Deliver R3	Pause	400
55	Argon dry	Pause	80
56	Prep S3	Pause	6
57[a]	Deliver S3	Pause	12
58	Pause	Pause	20
59	Deliver S3	Pause	100
60	Argon dry	Pause	120

Total time for cycle: 65 min 53 sec

[a] Adjust time to midpoint between S3 reaching and leaving filter cavity.

[b] If using on-line PTH analysis, adjust time to 2 sec less than that required for liquid to just emerge from end of injector backpressure line.

1. Remove cartridge blocks.
   - Loosen the upper and lower knurled knob fittings.
   - Unscrew and remove the upper locking ring.
   - Remove the Belleville washer and cartridge key.
   - Lift the cartridge blocks from the holder.

2. Clean cartridge blocks.
   - Discard the cartridge seal and used filter.
   - Clean the inner surfaces of both cartridge blocks, first with methanol and then with water.

3. Replace the lower cartridge block.
   - Put the lower cartridge block in the holder with "lip" side up.
   - Place a new cartridge seal on top of the block.
   - Close the cartridge door.

4. Fit a new filter into the upper cartridge block.
   - Center the filter in the cartridge recess.
   - Press the filter *gently* in place with the tamper tool. An off-center filter may cause cartridge sealing problems.

5. Load the BioBrene® solution onto the filter.
   - Load liquid onto the center of the filter so that it spreads evenly across the filter.

6. Dry the filter.
   - Place the upper cartridge block with filter in the cartridge chamber and close the cartridge door. The recirculating warm air will facilitate drying.

7. Reassemble the cartridge blocks.
   - Place the upper cartridge block, with the filter side down, on top of the lower block and seal.
   - Fit the cartridge key and Belleville washer on top of the blocks. Check that both knurled knob fittings are unscrewed sufficiently so that the tubing ferrules will not touch the glass blocks. Otherwise, excess pressure can be exerted on the ferrules while the upper locking ring is tightened.
   - Screw down the upper locking ring *hand tight*.
   - Tighten both knurled knob fittings *finger tight*.

8. Load the fraction collector tray.
   - Place vials in at least the first four positions.

9. Start the preconditioning run.
   - From the MAIN MENU, select the RUN SEQUENCE option.
   - From the RUNS AVAILABLE MENU, select Run 03RPRE.
   - From the RUN SEQUENCE MENU, select the BEGIN RUN option.
   - When the run program has been loaded from the tape, the reaction chambers have reached operating temperature (or you have *jumped to Step 1*), and Step 1 has started, set the Model 120A PTH analyzer control for three runs if on-line PTH analysis is being used.

2. Sample Loading

Once the preconditioning run has finished, move to the RUN STATUS MENU on the controller display and select the ABORT, RETURN TO MAIN MENU option. Then proceed with loading and sequencing of the sample as described below.

1. Remove cartridge blocks.
   - Loosen the upper and lower knurled knob fittings.
   - Unscrew and remove the upper locking ring.
   - Remove the Belleville washer and cartridge key.
   - Lift the cartridge blocks from the holder.
   - Discard the used cartridge seal.

2. Replace the lower cartridge block.
   - Put the lower cartridge block in the holder with "lip" side up.
   - Place a new cartridge seal on top of the block.

3. Load the sample solution onto the filter.
   - Load liquid onto the center of the filter so that it spreads evenly across the filter. Maximum liquid capacity of the filter is about 30 μℓ. When necessary, additional loadings can be made if the filter is allowed to dry in-between.

4. Dry the filter.
   - Place the upper cartridge block with filter in the cartridge chamber and close the cartridge door. The recirculating warm air will facilitate drying.

5. Load saline solution onto the filter.
   - Load 30 μℓ of 0.1 *M* NaCl onto the filter (this promotes dispersion of the peptide in the BioBrene® film).
   - Dry the filter as in step 4 above.

6. Reassemble the cartridge blocks.
   - Place the upper cartridge block, with the filter side down, on top of the lower block and seal.
   - Fit the cartridge key and Belleville washer on top of the blocks. Check that both knurled knob fittings are unscrewed sufficiently so that the tubing ferrules are unscrewed sufficiently so that the tubing ferrules will not touch the glass blocks. Otherwise, excess pressure can be exerted on the ferrules while the upper locking ring is tightened.
   - Screw down the upper locking ring *hand tight*.
   - Tighten both knurled knob fittings *finger tight*.
   - Close the cartridge door.

7. Load the fraction collector tray.
   - Remove the first three vials (those used during the conditioning run) from the tray.
   - Fill the tray with clean vials.

8. Start the sequencing run.
   - From the MAIN MENU, select the RUN SEQUENCE option.
   - From the RUNS AVAILABLE MENU, select Run 03RPTH (or any other desired, available run).
   - Set the Model 120A PTH Analyzer control for at least two runs more that the number of amino acid degradation cycles you intend to perform on the sample if on-line PTH analysis is being used.
   - From the RUN SEQUENCE MENU, select the BEGIN RUN option.
   - When the run program has been loaded from the tape and the reaction chambers have reached operating temperature (or you have *jumped to Step 1*), the sequencing run will commence.

III. PTH AMINO ACID ANALYSIS

A. Analytical Parameters

1. Resolution

The first requirement of a PTH analysis protocol is to resolve the PTH amino acids derived from the sequencer and any contaminating compounds with similar spectral properties. Fulfilling this requirement can be difficult; there are likely to be 20 to 30 chemically similar compounds that must be separated. Particular attention must be paid to providing good resolution of the PTH amino acids from any sequencer-derived contaminants. When small amounts of peptide are being sequenced, these contaminants may be present in much greater amounts than the PTH amino acids. Any overlap between a contaminant and a PTH amino acid can interfere with quantitation or identification of the PTH amino acid itself.

2. Sensitivity

The second requirement is to provide adequate detection sensitivity for the amount of sample being sequenced. The recoveries of certain PTH amino acids from the sequencer are typically low; the recoveries of all PTH amino acids decrease as the Edman degradation proceeds through the peptide. Thus, fulfilling this requirement generally means reaching a detection limit that is <10% of the starting amount of peptide.

Detection sensitivity is limited by a variety of factors:

1. Intrinsic noise level of the UV absorbance detector used to monitor the elution of compounds from the HPLC column: the best UV monitors have operating noise levels of <0.00002 absorbance units (AU), but others may exhibit noise levels substantially above this.

2. Elution volume of the PTH amino acids: smaller elution volumes mean higher concentration and, hence, higher detection sensitivity. This is determined by:
   - The particle size of the column packing (smaller particles generally give smaller volumes)
   - The uniformity of the packing (columns with more theoretical plates per unit length give smaller volumes)
   - Column size (shorter, narrower bore columns give smaller volumes)
   - Extracolumn effects (improper tubing connections and excessive tubing lengths between either the injector and the column or between the column and the detector give larger volumes)
   - Elution mode — with isocratic elution, the retention time of a particular component also influences the elution volume (longer elution times give larger volumes), while with gradient elution the gradient steepness influences the elution volume (steeper gradients give smaller volumes).

3. Detection wavelength: the local UV absorbance maximum for most PTH amino acids is between 266 and 270 nm. Variable wavelength detectors that can be set in this range will give a 40 to 50% signal increase over that obtained with a fixed wavelength detector operating from a mercury emission line at 254 nm. This provides increased sensitivity as long as the noise level of the variable detector is equivalent to that of the fixed wavelength detector.

4. Chromatography artifacts: random or periodic fluctuations in pump output or incomplete solvent mixing can produce baseline fluctuations that increase the effective detector noise. Contaminated solvents, column inlet frits, or column packings can give excessive baseline drift or produce artifact peaks that also increase the system noise level.

5. Sequencer artifacts: by-products of the Edman chemistry can coelute with some of the PTHs, thereby increasing the amount of PTHs required for reliable detection and quantitation.

6. Protein/peptide artifacts: contaminants in the peptide sample being sequenced can produce chromatography background peaks that decrease the effective sensitivity. Contaminants that are also peptides are particularly troublesome. They will undergo Edman degradation themselves, thus producing PTHs that interfere with identification of those produced by the primary sample.

3. Reliability

Many peptides are being sequenced in such small amounts that there is only enough PTH sample for one injection on the HPLC. Thus, reliability of the analysis system is crucial. With an automated injection system, malfunctions in the pump, injector, or detector can be devastating, because they may go undetected while a number of cycles are analyzed.

Retention time reproducibility is also important, especially if the elution of the several PTHs is closely spaced. Since there are numerous PTHs that share common chemical structures, this is almost always the case. Moreover, the occasional presence of chemically modified amino acids that elute close to the common ones can often be detected if retention times are quite reproducible. Generally, the relative standard deviation (RSD) for elution times should be less than 0.3%.

Analysis of reproducible portions of each cycle's PTH product is important for accurate quantitation. Repeated treatment of the protein sample by the cleavage acid during the Edman degradation gradually fragments the sample into a series of smaller peptides. This process tends to produce a steadily increasing background level of PTHs from the fragments which superimposed over the specific PTH released from the intact protein. Thus, the specific PTH, especially later in sequence runs, can only be identified by a quantitative distinction between the level of one PTH at a given cycle vs. the background level of that PTH at preceding and subsequent cycles.

Injection reproducibility is affected partly by the HPLC system, chiefly injector operation, and partly by the sample itself. In conventional PTH analysis, the PTHs must be quantitatively transferred from the sequencer and reconstituted in a reproducible volume of injection solvent. Furthermore, PTHs must remain stable between their formation and their injection onto the HPLC column. Practical limitations of these processes typically give a cycle-to-cycle variability (RSD) of 15 to 20%. With direct, on-line transfer of PTHs from the sequencer to the HPLC, however, PTH instability can be minimized and an RSD of only 2% is possible.[7]

4. Columns

A variety of HPLC column packings have been used for PTH analysis. Although the octyldecylsilyl[7,8] and cyanopropylsilyl[9] silicas are currently the most widely used, octylsilyl,[10] phenylsilyl,[11] and mixed[12] supports have also been used successfully. However, selectivity of column packings of the same nominal type vary substantially from one manufacturer to another, and even from one batch to another from the same manufacturer. The reverse-phase loading, residual silanol level, nature of any end-capping, particle diameter, and particle pore size all affect the selectivity and resolution of the packed column. The length and internal diameter of the column also affect resolution and detection sensitivity, with smaller diameter columns capable of providing higher sensitivity if used with suitable HPLC equipment.

5. Isocratic vs. Gradient Elution

Both isocratic[8,13,14] and gradient[7-12] elution systems have been used successfully for PTH analysis. Isocratic systems are generally simpler, less expensive, and more easily transported from one set of HPLC equipment to another. They also place less stringent requirements on the purity of the mobile phase and the function of the solvent pump and mixing systems. However, they also have many drawbacks.

When used with complex samples containing many compounds that have widely differing retention behaviors, they typically exhibit the general elution problem.[15] Early-eluting peaks in the chromatogram are easily seen, but are not well separated. Later-eluting peaks are separated well, but they come off the column too slowly and are difficult to detect because of excessive band broadening. Precolumn dead volume, either in the injector, the tubing between the injector and the column, or the column inlet, tends to broaden peak elution volumes in isocratic elution, thereby decreasing both resolution and sensitivity. Use of larger sample volumes also tends to broaden peak elution volumes, particularly with smaller bore columns. The necessity of using smaller sample volumes to obtain good resolution may limit the effective sensitivity of samples that require larger volumes for proper dissolution and transfer onto the column. Finally, highly retained compounds injected with one sample may elute as "ghost" peaks after injection of a second or third sample and interfere with peak identification in those samples.

Gradient elution systems are more complex and expensive, but they provide more uniform sensitivity as well as flexibility in adjusting the parameters necessary for good resolution.

They also permit injection of larger sample volumes without peak broadening of early-eluting components, and precolumn void volumes are relatively insignificant. As a result, they are well suited for use with smaller bore columns. Mobile phase purity is crucial to good results, since impurities can be concentrated on the column under starting solvent conditions and elute as artifact peaks or sharply rising baselines during gradient development.

6. Conventional vs. On-Line Analysis

With conventional HPLC analysis, the PTH samples are left in the sequencer either in solution or as a dried residue in fraction collector vials. Samples must be removed from the collector and manually prepared for analysis. Virtually any type of HPLC system can then be used, although a good HPLC autosampler is essential for reproducible results. One of the primary drawbacks of this technique is the significant delay between the Edman chemistry and obtaining the analytical results. This delay reduces the efficiency of the sequencer by as much as half. Moreover, the manual sample preparation often causes contamination, sample loss, and PTH degradation — problems that are particularly damaging to microsequencing results.

With on-line HPLC analysis, as each PTH sample is produced by the sequencer, a portion is automatically transferred into the HPLC system and analyzed. This provides rapid acquisition of results with minimal user intervention and none of the sample work-up problems inherent in conventional PTH analysis. An interface between the sequencer and the HPLC system is required to provide efficient and reproducible sample transfer and sampling as well as operational synchronization of the two units.

B. Instrumentation

The instrumentation used in the PTH separation protocol described in the next section is the Applied Biosystems Model 120A PTH Analyzer. This is a microbore, gradient HPLC system equipped with a high sensitivity, variable wavelength UV detector. It is designed specifically for use as an on-line PTH analysis system in conjunction with the Applied Biosystems Model 470A Gas Phase Protein Sequencer.

The basic components of the Model 120A PTH Analyzer are depicted schematically in Figure 5. Two solvents, labeled A and B, are pumped through the Model 120A by the micropump. The mixed solvents flowing from the pump pass through the injector to the analytical column housed in the temperature-controlled column oven. Solutions of PTH amino acid samples injected into the flow are resolved on the column. The separated PTHs eluted from the column pass through the flow cell of a variable wavelength UV detector where their absorbance at 270 nm is measured. The analog absorbance values, related to the concentration of the PTHs, are output to optional strip chart recorders, plotter-integrators, or computers for identification and quantitation. Model 120A operations are governed through the control panel located on the front of the instrument. The keyboards and displays on this panel allow user control of pump operation, injector actuation, oven temperature, and detector wavelength and output range.

1. Micropump

The MPLC® MicroPump (Figure 6) is a programmable, dual-syringe pump designed for optimum performance in both gradient and isocratic microbore chromatography. The syringe pumps are scaled to do one complete analysis with the volume of liquid contained in the syringe chambers (10 mℓ each) and then refill before the next analysis. This batch processing minimizes the problems associated with solvent compressibility at microbore flow rates, since the compression occurs only once per analysis, at the beginning when each syringe is full. Flow direction control (i.e., whether the syringes are pumping to the column or refilling) is controlled, not by check valves, but by a high-performance, motor-driven, switching valve that must also be cycled only once per analysis.

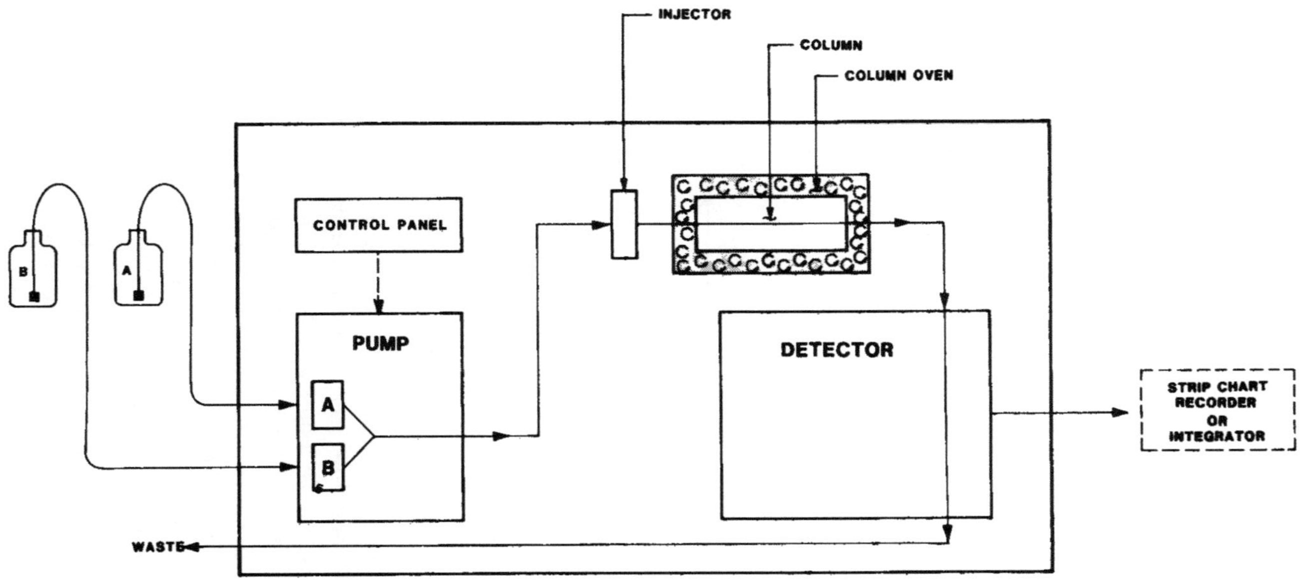

FIGURE 5. Schematic diagram of Model 120A PTH analyzer.

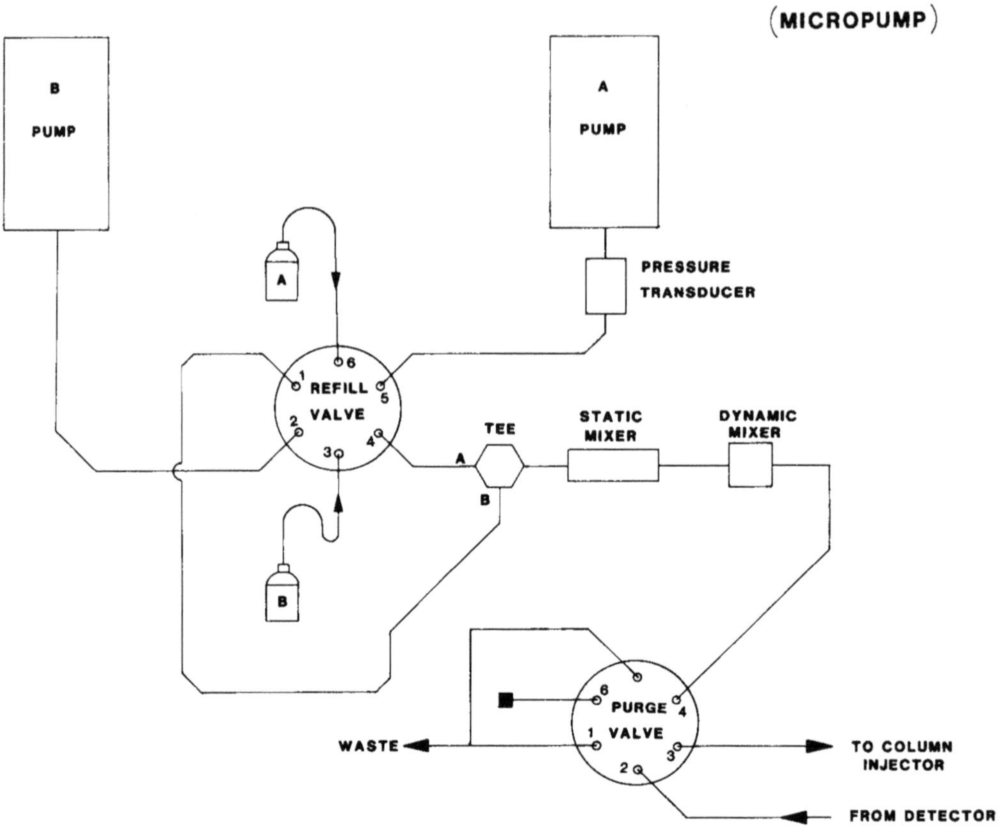

FIGURE 6. PTH analyzer MPLC MicroPump.

2. Injector

The pump output flows through the injector switching valve (Figure 7) before reaching the column. The injector directs the flow of solvent to the column, sample from the injection loop to the column, and residual sample from the Model 470A conversion flask back to its fraction collector.

The injector is toggled between the "sample load" and "sample inject" positions by energizing the appropriate solenoid valve controlling gas pressure inside the injector actuator. When the load solenoid is energized, the sample loop (located between Valve Ports 1 and 4) is isolated from the pump flow (coming from Port 3) going to the column (through Port 2). This allows the sample loop to be filled with sample solution from the syringe port located at the upper, right, front corner of the Model 120A. The back end of the sample loop is vented to atmosphere through a short length of 0.007-in. I.D. stainless steel tubing connected to injector Port 6. This tubing serves as a pressure regulator (restrictor) during on-line transfer of PTH solution from the Model 470A Sequencer. Its outlet is positioned above a small waste receptacle located on the right panel of the Model 120A.

When the inject solenoid is energized, the pump flow is directed through the sample loop to the analytical column, thus transferring the sample to the column for analysis. The syringe port is now connected to Injector Port 5 and from there through a length of 0.012-in. I.D. Teflon® tubing to the fraction collector spigot of the Model 470A Sequencer.

C. PTH Separation

The following protocol employs a gradient elution system that is suitable for PTH analysis using the on-line HPLC unit. It has been optimized to provide >95% separation of the

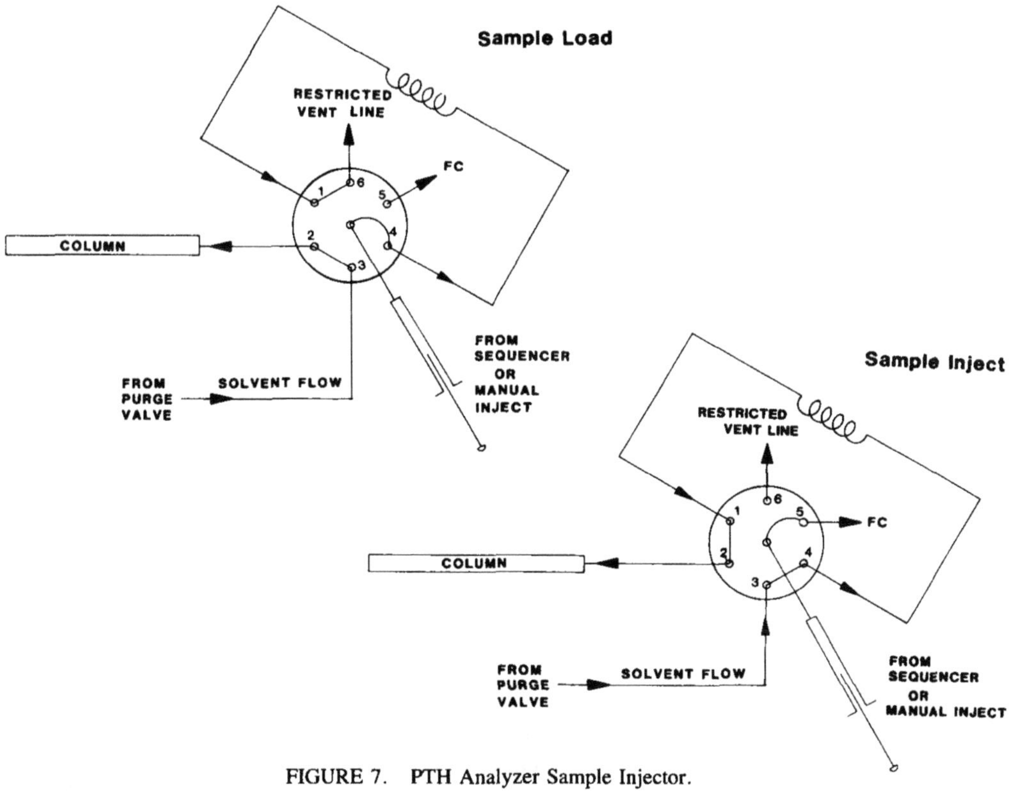

FIGURE 7. PTH Analyzer Sample Injector.

common PTH amino acids and the three primary artifacts from the gas phase sequencer. It uses the MPLC concept of interchangeable cartridges and a reusable holder that allows easy replacement of the analytical cartridge with finger-tightened connections.

Each PTH-C18 cartridge (Applied Biosystems Part Number 400293) is slurry packed with a porous, 5-μm octyldecylsilyl-type sorbent specially selected for PTH analysis. The cartridges are 22 cm in length with 2.1-mm I.D. The integral frits sealed in Tefzel® at each end of the cartridge retain the sorbent and filter solids from the sample. The cartridges are shipped containing the mobile phase (40% aqueous acetonitrile) with which the columns were tested.

1. Column Installation and Maintenance

Use clean solvents and only those which are compatible with the column packing material and Type 316 stainless steel. *Halogen acids and salts can corrode stainless steel.* The acceptable pH range is 2.2 to 8.0. All solvents should be free of particulate matter which might plug the 2-μm entrance frit in the cartridge. An in-line filter placed just before the injection system is also highly recommended.

The cleanliness of the injected sample greatly affects the column life. Samples containing components which are not eluted during the analysis cause a loss in efficiency and increase column back pressure. Column life can be extended by preparing samples free from retained components or by injecting smaller quantities. If performance loss due to retained sample components is suspected, reversing the column flow and pumping pure acetonitrile through the column may help restore performance.

High back pressure may be an indication of plugged frits. A plugged column can often be unplugged by reversing the direction of flow through the column (the column outlet should be temporarily disconnected from the detector cell to prevent particulates from lodging there).

To install the column holder, connect the female compression fittings on each of the holder end nuts to the HPLC unit using the male fittings provided with the holder and $^1/_{16}$-in. O.D. tubing. To install a cartridge, unscrew the holder end nuts from the cartridge jacket and carefully insert the cartridge. The preferred flow through the cartridge is from left to right as you read the label, but the design is symmetrical and the column may be backflushed. Reseal the holder to finger tightness using both end nuts. *Never use tools to tighten the end nuts to the jacket.* Once the cartridge is installed, leave it in the 55°C column heater or oven until temperature equilibrium is reached. Then, retighten the column holder end nuts *before flowing liquid through the column.* Loose end nuts can result in leakage of the column packing from the outlet end of the column. Leaked packing material can plug the column outlet line and/or the detector flow cell.

If the cartridge leaks at less than its rated pressure limit (7000 psi), do not try to force a seal with tools. A leak indicates that the sealing surface is either dirty, scratched, or deformed. Replace the cartridge with another and check again for leaks. If the leak persists, the high pressure seals inside the column holder end nuts must be replaced. Use the Snap Ring Seal Replacement Kit (Brownlee Part Number 140-260) for seal replacement.

To remove a cartridge, unscrew both holder end nuts. Then, using your thumb and forefinger together, pull the cartridge out. If you remove the cartridge from the holder and intend to reuse it later, store it with its ends wrapped with Parafilm® or a similar material to protect the Tefzel® frit assemblies from dirt and scratches.

2. Column Flushing

The column must be purged of the packing, testing, and shipping solvents before use for PTH analysis. This flush procedure will shorten the time required for optimizing the PTH separation and will significantly improve the peak shape of the charged PTH amino acids — Asp, Glu, His, and Arg. *This purge procedure should always be performed when installing a new column.*

Use only HPLC-quality water, tetrahydrofuran (THF), acetonitrile, sodium acetate, and acetic acid and reagent grade phosphoric acid. The HPLC grade acetonitrile should have a UV cutoff value of 188 nm or less. THF should be free of peroxides that destroy low levels of PTH amino acids and other UV-absorbing impurities that give high baselines. A 5% aqueous THF solution is more stable than neat THF.

All glassware used with the solutions must be scrupulously clean! Rinse it thoroughly with pure water or solvent, as appropriate, before using. Common laboratory detergents tend to leave deposits on the glassware that result in high gradient baselines or specific peaks in the chromatogram. Since this residue is difficult to remove by simple rinsing, *you should not generally have HPLC glassware cleaned with detergent.* Instead, have glassware dedicated only for HPLC use, rinse the glassware with water or solvent after use, and cover with aluminum foil during storage.

Start the purging process with 0.2% phosphoric acid as Solvent A and acetonitrile as Solvent B. Set the HPLC to deliver 50% A/50% B 0.4 mℓ/min for 2 hr. Discard the phosphoric acid solution, rinse the container with water, fill the container with water, and continue the purge for 20 min at 50% B. Discard the water, rinse the container with fresh water, fill the container with water, and continue the purge for another 20 min at 50% B. The system is now ready to install the separation solvents for PTH analysis.

3. Separation Solvents

A solvent kit for PTH separation is available from Applied Biosystems (Part Number 400321). It includes two solvents (acetonitrile and 5% aqueous THF), two 3-*M* sodium acetate buffer concentrates (pH 3.8 and 4.6), and an oxidant scavenger (*N,N*-dimethyl-*N'*-phenylthiourea, DMPTU). These reagents are used to form the mobile phases required for

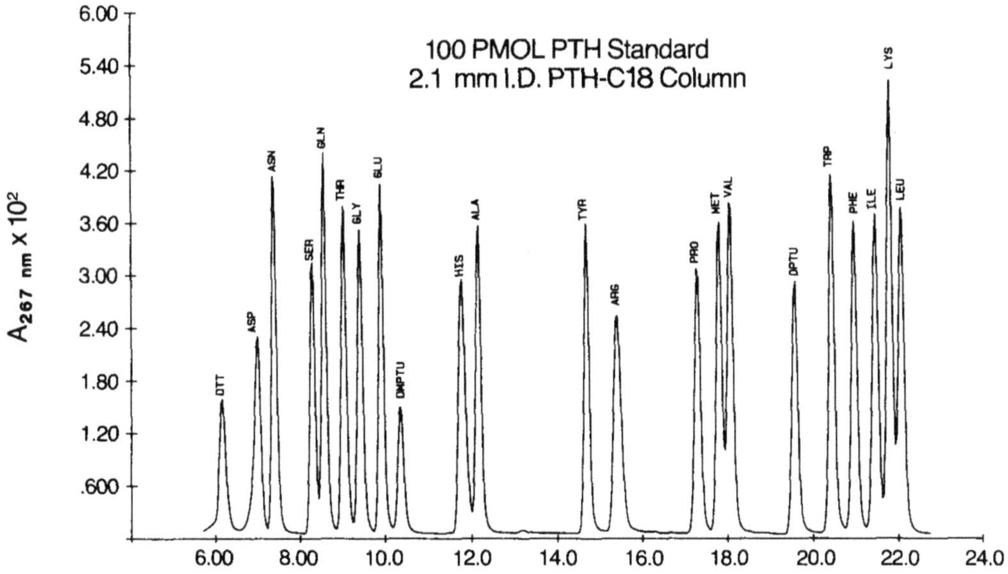

FIGURE 8. Separation of PTH amino acids on PTH-C18 column.

the gradient elution of PTHs from the analytical column. The solvents should be stored in a cool, dark place. The buffers should be stored at 4°C and the DMPTU should be stored at −20°C.

A standard PTH mixture must include the PTHs and common sequencer-derived contaminants. The Applied Biosystems PTH Standard (Part Number 400316) includes 19 PTH amino acids (no cysteine derivative), *N,N′*-diphenylthiourea (DPTU), dithiothreitol (DDT), and DMPTU. Vials containing dried PTH residues or stock solutions in acetonitrile should be stored at −20°C and protected from contamination with water.

4. Separation Optimization

Start with the following procedure and optimize it according to the accompanying stepwise procedures described below until you have obtained the correct separation. This optimization may include changes in the %B gradient, buffer pH, buffer concentration, and column temperature. Typical PTH separations obtained at Applied Biosystems are shown in Figure 8.

NOMINAL STARTING SOLVENT COMPOSITIONS

Solvent A (per liter)	5% Aqueous THF 30 mℓ pH 3.8 buffer 7 mℓ pH 4.6 buffer
Solvent B (per liter)	Acetonitrile 500 nmol DMPTU

NOMINAL HPLC PARAMETERS

Flow rate	0.2 mℓ/min
Temperature	55°C
Detector wavelength	270 nm
Nominal gradient	10% B at time 0 min 14% B at time 2 min 40% B at time 20 min 60% B at time 25 min 10% B at time 25.1 min

Two stages of optimization of the elution protocol are required. First, the gradient of increasing acetonitrile concentration must be adjusted to provide satisfactory resolution of all of the neutral PTHs. This adjustment may also include some adjustment of the column oven temperature. Second, the absolute and relative concentrations of the two buffers that are added to Solvent A must be adjusted to position the elution times of the charged PTHs.

Only two of the neutral PTH amino acids, PTH-Gln and PTH-Lys, are likely to require repositioning because of overlap with other, closely eluting PTHs. PTH-Gln elutes between PTH-Ser and PTH-Thr. It can be positioned away from PTH-Ser (and towards PTH-Thr) by lowering the %B increase during the initial 2 min of the gradient. Raising the %B increase moves it towards PTH-Ser.

PTH-Lys elutes between PTH-Ile and PTH-Leu. It can be positioned away from PTH-Ile (and towards PTH-Leu) by lowering the %B increase during the next gradient step. Raising the %B increase moves it towards PTH-Ile. With both PTH-Gln and PTH-Lys, small changes (a few %B) should be sufficient to provide the required separation, although HPLC units with very large or very small mixer dead volumes may require larger changes.

Once all of the neutral PTHs are separated from each other, the positions of PTH-Asp and PTH-Glu should be adjusted. Their positions are determined primarily by the pH of the Solvent A buffer. Lowering the pH causes both PTHs to elute later; raising the pH causes both PTHs to elute earlier. PTH-Asp should be positioned just before PTH-Asn and well after oxidized DDT, which will be present in both sequencer samples and the PTH standard. If there is insufficient spacing between DDT and PTH-Asn or PTH-Asp, then either decrease the column oven temperature by a few degrees or reduce the %B at time 0.0 of the gradient to provide the required spacing. These latter parameters may also affect the positioning of PTH-Gln and PTH-Lys.

Finally, position PTH-His and PTH-Arg by adjusting the Solvent A buffer concentration. Increasing the buffer concentration causes both PTHs to elute earlier; decreasing it causes both to elute later. They should be positioned somewhere between DMPTU and PTH-Pro so that they do not coelute with PTH-Ala, PTH-Tyr, or the DTT adduct of PTH-Ser (which elutes midway between PTH-Ala and PTH-Tyr). Ideal positions are just after PTH-Ala for PTH-His and just after PTH-Tyr for PTH-Arg.

5. Sample Considerations

The recommended solvent for sample injections is 10 to 20% aqueous acetonitrile. If higher solvent concentrations are used, especially with large injection volumes on the 2.1-mm I.D. column, early eluting PTHs may show peak broadening and poor resolution. Some PTHs, notably PTH-Ser and PTH-Thr, are unstable in aqueous acetonitrile. Standards for manual injection should be made fresh daily by diluting with water a stock solution made

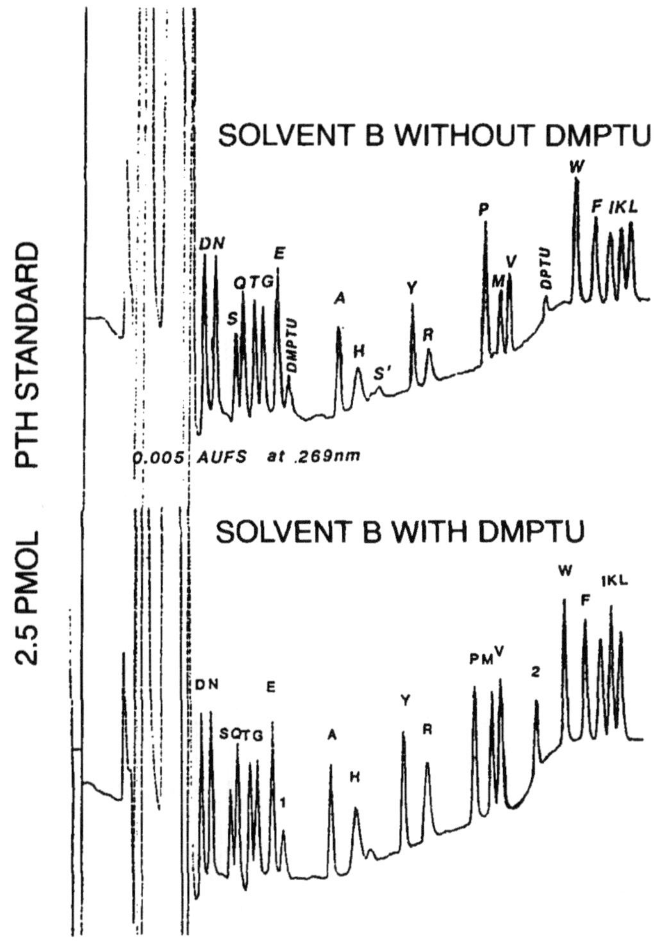

FIGURE 9. Effect of solvent scavenger addition on recoveries of low amounts of PTH amino acids from PTH-C18 column.

up in neat acetonitrile containing 0.001% DTT. Standards stored in the sequencer for on-line analysis should be dissolved in neat acetonitrile containing 0.001% DTT.

Dissolution of the PTHs in the solvents is not instantaneous. As long as 20 to 30 min may be required. Failure to wait may result in apparent low recoveries of the PTHs during the HPLC analysis. Mixing the vial contents by vortexing periodically during this time can aid in dissolution.

When the total amount of all the PTHs in a mixture injected for analysis is less than a few hundred picomoles, the recovery of several of the PTHs from the column may be low (Figure 9). This loss of sample on the column can be minimized by adding DMPTU to the Solvent B reservoir at 500 nmol DMPTU per liter of Solvent B. It serves as a scavenger for elements on the column packing surface or in the mobile phases that might otherwise destroy the PTHs. This addition to Solvent B will cause a very slight increase (about 0.001 AU) in the baseline absorbance level at the end of the gradient.

D. Amino Acid Assignment

Figure 10 shows typical chromatographic data from the first portion of a sequence analysis of a small protein. The data were obtained using on-line PTH analysis in conjunction with the gas phase protein sequencer. In data such as this, the assignment of amino acid sequence

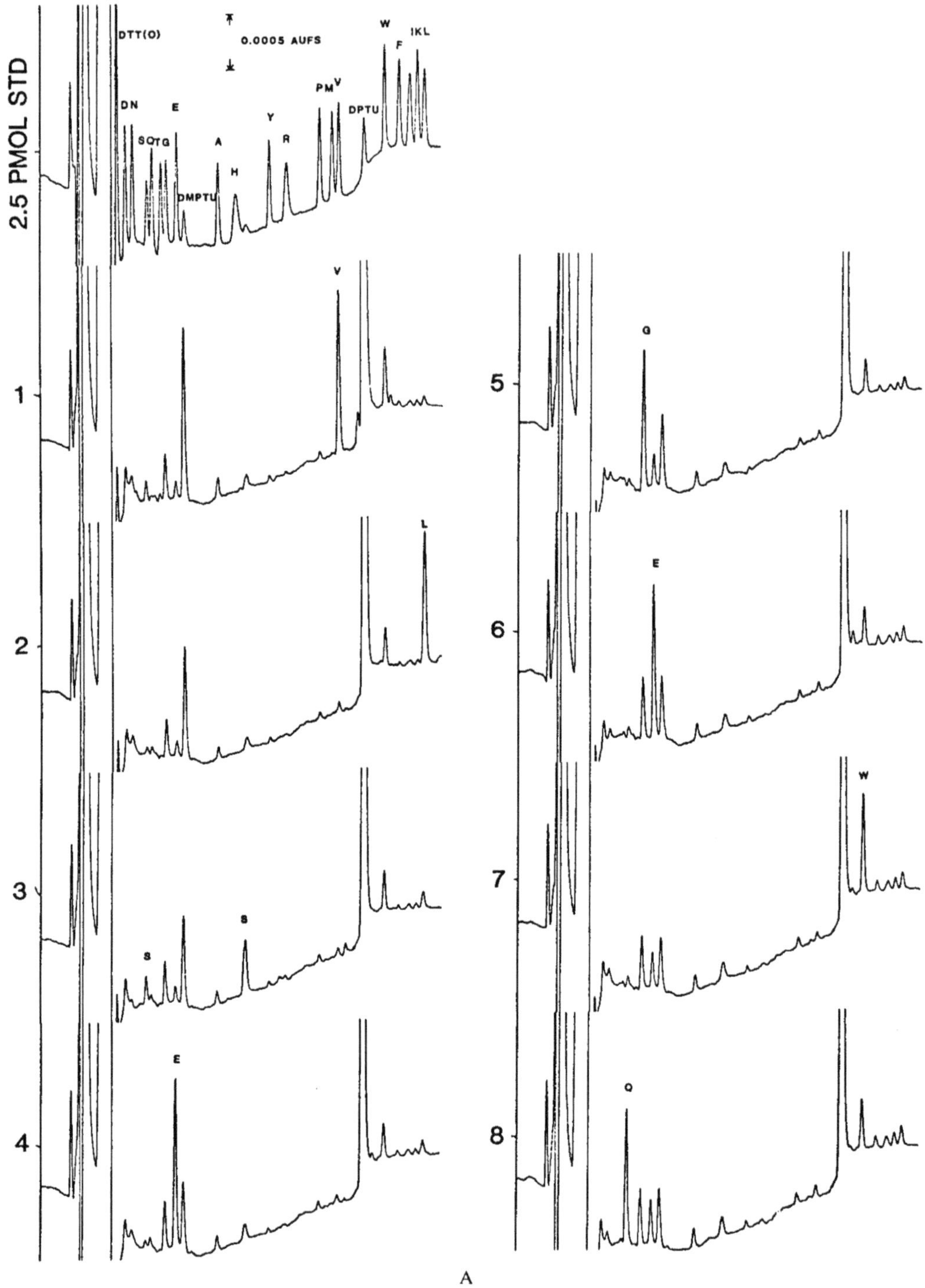

FIGURE 10. On-line PTH analysis data. Sperm whale apomyoglobin (20 pmol) was sequenced on a Model 470A Gas Phase Protein/Peptide Sequencer using Run Program 03RPTH. Portions (40%) of the PTH solution produced at each of the first 24 cycles of the degradation were automatically transferred into a Model 120A PTH Analyzer and chromatographed.

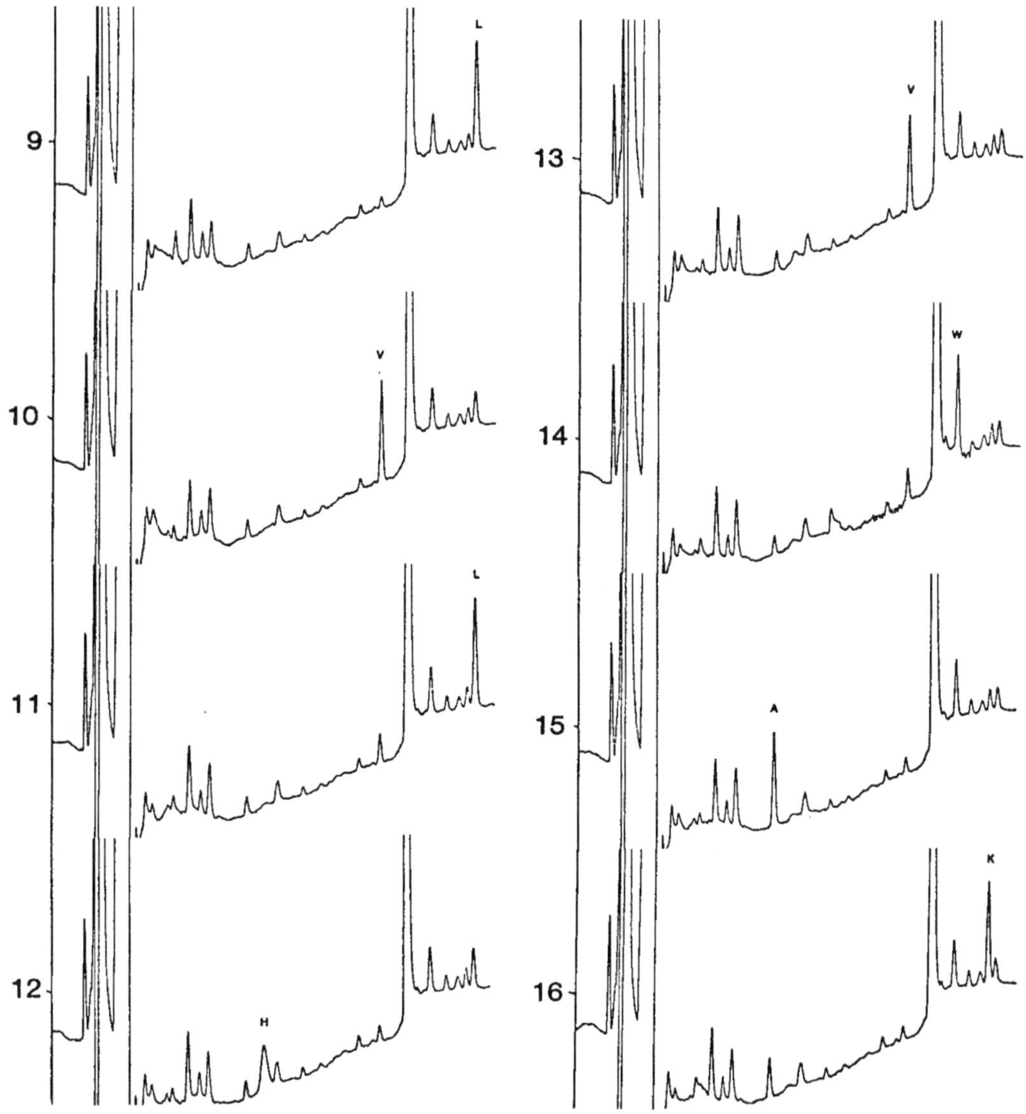

FIGURE 10B

at each cycle must be made in relation to both the background level of amino acids present and to the carryover from the preceding cycles due to incomplete degradations. The simplest method for making qualitative amino acid assignment is to overlay chromatograms from succeeding cycles on a light box and compare the increase and decrease in the heights of specific peaks. The carryover of signal in cycles immediately after that in which a signal increase above background occurs can be used to confirm the assignment of the amino acid residue at that cycle.

Quantitation can be made either by peak height or peak area measurements. Manual peak height analysis is simple and requires no on-line computer system. It is probably more accurate than peak area analysis when measuring amounts of PTH amino acids near the limits of the HPLC system's sensitivity. It is, however, quite time consuming. Peak area analysis generally requires an automated integrator that will increase the cost of the HPLC system. However, some of the more sophisticated data systems greatly simplify and speed

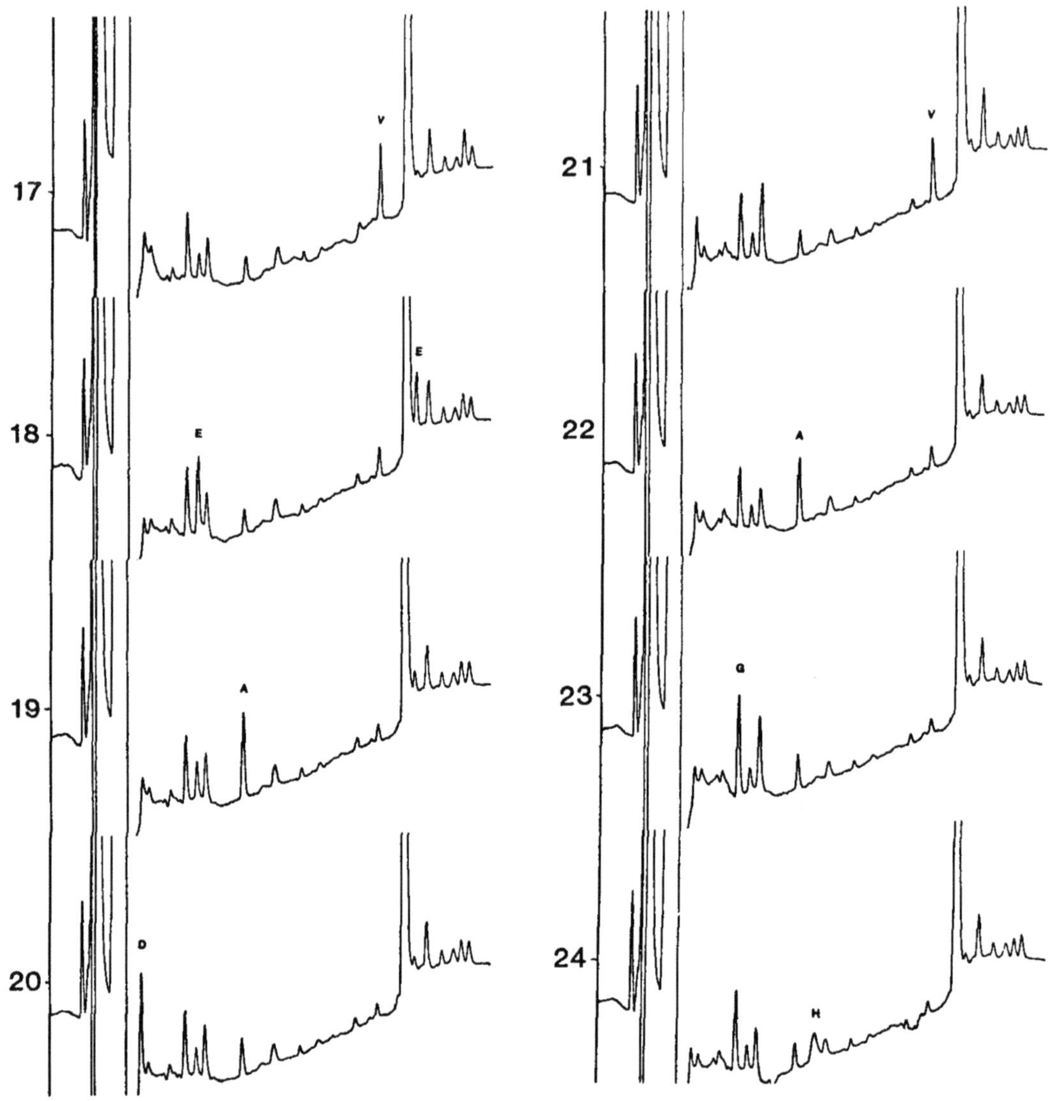

FIGURE 10C

calculations, particularly those such as baseline subtraction, correction of injection volume variability using internal standards, and conversion of peak areas to molar quantities using external standards.

Many amino acids yield more than one PTH derivative during sequencing. All of these derivatives should be used to confirm the sequence assignment based on the primary PTH derivative. Those secondary derivatives that we have identified are described below. The elution positions noted are for gradient elution using the PTH-C18 column and separation protocol described above.

1. Aspartic Acid

An unidentified derivative forms upon exposure of aspartyl residues to the coupling reagent and base during the Edman chemistry. The amount of this derivative increases through the sequencing run at the expense of PTH-Asp. However, only a few percent of the total aspartic acid appears as this derivative even after 40 or more degradation cycles in the gas phase sequencer. It elutes just before DPTU.

2. Asparagine

About 10% of PTH-Asn is degraded by deamination to yield PTH-Asp in the conversion flask under typical conversion conditions. Additional PTH-Asp can result from deamination of asparaginyl residues during purification or handling of the protein prior to sequencing.

Asparaginyl residues N-linked to complex carbohydrate moieties produce ATZ derivatives that are insoluble in Applied Biosystems Sequencer Solvent S3 (1-chlorobutane). However, removal of all but the directly linked *N*-acetylgalactosamine (AGA), by treatment of the protein or peptide prior to sequencing with endoglycosidase H, results in formation of an S3-soluble ATZ. PTH-AGAAsn elutes between oxidized DTT and PTH-Asp.[16]

3. Serine

The seryl hydroxyl group is esterified by trifluoroacetic acid during the cleavage reaction of the Edman chemistry. During subsequent conversion of the ATZ, loss of the trifluoroacetyl group gives PTH-dehydroalanine. This derivative has frequently been used for identification of serine in Edman sequencing, although it is very reactive and unstable. It elutes near PTH-Tyr and can be monitored by its absorbance at 313 nm.

The standard gas phase sequencer programs are designed to trap and stabilize the dehydro product with DTT delivered to the conversion flask just before transfer of the ATZ from the cartridge. The DTT-trapped derivative elutes midway between PTH-Ala and PTH-Tyr and can be monitored by its absorbance at 254 to 270 nm. Recoveries of this derivative are typically 40 to 60%. Some authentic PTH-Ser, usually 10 to 20%, is also recovered.

4. Glutamine

About 10% of PTH-Gln is degraded by deamination to yield PTH-Glu in the conversion flask under typical conditions. Additional PTH-Glu can result from deamination of glutaminyl residues during purification or handling of the protein prior to sequencing.

5. Threonine

The threonyl hydroxyl group is esterified by trifluoroacetic acid during the cleavage reaction of the Edman chemistry. During subsequent conversion of the ATZ, loss of the trifluoroacetyl group gives PTH-dehydro-α-aminoisobutyric acid. This derivative has frequently been used for identification of threonine in Edman sequencing, although it is only moderately stable. It elutes near PTH-Pro and can be monitored by its absorbance at 313 nm.

As noted with serine above, the standard gas phase sequencer programs trap and stabilize some of the dehydro product with DTT delivered to the conversion flask just before transfer of the ATZ from the cartridge. The DTT-trapped derivative elutes as two or four peaks midway between PTH-Tyr and PTH-Pro and can be monitored by their absorbance at 254 to 270 nm. Recoveries of these derivatives are typically 5% each. Some authentic PTH-Thr, typically 20 to 30%, is also recovered.

6. Glycine

ATZ-Gly converts to PTH-Gly somewhat slowly, the reaction being only 80 to 85% complete during the standard conversion conditions in the gas phase sequencer. The remaining 15 to 20% is observed as phenylthiocarbamylglycine (PTC-Gly), which elutes near the end of the solvent front.

7. Glutamic Acid

An unidentified derivative forms upon exposure of glutamyl residues to the coupling reagent and base during the Edman chemistry. The amount of this derivative increases through the sequencing run at the expense of PTH-Glu. After 40 or more degradation cycles in the gas phase sequencer, a substantial portion of the glutamic acid is represented by this compound. It elutes midway between DPTU and PTH-Trp.

8. Proline

Hydroxyproline (HYDPro), if present in the protein or peptide, produces PTH-HYDPro that elutes as two peaks, one just before and one just after PTH-Ala.

9. Tryptophan

An unidentified derivative of tryptophan that elutes midway between DPTU and PTH-Trp can be observed in many samples of tryptophan-containing proteins and peptides. Presumably, it is the result of modification of the tryptophanyl indole ring during sample purification or handling prior to sequencing. It may account for 0 to 100% of the tryptophan signal, but it is generally only stable enough to be seen with on-line PTH analysis.

10. Lysine

Hydroxylysine (HYDLys), if present in the protein or peptide, produces PTH-HYDLys that elutes just after PTH-Val. Methyllysine (METLys), if present, produces PTH-METLys that elutes just after PTH-Leu, with its exact position being sensitive to the chromatography buffer concentration. Succinyllysine (SUCLys), if present, produces PTH-SUCLys that elutes midway between DMPTU and PTH-Ala, with its exact position being sensitive to the chromatography buffer pH.

11. Cysteine

Authentic PTH-Cys is not usually recovered in sufficient yield to be seen. PTH-dehydroalanine, generated by loss of H_2S from the side chain, can be observed directly by monitoring at 313 nm or indirectly as its DTT derivative by monitoring at 254 to 270 nm, although the recovery of this compound is generally less with cysteine than with serine (see discussion of serine above).

Cysteine is easily identified after modification of its side chain to give a form more stable to the Edman chemistry, and a variety of modification methods have been used. Alkylation with 4-vinylpyridine to give *S*-β-(4-pyridylethyl)cysteine (PECCys) is the ideal method.[17] PTH-PECCys, which is positively charged at pH 4, can be positioned to elute midway between PTH-Val and DPTU by adjusting the chromatography buffer concentration.

Alkylation of the protein with iodoacetic acid gives *S*-carboxymethylcysteine (CMCys). PTH-CMCys elutes near PTH-Ser and PTH-Gln, with the exact position being sensitive to the chromatography buffer pH.

Alkylation of the protein with iodoacetamide gives *S*-carboxamidomethylcysteine (CAMCys). PTH-CAMCys elutes just before DMPTU. About 50% of PTH-CAMCys is degraded by deamination to yield PTH-CMCys in the conversion flask under typical conversion conditions. Additional PTH-CMCys can result from deamination of CM-cysteinyl residues during purification or handling of the protein prior to sequencing.

Oxidation of the protein with performic acid gives cysteic acid (CysA). PTH-CysA elutes near the end of the solvent front.

REFERENCES

1. **Edman, P.,** Method for determination of the amino acid sequence in peptides, *Acta Chem. Scand.*, 4, 283, 1950.
2. **Edman, P. and Begg, G.,** A protein sequenator, *Eur. J. Biochem.*, 1, 80, 1967.

3. **Wittmann-Liebold, B., Graffunder, H., and Kohls, H.,** Amino acid sequence studies on ten ribosomal proteins of *Escherichia coli* with an improved sequenator equipped with an automatic conversion device, *Hoppe-Seyler's Z. Physiol. Chem.*, 354, 1415, 1973.
4. **Hewick, R. M., Hunkapiller, M. W., Hood, L. E., and Dreyer, W. J.,** A gas-liquid-solid phase peptide and protein sequenator, *J. Biol. Chem.*, 256, 7990, 1981.
5. **Rodriguez, H., Kohr, W. J., and Harkins, R. M.,** Design and operation of a completely automated Beckman microsequencer, *Anal. Biochem.*, 140, 538, 1984.
6. **Machleidt, W. and Hofner, H.,** Fully automated solid phase sequencing with on-line identification of PTHs by high pressure liquid chromatography, in *Methods in Peptide and Protein Sequence Analysis*, Birr, C., Ed., Elsevier/North-Holland, Amsterdam, 1980, 35.
7. **Hunkapiller, M. W.,** PTH amino acid analysis, in *Proc. Symp. of American Protein Chemists*, L'Italien, J., Ed., Plenum Press, New York, in press.
8. **Zimmerman, C. L., Apella, E., and Pisano, J. J.,** Rapid analysis of amino acid phenylthiohydantoins by high-performance liquid chromatography, *Anal. Biochem.*, 77, 569, 1977.
9. **Johnson, N. D., Hunkapiller, M. W., and Hood, L. E.,** Analysis of phenylthiohydantoin amino acids by high-performance liquid chromatography on DuPont Zorbax cyanopropylsilane columns, *Anal. Biochem.*, 100, 335, 1979.
10. **Wittmann-Liebold, B.,** An evaluation of the current status of protein sequencing, in *Methods in Protein Sequence Analysis*, Elzinga, M., Ed., Humana Press, Clifton, N.J., 1982, 27.
11. **Henderson, L. E., Copeland, T. D., and Oroszlan, S.,** Separation of amino acid phenylthiohydantoins by high-performance liquid chromatography on phenylalkyl support, *Anal. Biochem.*, 102, 1, 1980.
12. **Cunico, R. L., Simpson, R., Correia, L., and Wehr, C. T.,** High-sensitivity phenylthiohydantoin amino acid analysis using conventional and microbore chromatography, *J. Chromatogr.*, 336, 105, 1984.
13. **Lottspeich, F.,** Identification of the phenylthiohydantoin derivatives of amino acids by high pressure liquid chromatography, using a ternary, isocratic solvent system, *Hoppe-Seyler's Z. Physiol. Chem.*, 361, 1829, 1980.
14. **Tarr, G. E.,** Rapid separation of amino acid phenylthiohydantoins by isocratic high-performance liquid chromatography, *Anal. Biochem.*, 111, 27, 1981.
15. **Glach, J. L.,** Gradient elution in liquid chromatography, *LC Mag.*, 2, 746, 1984.
16. **Paxton, R. J. and Shively, J. E.,** Structural analysis of carcinoembryonic antigen (CEA) and a related tumor associated antigen (TEX), in *Proc. Symp. of American Protein Chemists*, L'Italien, J., Ed., Plenum Press, New York, in press.
17. **Fullmer, C. S.,** Identification of cysteine-containing peptides in protein digests by high-performance liquid chromatography, *Anal. Biochem.*, 142, 336, 1984.

Chapter 6

HOMOLOGOUS EDMAN-TYPE REAGENTS IN MICROSEQUENCE ANALYSIS OF POLYPEPTIDES

B. Wittmann-Liebold, J. Shan-Wei, and J. Salnikow

TABLE OF CONTENTS

I. INTRODUCTION

Since the introduction of phenylisothiocyanate (PITC) for the sequential degradation of proteins and peptides by Edman,[1] this reagent has found widest application in the structure analysis of polypeptides. This was further facilitated by the design of sequencers which automatized the process and allowed to perform the degradation with PITC either in the liquid, solid, or gas phase mode.[2-4] The application of this reagent essentially increased the knowledge of peptide or protein structures in the past two to three decades. At present, the Edman chemistry using PITC is still one of the main basic techniques to elucidate the structure of these macromolecules and to gain insight in their organization and interactions at a molecular level. It complements DNA and RNA sequence analysis and allows the location of expressed genes and their initiation sites. Thus, since the rapid development of gene technology in the last few years, protein sequence analysis has become even more important.

In the beginning of protein sequencing, the Edman degradation was performed with considerably high amounts of polypeptides, e.g., in the range of milli- to micromole quantities, then the sensitivity was enhanced to 50 to 100 nmol of substance. More sophisticated versions of the degradation enabled microsequencing, facilitated by the application of higher purity grade chemicals, by miniaturization of the main sequencer parts, by decreasing reagent and solvent consumption, and by advanced HPLC detection of the released phenylthiohydantoins (for review, see Wittmann-Liebold[5]). Also, most microtechniques still use PITC; degradation with this reagent is the best studied so far, with repetitive degradation yields approaching optimal values. The question arises whether reagents homologous to PITC can be used which would allow an enhanced sensitivity in polypeptide sequencing and guarantee safe sequencing in the low picomole to femtomole range. Such Edman-type reagents were introduced in the past, like 4-*N,N*-dimethylaminoazobenzene-4′-phenylisothiocyanate (DABITC) proposed by Chang and Creaser,[6] which yields amino acid derivatives visible as red-colored products in low picomole quantities.[7] Fluorescein-isothiocyanate (FITC), introduced for sequencing by Maeda and Kawauchi,[8] forms fluorescent amino acid derivatives which are detectable in femtomole quantities.[9] Another class of isothiocyanates carrying a boc-protected amino group was proposed by L'Italien and Kent,[10] such as Boc-aminomethylphenylisothiocyanate (BAMPITC) and its homologues with extended side chains. These reagents were presumed to yield amino acid hydantoins with a free amino function upon cleavage, allowing sensitive detection with fluorescent reagents after the degradation.

In this chapter, experiences with these different reagents are summarized which have been extensively tested in our laboratories. Their applications for diverse microsequencing techniques are discussed. Further, we present a newly synthesized reagent, proposed by Jin et al.,[11] namely, dimethylaminonaphtylsulfonylamino-phenylisothiocyanate (DNSAPITC), and compare it with PITC and the other homologues for its use in microsequence analysis. Finally, we draw conclusions about future design of Edman-type reagents for advanced manual and the different automatic sequencing techniques.

II. DABITC DEGRADATION

A. Merits of the Reagent

DABITC has found much application in microsequence analysis of peptides and proteins, especially for manual sequencing, since the DABITC/PITC double coupling method was introduced.[12] It replaced the more recent dansyl Edman technique and enabled sequencing in the low nanomole range. In the "dansyl-Edman" technique,[13] the polypeptide is degraded with PITC and the identification of the next N-terminal residue is made by labeling of a peptide aliquot with dansyl chloride, dimethylaminonaphthylsulfonyl chloride.[14,15] In comparison with this technique, the DABITC/PITC method has four main advantages:

1. The degraded amino acid can directly be identified from the released derivative. With minute amounts of peptide present, the degraded residues equivalent to the total peptide sample can be used for identification. In contrast, the dansyl amino acid originates from an aliquot of the polypeptide that has to be withdrawn from the residual sample after the degradation for the reaction with dansyl chloride; this portion is then hydrolyzed and lost for further sequencing. With initial amounts of peptide of about 50 nmol, these peptide losses can be neglected, but in sequencing minute amounts of material, the fractions that have to be withdrawn for the dansylation are proportionally greater and rapidly reduce the amount of peptide available for sequencing in the following steps.
2. The direct identification of the released DABTH derivatives allows an easy differentiation of the acidic residues and their amides, which was not possible by the dansyl method due to the deamidation of the dansyl-peptide during acid hydrolysis. Identification of tryptophan residues poses a problem in the dansyl technique due to the acid lability of the indole side chain.
3. The released DABTH-amino acid derivatives are visible by eye as red-colored products after exposure to HCl gas. Therefore, their chromatographic detection on small-sized polyamide thin-layer sheets is very sensitive. By visual inspection, on these sheets, 20 pmol are easily detectable. With HPLC methods, the detection is more sensitive; quantitative determination is performed at 436 nm, either by isocratic[16] or gradient elution[17] employing C-18 or CN-derivatized reversed-phase columns.
4. The degradation with DABITC can be performed, not only with peptides, but also quite efficiently with proteins as well. In contrast, the dansyl-Edman degradation of proteins is often hampered by lower labeling efficiency of the dansyl fluorophor.

The main advantage of the reaction with DABITC is that all UV-active side products of the Edman degradation, such as diphenylurea (DPU) and diphenylthiourea (DPTU), do not obscure the identification of the released phenylthiohydantoin derivatives. Their blue color is easily distinguished from the red-colored DABTH derivatives on polyamide sheets; they also do not interfere with the detection at 436 nm.

B. Application

1. Manual DABITC Methods

DABITC is most frequently applied as DABITC/PITC double coupling technique in the manual microsequencing mode described in detail elsewhere.[12,18,19] This method is applied for peptides and proteins in the range of 0.5 to 5 nmol. Higher amounts, e.g., 10 to 20 nmol, do not substantially improve the result. With more substance present, relatively more peptide is lost during the washes (e.g., after the coupling). Small amounts of polypeptide, 500 pmol to 2.5 nmol, are best degraded with DABITC in small-sized tubes, as described by Wittmann-Liebold et al.[19] Accordingly, the applied reagent and solvent volumes are reduced by a factor of 2 to 4, resulting in decreased extraction losses of the peptides.

Very short hydrophobic peptides are covalently attached to solid support following one of the three main procedures, namely, (1) by attaching the peptide via its N-terminal and ε-amino groups of the lysine residues;[20] (2) by attaching the homoserine lactone of cyanogen bromide cleaved peptides;[21] or (3) by activating the C-terminal carboxyl group.[22] The attachment to aminopropyl glass (APG)[23] and (*p*-phenylenediisothiocyanate, DITC) DITC-activated APG is preferred to the attachment of peptides to aminopolystyrene (APS) or tetraethylenetetraminepolystyrene (TETA-resin) used earlier, since styrene supports lead to extensive reagent background due to strong adsorption of the DABITC chromophore to the aromatic support. The solid phase attachment methods have been described in detail[18,20,24] and were reviewed recently.[25,26] The former also includes necessary purification procedures of reagents and solvents for microsequencing.[25]

2. Automatic DABITC Methods

The DABITC/PITC double coupling method has been applied successfully in solid phase sequencers.[24,25] Here, the colored reagent allows a precise adaptation of the program: the color change of the APG-bound peptide to yellow/orange at the coupling and the drastic change to red upon cleavage indicate the proper timing for the reaction. Extraction of the excess DABITC reagent after the coupling reaction can be completed by thorough rinsing of the column with methanol. Usually, no losses of peptide occur at the wash stages of the program. In contrast, this cannot be avoided in either of the two other sequencer types, the liquid or gas phase machines, even if polybrene is used as a carrier.

Small-sized peptides can be degraded after attachment via the lysines or carboxyl termini, depending on the amino acid composition of the fragment. Larger peptides and proteins are preferably attached to DITC-activated APG via the lysines.[25,26] Recently, the on-column attachment of the proteins via the lysine residues, after filling of the solid phase column with DITC-activated APG, was described.[26] This method enables a higher sensitivity of the method, as the polypeptide can be attached in a closed system, and due to the recycling, higher initial yields of the released amino acid derivatives are obtained.

It is interesting to note that salt contaminations and the presence of SDS do not impair the attachment to a solid support. Up to 2% of SDS can be present or added to the polypeptide to increase its solubility during the attachment. Also, a short TFA pretreatment is recommended. Ammonium ions and buffers containing amino groups like Tris buffer have to be removed, since they will reduce attachment yields by competitive reaction with the activated support.

In most cases, automatic solid phase sequencing with DABITC/PITC allows complete degradation of polypeptides to the C-terminal residue, up to about 40 to 45 steps. By this means, complete sequences of many cyanogen bromide-generated peptides were determined.[18] Almost all purified cyanogen bromide fragments derived from ribosomal proteins of different sources (e.g., *Escherichia coli, Bacillus subtilis, B. stearothermophilus*, and of Halobacterium and other Archaebacteria [recent work in our laboratories]) were sequenced by this method.[27-31]

Further, synthetic peptides can easily be controlled in their sequence with the solid phase DABITC technique after attachment to APG. Synthetic peptides meet an increasing interest in gene technology; therefore, their sequencing at the solid support has a wide future application.

C. Limitations

DABITC has the following disadvantages when compared to PITC:

1. This reagent is not volatile; its excess after the coupling has to be extracted with sufficient amounts of solvent, which is no problem if the peptide is covalently attached to a solid support. In the liquid phase sequencer, it is retained in the Teflon® tubings which causes yellow staining (memory effect). Therefore, the extraction time has to be prolonged, which causes greater polypeptide losses.
2. In the manual sequencing techniques, hydrophobic peptides are easily lost during the extraction after the coupling. A change in the heptane/ethylacetate composition of the solvent to higher concentrations in heptane reduces these losses. But, in such cases, the attachment to a solid support and subsequent microsequencing either performed manually[18] or automatically in a solid phase sequencer is more recommended.
3. Usually, the DABITC reagent is less suitable for the identification of lysines, serines, and threonines caused by the formation of additional products.[12,18] Peptide amounts of 5 nmol for sequencing, however, make their positive identification possible.

Employing 1 *M* HCl or HCl gas-saturated acetic acid for the conversion improves the

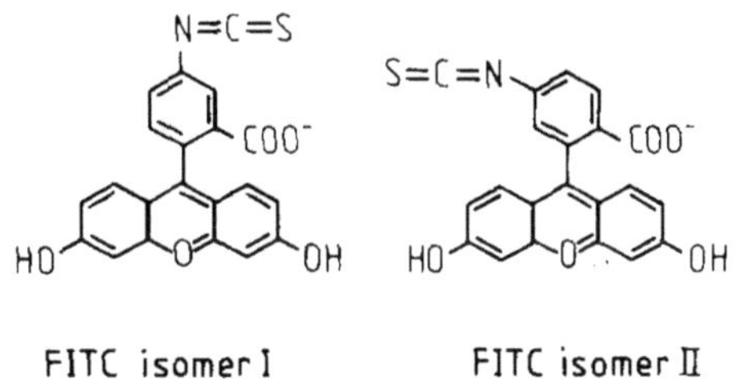

FIGURE 1. Fluorescein isothiocyanate (FITC) isomer I and II as used for microsequencing.

DABTH-serine and -threonine yields, but aqueous trifluoroacetic acid is more suitable for the detection of the acid amides. By HPLC separation, the serine and threonine derivatives can be differentiated from all other DABTH-amino acids; further, DABTH-isoleucine and DABTH-leucine separate well.[16,17]

III. FITC DEGRADATION

A. Merits of the Reagent

Recent advances in the isolation and purification of trace amounts of biologically important peptides require the development of more sensitive analysis methods. In general, there is a tendency towards the use of fluorescent compounds for microsequencing and sensitive amino acid analysis. The reagent FITC has been shown useful for adding a fluorescent chromophore to the amino groups of amino acids, peptides, and proteins.[8,32] However, the commercially available FITC dyes have shown great variation in purity. They should be checked by infrared spectroscopy, where isothiocyanate forms a strong isothiocyanate absorption peak with a maximum at 2080 cm^{-1} in dry dimethylformamide.[33,34]

FITC is available in two isomeric forms (FITC I and FITC II), as shown in Figure 1. According to the spectroscopic measurements, isomer I might be superior for fluorescent detection. In our laboratory, both isomers were tested for their ability to couple to polypeptides, and isomer II gave better results in microsequencing.[49]

Detection of the released FTH-amino acid derivatives can be done by two-dimensional polyamide thin-layer chromatography[32,35] or by HPLC separation.[9] Micropolyamide thin-layer chromatography was performed on 5 × 5 cm sheets using the following three different solvent systems:[36]

- System I: acetone/benzene/glacial acetic acid (4:2:1, v/v)
- System II: water/pyridine/glacial acetic acid (10:8:1, v/v)
- System III: hexane/methanol/pyridine (10:3:2, v/v)

We performed the TLC identification on 5 × 5 or 2.5 × 2.5 cm sheets with Solvent Systems I and II, but changed the composition of II to 10:7.5:1.5, v/v. Detectable are 10^{-9} mol of FTH-amino acids as yellowish spots and 10^{-12} mol as their greenish-yellow fluorescence under UV light (with increased fluorescence after exposure to ammonia gas).

After HPLC separation on C18 reversed-phase columns with a gradient of 10 m*M* sodium phosphate buffer, pH 7.0, and acetone at 60°C, the fluorometric determination of the FTH-amino acids was made at 515 nm (maximum emission wavelength is at 518 nm) with the excitation wavelength at 470/480 nm (or excitation with the 360-nm line of a Hg lamp).[9] Detection limit was less than 0.5 pmol for a single component.

Table 1
DETERMINATION OF INCOMPLETE REACTION OF FITC WITH INSULIN B CHAIN (OXIDIZED)

Cycle	a	b	c	d	e
1	*FITC*	PITC	PITC	PITC	PITC
2	PITC	*FITC*	PITC	PITC	PITC
3		PITC	*FITC*	PITC	PITC
4			PITC	*FITC*	PITC
5				PITC	PITC

Note: In parallel experiments a to e, 5 nmol each of insulin B were degraded, applying one coupling each with the indicated reagent. The degradation with PITC was done at 55°C for 50 min; that with FITC at 55°C for 60 min.
The FTH-amino acid derivatives were identified by two-dimensional polyamide thin-layer chromatography (see text) on 5 × 5 cm sheets. Overlaps caused by the incomplete reaction with FITC were calculated from the reduced PTH-amino acid liberation in the following step in comparison to the reaction with PITC. PTH-amino acids were determined by isocratic HPLC separation on reversed phase C_8-columns.[43]

B. Application

Kawauchi et al.[37-42] have determined the sequences of some short peptide hormones and some proteins using FITC in the manual liquid phase sequencing mode. But since FITC carries functionally active hydrophilic groups (–COOH or lactone and –OH), more byproducts are formed at the coupling with this reagent than with PITC or DABITC. These interfere with the FTH-amino acid identification. Thorough extraction after the coupling stages is advisable, which can be performed more easily if the peptide is covalently bound to a solid support. Further, this would enable a greater excess of the reagent to be applied in order to increase the coupling yield.

Thus, we tested the FITC isomers I and II with oxidized insulin B chain, both by the manual liquid and by the solid phase method, and determined the rate of overlapping by replacing the FITC reagent by PITC alternately in successive cycles (degradation procedure as given in Section III.C, but only one coupling, each omitting the second coupling with PITC) in five parallel experiments as follows (see, also, Table 1):

- Cycle 1: 1a with FITC; 1b to 1e with PITC
- Cycle 2: 2a with PITC; 2b with FITC; 2c to 2e with PITC
- Cycle 3: 3a and 3b with PITC; 3c with FITC; 3d and 3e with PITC
- Cycle 4: 4a to 4c with PITC; 4d with FITC; 4e with PITC
- Cycle 5: 5a to 5e with PITC

The rates of overlapping were estimated by TLC identification of the released FTH-amino acid derivatives and by quantitative determination of the PTH-amino acids on isocratic reversed-phase HPLC.[43]

C. Solid Phase Sequencing with FITC

1. Attachment of the Peptide to DITC-APG

Ten to fifty nanomoles of peptide were dissolved in 50 μℓ 50% pyridine and transferred to a glass-fritted stopper tube and dried *in vacuo* 0.8 to 1.5 mℓ of 0.2 *M* $NaHCO_3$/NaOH buffer of pH 9.0 and 20 to 50 mg of DITC-APG (see Wittmann-Liebold[25]) were added under nitrogen and the glass kept at 55°C for 1 hr with gentle stirring; 20 to 50 μℓ of ethanolamine were added and stirred for 15 min at 55°C under nitrogen (saturation of excess of DITC groups). The support was washed twice with 2 mℓ of bidistilled water, centrifuged, and rewashed twice with 2 mℓ methanol. After centrifugation, the upper layers were discarded and the glass dried *in vacuo*.

2. Solid Phase Degradation with FITC/PITC Double Coupling Method

First cycle with PITC — To 1 to 10 nmol peptide (0.5 to 1 nmol/mg DITC glass) 600 μℓ of 67% pyridine in water and 50 μℓ of PITC were added under nitrogen. The mixture was kept at 55°C for 30 min with gentle stirring, washed twice with 2 mℓ methanol, centrifuged, the upper layer discarded, and the glass dried *in vacuo*; 200 μℓ anhydrous TFA were added under nitrogen and kept at 55°C for 10 min with stirring. The phenylthiohydantoin was extracted twice with 2 mℓ methanol and the extracts dried *in vacuo*.

Second cycle with FITC/PITC — Four hundred microliters of 50% pyridine in water and 200 μℓ of FITC solution (5 μmol/mℓ in pyridine) were added to DITC glass-attached peptide. The tube was flushed with nitrogen and incubated at 55°C for 30 min. A further 200 μℓ of FITC solution was added, flushed, and reincubated at 55°C for 30 min. Then 50 μℓ of PITC was added, the tube flushed with gas and incubated for 30 min at 55°C. The mixture was centrifuged and the supernatant removed, the beads washed three times with acetone and methanol, the glass support dried *in vacuo* and 200 μℓ of anhydrous TFA added. The incubation was made at 45°C for 10 min (for serine, threonine, arginine, and histidine residues), 30 min (for valine, isoleucine, phenylalanine, and leucine), or for 15 min (all others). After centrifugation the TFA layer was transferred to another tube and dried *in vacuo*. The dried thiazolinone was converted in 50 μℓ 40% TFA in water for 15 to 30 min at 45°C under nitrogen (Ser, Thr, Arg, His — 15 min; other amino acids — 20 min; Gly — 30 min). The FTH-amino acids were dried *in vacuo*. The residual glass-attached peptide was washed twice with 2 mℓ acetone and dried *in vacuo*, then subjected to further degradation cycles.

3. Results of the Degradations

FITC is more suitable for solid phase sequencing of the peptides tested. In the manual liquid mode, peptides were lost more readily at the wash procedures. By this method more spots were visible on TLC than after attachment to solid support. In comparison with the DABITC/PITC double coupling method, no increased sensitivity could be reached applying the FITC/PITC technique, since the coupling with FITC was decreased to 30 to 40%. The increased background contaminants cause less safe identification of the released FTH-amino acids as compared to DABITC and do not allow extended degradations of more than 15 residues. Although the glass beads were thoroughly washed, the excess of the reagent could not be removed completely.

D. Limitations

1. Degradation of polypeptides with FITC is less suitable for the liquid phase method. Due to fewer contaminants, sequencing in the solid phase manner was more efficient than by the liquid phase manual method.
2. Comparison of the FITC reagent with DABITC showed no increased sensitivity: the higher detection sensitivity of the fluorescent isothiocyanate is compensated by the lower coupling yield of this reagent, probably caused by steric hindrance of the bulky chromophore. Especially the N-terminal residues react poorly, which may be caused by repulsion of the negative groups between the dye and the peptide.

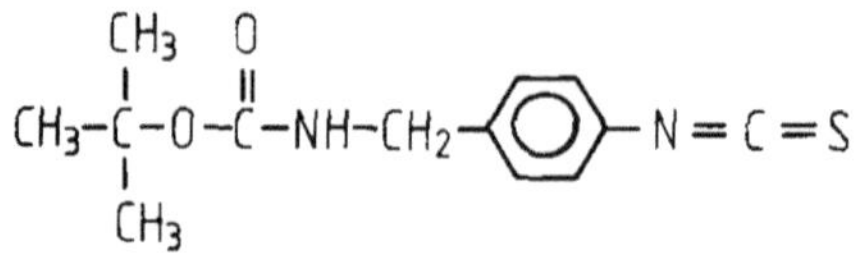

FIGURE 2. Tert.-Boc-aminomethyl-phenylisothiocyanate (BAMPITC) as proposed for microsequencing[10] and synthesized according to Palasz et al.[45]

3. FITC may decompose during the degradation. The reagent carries many functional side groups that can give rise to many side reactions. Consequently, many extra spots are visible on the thin-layer sheets which obscure the degradation result. Therefore, higher quantities of the peptide must be subjected to the sequence analysis than with DABITC or PITC.
4. Removal of the residual FITC reagent after the reaction is more difficult, even with extensive wash cycles.
5. Separation of all FTH-amino acids by two-dimensional thin-layer chromatography on polyamide sheets is not possible: FTH-Met/Phe/Val and FTH-Ile/Leu are poorly resolved. If the isomer II is applied instead of isomer I, FTH-Met/Phe/Val are better resolved; the reverse is true for the separation of FTH-Ile/Leu.
6. Noncovalent adsorption of the FITC fluorochrome to proteins cannot be neglected. Sequencing of proteins is hampered by this effect. Further, adsorption of FITC to the glass support seems possible.

IV. BAMPITC DEGRADATION

A. Design of the Reagent

A novel Edman-type class of reagents was proposed by L'Italien and Kent[10] which carries a masked amine of the general structure tert.-Boc-NH-(CH2)n-PITC (see Figure 2). The feature of these compounds would be the unmasking of a primary amino group during the acid cleavage step of the degradation of the polypeptide. This free amino function would be available for any fluorescent postdegradation detection of the released amino-(CH2)*n*-phenylthiohydantoin, e.g., by reaction with ortho-phthaldialdehyde (OPA)[44] or other reagents which generate fluorescing products. In principle, this type of reagent should yield at least a 2.5-fold increase in sensitivity over the usual 254-nm UV detection of the PTH-amino acids released. Therefore, the application of this class of reagents would provide a general method for sequencing in the low picomolar to subpicomolar range.

As this type of reagent seemed very promising for future high sensitivity microsequencing, we synthesized the reagent BAMPITC, tert.-Boc-aminomethyl-phenylisothiocyanate,[45] and tested its sequencing abilities. As has been stated in previous papers,[10,45] the degradation of peptides or proteins with this reagent is nearly as complete as with PITC. We examined the degradation of oxidized insulin B chain, myoglobin, and of ribonuclease A with this reagent and determined the amount of carryover after one, two, three, and four cycles of BAMPITC degradation by reaction with PITC thereafter and PTH-amino acid HPLC analysis. In parallel experiments, the carryover was determined by replacing BAMPITC by (1) PITC or (2) DABITC, respectively, as given for the experiments with FITC in Table 1. By sequencing manually under liquid phase conditions, the BAMPITC coupling yield ranges from 80 to 95%, depending on the amino acid residue under examination. This is more complete than the reaction with DABITC and, therefore, a second coupling with PITC to trap the unreacted free amino groups of the polypeptide by degradations with BAMPITC is not always mandatory. However, for hydrophobic and basic residues, a short incubation with PITC following the reaction with BAMPITC is advisable.

In the case of sequencing with BAMPITC in a solid phase sequencer (conditions as employed for the degradation with DABITC, see Salnikow[24]), a second coupling with PITC thereafter is necessary, as coupling yields of only 50% were obtained similar to those reported for the manually performed solid phase degradation.[45]

B. Identification of the Released Derivatives and Coupling Yields

We tried to identify the amino acid derivatives released by degradation with BAMPITC by (1) direct UV detection in an isocratic PTH-amino acid system; (2) by thin-layer chromatography on polyamide sheets after reaction of the released thiohydantoins with dabsyl chloride (dimethylaminoazobenzenesulfonyl chloride);[46] (3) by HPLC analysis of the OPA-reacted derivatives; and (4) by mass spectrometry of the released thiohydantoins or the dabsyl derivatives.

Direct UV detection of the released derivatives after degradation with BAMPITC was done by HPLC in the isocratic 19.5% 2-propanol/1% tetrahydrofuran system described for PTH-amino acids,[43] which yielded specific derivatives for the different amino acids in each of the degradation cycles eluting earlier in the chromatogram than the PTH-amino acids, which is expected due to the increased hydrophilicity caused by the additional amino group in these molecules. For quantitative measurements the content of 2-propanol in the HPLC solvent mixture was reduced to 15% to obtain a higher resolution of the hydrophilic derivatives. Oxidized insulin B chain was covalently bound to aminopropyl glass and degraded in a solid phase sequencer once by performing a BAMPITC/PITC double coupling (conditions as for the DABITC/PITC method and 0.5% BAMPITC dissolved in DMF) and once only employing the BAMPITC reagent. Under the conditions of the double coupling method the degradation was complete, but with BAMPITC alone approximately 50% overlaps were noticed.

We further degraded the di-peptide Phe-Pro covalently attached to aminopolystyrene (via the carboxyl-groups) with BAMPITC and investigated the thiazolinone released in the first cycle by TFA treatment after evaporation of the TFA by mass spectrometry (E.I. mass spectrum at 70 eV accelerating voltage and 300°C emitter temperature) and obtained the expected mass ion of the amino-methyl-phenylthiazolinone of phenylalanine of 311.[50]

On the other hand, thin-layer chromatography of the released thiohydantoins of insulin degradations yielded no specific derivatives characteristic for each individual amino acid after the reaction with dabsyl chloride, nor were they found after OPA precolumn derivatization in a HPLC amino acid analyzer in the expected amounts. The dabsyl reaction products were further investigated by mass spectrometry which could not prove the formation of the expected dabsylamino-phenylthiohydantoins. Obviously, the degradation with BAMPITC and the reaction thereafter to form fluorescent amino acid derivatives causes partial decomposition and additional side reaction(s) which lower the expected yields of the amino acid derivatives suitable for any fluorescent detection.

Therefore, the application of BAMPITC for the degradation of minute polypeptide quantities to obtain highly sensitive fluorescent amino acid derivatives is hampered by (1) the decreased coupling yields of residues with bulky side chains, (2) the insufficient formation of the wanted fluorescent products, and (3) by the need to perform an additional reaction following the degradation causing additional side reactions. On the other hand, the direct detection of the released BAMPITC-reacted amino acids by UV offers no advantage over the usual PITC degradation, since no enhanced sensitivity is obtainable.

Concerning the other reagents proposed of the type tert-Boc-(CH2)-phenylisothiocyanate, it is likely that also the reagents with elongated carbon side chains may behave similarly and, thus, would be useless for an enhanced fluorescence identification of the degraded amino acid residues of unknown polypeptides. Whether the reagent tert-Boc-amino-phenylisothiocyanate lacking the spacer methyl group would be superior has not been examined

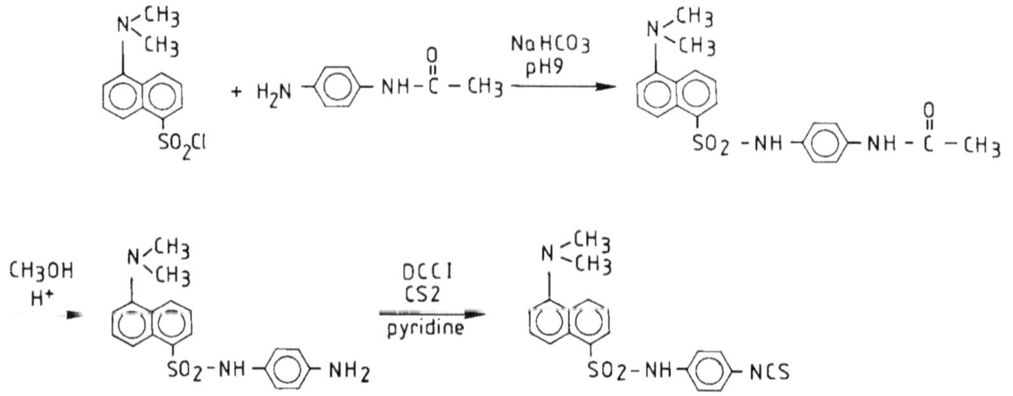

FIGURE 3. Synthesis scheme of 4,-*N,N*-dimethylamino-naphthyl-5-sulfonyl-amino-phenylisothiocyanate (DNSAPITC). (From Jin, S.-W., Chen, G.-X., Palacz, Z., and Wittmann-Liebold, B., *FEBS Lett.*, 198, 150, 1986. With permission.)

yet in our laboratories. Instead, it seems more feasible to employ Edman-type reagents which carry directly a suitable fluorophor for sensitive microsequencing. Although then it might be necessary to carry out a second coupling with PITC to complete the reaction, the degradation employing such a reagent would be far simpler in its use.

V. DNSAPITC DEGRADATION

A. Properties of the Reagent

A novel fluorescent PITC homologous Edman-type reagent, the DNSAPITC (4-*N,N*-dimethylaminonaphthyl-5-sulfonylamino-phenylisothiocyanate), was proposed by Jin et al.[11] following a simple three-step synthetic route as shown in Figure 3. This reagent should react more readily with the free amino groups of polypeptides than does FITC, comparable in its coupling abilities with that of dansyl chloride. The new isothiocyanate reagent combines the useful properties of the fluorochromophore with that of a hydrophilic sulfonylamine spacer, which provides an increased solubility and less steric hindrance of the chromophore for the attachment of the polypeptide to the isothiocyanate group. The reagent was crystallized and characterized by thin-layer chromatography, I.R., and mass spectrometry.[11]

B. Manual Microsequencing with DNSAPITC

Manual degradations with the so-called "Jin" reagent were studied by degrading the oxidized insulin chains employing a double coupling method similar to that described for the reaction with DABITC and identification of the released DNSAPTH-amino acid derivatives by two-dimensional thin-layer chromatography as detailed.[11] More recently, a variety of synthetic peptides and standard proteins were studied with this reagent, and the procedure for the manual liquid phase degradation, mainly of the extraction of the excess reagent by employing ethylacetate/benzene (1:1, v/v) or benzene/butylacetate (1:1, v/v) as solvent mixture, was improved.[47]

C. Automatic Solid Phase Sequencing with DNSAPITC[48]

For solid phase sequencing in the automated mode, 0.5 to 1 nmol oxidized insulin B chain was coupled to 5 to 10 mg aminopropyl glass (lysine attachment) and degraded in a modified LKB solid phase sequencer using PITC for the first and a double coupling with DNSAPITC and PITC for all other steps; in parallel, one cycle each was reacted with DNSAPITC alone and the overlap produced measured in the following step by the reaction

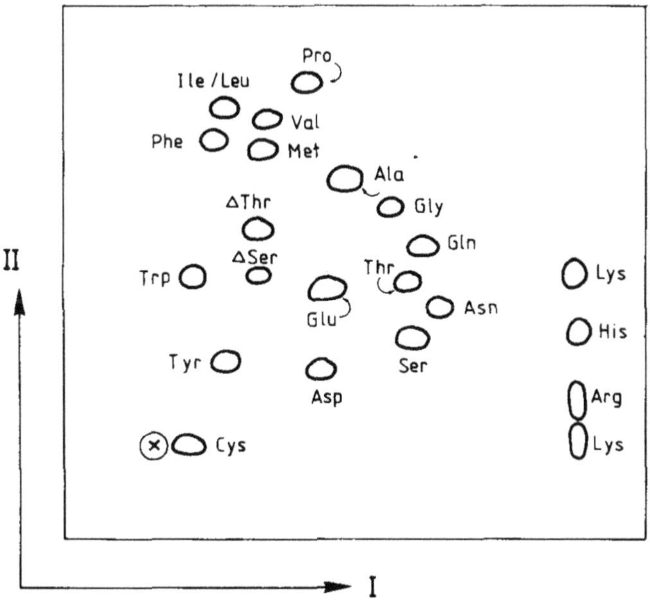

FIGURE 4. Two-dimensional polyamide thin-layer chromatography of standard DNSA-PTH-amino acid derivatives (for details see text); detection at 366 nm.

with PITC. The sequencing program contained a total of 30 min coupling time at 50°C in 20% 0.4 *M N*-ethyl morpholine trifluoroacetate, pH 9/80% pyridine as coupling buffer, and a solution of 1% reagent in DMF (reaching a final reagent concentration of 0.3 to 0.4% and of 53% pyridine in the sequencing column). The wash procedures and the cleavage were performed as given for DABITC/PITC degradations.[24]

D. Two-Dimensional Thin-Layer Chromatography

Standard DNSA-PTH-amino acids were prepared and appropriate solvent mixtures for the two-dimensional chromatography on 5 × 5 or 2.5 × 2.5 cm polyamide sheets (Schleicher/Schuell, F1700) developed.[11] Best separation of all derivatives (except DNSA-PTH-Ile/Leu) was achieved using the following solvents (see Figure 4):

- System I: water/acetonitrile/formic acid (5:1:0.5, v/v)
- System II: benzene/acetic acid (2:1, v/v)

On these polyamide sheets, 1 to 5 pmol of the DNSA-PTH-amino acid derivatives are visible on exposure at 366 nm.

The thin-layer identification, however, is obscured to some extent (above the origin and in the area of the basic derivatives) by additional fluorescing spots caused by side reactions of the degradation and obviously even more by traces of fluorescence-positive contaminants of the sheets. It was found that the quality of the thin-layer chromatogram can largely be improved by prechromatography of the sheets prior to sample application in the second-dimension solvent mixture and in the direction of the second dimension and then followed by application of the DNSA-derivatives and performing the two-dimensional chromatography. In this manner, clear spots of individual degradation cycles can be obtained, thus rendering this technique useful for manual microsequence analysis. When performing the DNSAPITC degradations in shortened "dansyl glass" tubes, the method can be applied for 100 to 500 pmol; simultaneously, up to 30 samples can be degraded.[47]

E. HPLC Separation of DNSA-PTH-Amino Acids

A gradient HPLC system was established for the separation of the DNSA-PTH-amino acid derivatives employing a Waters HPLC system equipped with a Shimadzu fluorescence HPLC monitor model RF-530 set at 340 nm (excitation wave length) and 540 nm (emission).[48] All but two pairs of derivatives (except Ala/Tyr, Val/Trp) were completely resolved on a 3-μm Spherisorb C8 column using a linear acetonitrile gradient from 45% towards 70%; they can be further examined by additional thin-layer chromatography (see above). The detection limit of the DNSA-PTH derivatives employing this fluorescence detector was at 100 to 200 fmol (highest setting of the detector), whereas 1 to 10 pmol were easily detectable at lower range settings.

F. Coupling Yields

Standard peptides and proteins were degraded with this reagent by the manual liquid and automated solid phase technique as described above. The rate of overlapping was determined by replacing PITC by the new reagent at the respective cycles and measuring the overlap in the following step by reaction with PITC (as exemplified in the scheme of Table 1). In the manual mode, coupling yields comparable to that of PITC and BAMPITC were obtained; in the solid phase mode 60 to 70% coupling yields were found. The performance of a second coupling is mandatory for all long degradations and if amino acids with bulky side chains are contained in the peptide. However, the reagent reacts faster as DABITC with the primary amino groups of the peptides and, consequently, a shorter second coupling with PITC following the first with the fluorescent reagent suffices (e.g., 10 min at 50°C).

G. Applications

The new reagent DNSAPITC is well suited for sensitive microsequencing of peptides and proteins, since the coupling yields are much better than those observed with DABITC or FITC. For short peptides, the application of a double coupling with DNSAPITC/PITC is not mandatory; for extended sequencing, e.g., of bigger peptides or proteins, a second, but short, incubation with PITC after the reaction with the fluorescent reagent is advisable to diminish overlap at certain residues.

The reagent is not volatile and, hence, it has to be extracted with organic solvents, e.g., mixtures of ethylacetate or butylacetate and benzene. In comparison to the reagent DABITC, the reaction time is shorter and the sensitivity of the fluorescent derivatives increased over the red-colored ones. On the other hand, the easy differentiation between the real red-colored DABTH-amino acids and the blue unwanted by-products of the degradation is not given in the case of the application of any fluorescent isothiocyanate. However, by the improved thin-layer identification of the DNSA-PTH-amino acids as described in this chapter, only two faint additional fluorescent spots are visible on the two-dimensional micro-TLC-sheet (probably caused by the usual thiourea derivates), and they serve as internal marker for the relative position of the newly released thiohydantoin amino acids in a similar manner, as was done by the aid of the two blue markers in the case of the DABTH-amino acid thin-layer method.

The sensitivity limit of the released DNSA-PTH-amino acids by analytical or microbore HPLC employing a fluorescent detector was found at 100 to 200 fmol. On the small polyamide thin-layer sheets 1 to 5 pmol of the derivatives are visible at 366 nm. In addition, work is in progress in this laboratory to allow to identify these derivatives by means of an isocratic HPLC separation system.

Further, at present, other DNSA-isothiocyanates have been synthesized, such as the aliphatic analogs, e.g., DNSA-pentyl-isothiocyanate and the methyl-homologue, replacing the rigid aryl ring system by a more flexible aliphatic spacer group. By sequencing with these reagents, it was observed that they present much reduced coupling capacities (in the range of only 10 to 20%), in comparison with the aromatic homologs DNSAPITC or PITC.[48]

VI. CONCLUSION

An increase in sensitivity for polypeptide sequencing is achieved if several parameters are refined, namely:

1. Sophisticated isolation and purification of the protein or peptide sample under conditions suitable for microsequence analysis thereafter
2. Sufficient purity grade of reagent and solvents to allow for high quality sample purification and degradation
3. Advanced technology for microsequencing and sensitive identification of the released amino acid derivatives
4. Adapted methods and suitable chemistry

Independent of any improvements of the above-mentioned first points, the adaptation of the most suitable chemistry in microsequence analysis is still an open question and leaves a wide field of possible strategies for further development. Many different isothiocyanate-type reagents have been applied in the literature and other reagents were proposed. But only the Edman degradation with the classical PITC reagent has been studied in much detail. Therefore, the synthesis of isothiocyanate homologues offers certain advantages for future development.

What are the criteria for selecting a promising reagent? We believe these are mainly the following:

1. The reagent should be available following simple synthesis routes which allow it to be purified easily to the highest possible grade.
2. The reagent should possibly be volatile; then the removal of the excess of reagent after the reaction would not cause losses of peptides or proteins in any of the different sequencing modes. So far, only PITC and methylisothiocyanate (MITC) meet this demand.
3. The coupling of the reagent to the free α-amino groups of the proteins should exceed 95%; the reaction must be as complete for hydrophilic and small residues as for those with bulky and hydrophobic side chains. This is only possible if the reactive group of the reagent is not sterically hindered.
4. The reagent should provide a chromophore prior to or after the degradation, which enables sensitive detection of the released amino acid derivatives.
5. Complete cleavage of the coupled residue from the polypeptide chain at the cleavage should not be sterically hindered by the chromophore part of the reagent.

At present, no Edman-type reagent which meets all these requirements is available. Yet, DNSAPITC appears most suitable for sequencing of minute polypeptide amounts; the reagent produces fluorescent PTH-amino acid derivatives which are easily detectable in low picomole to femtomole quantities employing HPLC separation and detection with a fluorescence detector. In addition, the application of small-sized polyamide thin-layer sheets for additional identification of 1 to 10 pmol is possible, which allows safe microsequencing of high reliability. The excellent fluorescent properties and stability of the derivatives add to the quality of this Edman-type reagent and serve as a possibility for further enhanced microsequencing at the low picomole level. Further, manual sequencing employing DNSAPITC can be performed on the 100- to 500-pmol level and with many samples at the same time. Hence, this reagent serves as a possible alternative to the classical PITC or DABITC/PITC sequencing approaches. In the automated mode, this new reagent has been tested in the solid phase sequencer where it yields fluorescent derivatives of high purity.

LIST OF ABBREVIATIONS

PITC	Phenylisothiocyanate
DABITC	4-N,N-Dimethylaminoazobenzene-4′-phenylisothiocyanate
FITC	Fluorescein-isothiocyanate
BAMPITC	Boc-aminomethyl-phenylisothiocyanate
DNSAPITC	Dimethylaminonaphthylsulfonylamino-phenylisothiocyanate
MITC	Methylisothiocyanate
DITC	p-Phenylenediisothiocyanate
APG	Aminopropyl glass
APS	Aminopolystyrene
TETA-resin	Tetraethylenetetramine-polystyrene resin

REFERENCES

1. **Edman, P.,** *Acta Chem. Scand.*, 4, 283, 1950.
2. **Edman, P. and Begg, G.,** A protein sequenator, *Eur. J. Biochem.*, 1, 80, 1967.
3. **Laursen, R. A.,** General description of original sequencer and preparation of polystyrene, *Eur. J. Biol. Chem.*, 20, 89, 1971.
4. **Hewick, R., Hunkapiller, M., Hood, L. E., and Dreyer, W. J.,** A gas-liquid solid phase peptide and protein sequenator, *J. Biol. Chem.*, 256, 7990, 1981.
5. **Wittmann-Liebold, B.,** Advanced automatic microsequencing of proteins and peptides, in *Modern Methods in Protein Chemistry*, Vol. 1, Tschesche, H., Ed., Walter de Gruyter, Berlin, 1984, 229.
6. **Chang, J. T. and Creaser, E. H.,** A novel manual method for protein sequence analysis, *Biochem. J.*, 157, 77, 1976.
7. **Chang, J. Y., Creaser, E. H., and Bentley, K. W.,** 4-N,N-Dimethylaminoazobenzene-4′-isothiocyanate, a chromophoric reagent for protein sequence analysis, *Biochem. J.*, 153, 607, 1976.
8. **Maeda, H. and Kawauchi, H.,** A new method for the determination of the N-terminus of peptide chains with a fluorescent isothiocyanate, *Biochem. Biophys. Res. Commun.*, 31, 188, 1968.
9. **Muramoto, K., Kamiya, H., and Kawauchi, H.,** The application of fluorescein isothiocyanate and high performance liquid chromatography for the microsequencing of proteins and peptides, *Anal. Biochem.*, 141, 446, 1984.
10. **L'Italien, J. J. and Kent, S. B. H.,** Protein microsequencing with post-column fluorescent phenylisothiocyanate analogues, *J. Chromatogr.*, 283, 149, 1984.
11. **Jin, S.-W., Chen, G.-X., Palacz, Z., and Wittmann-Liebold, B.,** A new sensitive Edman-type reagent: 4-(N-1-dimethylaminoaphthalene-5-sulfonylamino)-phenyl-isothiocyanate, *FEBS Lett.*, 198, 150, 1986.
12. **Chang, J. Y., Brauer, D., and Wittmann-Liebold, B.,** Micro-sequence analysis of peptides and proteins using 4-N,N-dimethylaminoazobenzene-4′-isothiocyanate/phenyl-isothiocyanate double coupling method, *FEBS Lett.*, 93, 205, 1978.
13. **Gray, W. R. and Hartley, B. S.,** A fluorescent endgroup reagent for proteins and peptides, *Biochem. J.*, 89, 59, 1963.
14. **Gray, W. R.,** Dansyl chloride procedure, in *Methods in Enzymology*, Vol. 11, Colowick, S. P. and Kaplan, N. O., Eds., Academic Press, New York, 1966, 139.
15. **Gray, W. R. and Hartley, B. S.,** The structure of a chymotrypsin peptide from *Pseudomonas* cytochrome c-551, *Biochem. J.*, 89, 379, 1963.
16. **Lehmann, A. and Wittmann-Liebold, B.,** Complete separation and quantitative determination of DABTH-amino acid derivatives by isocratic reversed-phase high performance liquid chromatography, *FEBS Lett.*, 176, 360, 1984.
17. **Knecht, R., Seemüller, U., Liersch, M., Fritz, H., Braun, D. G., and Chang, J. Y.,** Sequence determination of Eglin C using combined microtechniques of amino acid analysis, peptide isolation, and automatic Edman degradation, *Anal. Biochem.*, 130, 65, 1983.
18. **Wittmann-Liebold, B. and Kimura, M.,** Microsequencing of peptides and proteins with 4-N,N-dimethylaminoazobenzene-4′-isothiocyanate, in *Methods of Molecular Biology*, Vol. 1, Walker, J. H., Ed., Humana Press, Clifton, N. J., 1984, 221.

19. **Wittmann-Liebold, B., Hirano, H., and Kimura, M.,** Manual microsequence determination of proteins and peptides with the DABITC/PITC method, in *Advanced Methods in Protein Microsequence Analysis,* Wittmann-Liebold, B., Salnikow, J., and Erdmann, V. A., Eds., Springer-Verlag, New York, 1986, 77.
20. **Laursen, R. A.,** Coupling techniques in solid-phase sequencing, in *Methods in Enzymology,* Vol. 47, Colowick, S. P. and Kaplan, N. D., Eds., Academic Press, New York, 1977, 277.
21. **Horn, M. J. and Laursen, R. A.,** Preparation of TETA-polystyrene, *FEBS Lett.,* 36, 285, 1973.
22. **Wittmann-Liebold, B. and Lehmann, A.,** Comparison of various techniques applied to the amino acid sequence determination of ribosomal proteins, in *Solid Phase Methods in Protein Sequence Analysis,* Laursen, R. A., Ed., Pierce Chemical Corp., Rockford, Ill., 1975, 81.
23. **Machleidt, W., Wachter, E., Scheulen, M., and Otto, J.,** Coupling proteins to aminopropyl glass, *FEBS Lett.,* 37, 217, 1973.
24. **Salnikow, J., Lehmann, A., and Wittmann-Liebold, B.,** Improved automated solid-phase microsequencing of polypeptides using DABITC, *Anal. Biochem.,* 117, 433, 1981.
25. **Wittmann-Liebold, B.,** Practical methods for solid-phase sequence analysis, in *Practical Protein Chemistry — A Handbook,* Darbre, A., Ed., John Wiley & Sons, Chichester, 1986, 375.
26. **Salnikow, J.,** Automated solid-phase microsequencing using DABITC, on-column immobilization of proteins, in *Advanced Methods in Protein Microsequence Analysis,* Wittmann-Liebold, B., Salnikow, J., and Erdmann, V. A., Eds., Springer-Verlag, New York, 1986, 108.
27. **Kimura, M.,** Proteins of the Bacillus stearothermophilus ribosome, *J. Biol. Chem.,* 259, 1051, 1984.
28. **Kimura, M., Ernst, H., and Appelt, K.,** The primary structure of initiation factor IF3 from *Bacillus stearothermophilus, FEBS Lett.,* 160, 78, 1983.
29. **Kimura, M., Kimura, J., and Ashman, K.,** The complete primary structure of ribosomal proteins L1, L14, L15, L23, L24, and L29 from *Bacillus stearothermophilus, Eur. J. Biochem.,* 150, 491, 1985.
30. **Arndt, E., Breithaupt, G., and Kimura, M.,** The complete amino acid sequence of ribosomal protein H-S11 from the archaebacterium *Halobacterium marismortui, FEBS Lett.,* 194, 227, 1986.
31. **Kimura, M., Kimura, J., and Watanabe, K.,** The primary structure of ribosomal protein L2 from *Bacillus stearothermophilus, Eur. J. Biochem.,* 153, 289, 1985.
32. **Kawauchi, H., Tuzimura, K., Maeda, H., and Ishida, N.,** Reaction of fluorescein isothiocyanate with proteins and amino acids, *J. Biochem.,* 666, 783, 1969.
33. **McKinney, R. M., Spillane, J. T., and Pearce, G. W.,** Determination of purity of fluorescein isothiocyanate, *Anal. Biochem.,* 7, 74, 1964.
34. **Sinsheimer, J. E., Jagodic, V., and Burckhalter, J. H.,** Fluorescein isothiocyanate: improved synthesis and purity, spectral studies, *Anal. Biochem.,* 57, 227, 1974.
35. **Kawauchi, H., Kadooka, K., Tanaka, M., and Tuzimura, K.,** Micro-analysis of amino acids with fluorescein isothiocyanate, *Agric. Biol. Chem.,* 35, 1720, 1971.
36. **Muramoto, K., Kawauchi, H., and Tuzimura, K.,** Micropolyamide thin-layer chromatography of fluorescein-thiohydantoin-amino acids at picomole level, *Agric. Biol. Chem.,* 39, 2241, 1975.
37. **Muramato, K., Meguro, H., Tuzimura, K., Sakurai, A., and Tamura, S.,** Application of FTC-method to amino acid sequence analysis of the peptidyl factor in *Saccharomyces cerevisiae, Agric. Biol. Chem.,* 40, 2503, 1976.
38. **Muramato, K., Kawauchi, H., and Tuzimura, K.,** Sequence determination of a peptide by the combined use of fluorescein isothiocyanate, *Agric. Biol. Chem.,* 42, 1559, 1978.
39. **Kawauchi, H., Muramato, K., and Ramachandran, J.,** Isolation and primary structure of adrenocorticotropin from several species of whale, *Int. J. Peptide Protein Res.,* 12, 318, 1978.
40. **Kawauchi, H. and Tsubokawa, M.,** Isolation and characterization of fin whale prolactin, *Int. J. Peptide Protein Res.,* 13, 229, 1979.
41. **Kawauchi, H. and Muramato, K.,** Isolation and primary structure of melanotropins from salmon pituitary glands, *Int. J. Peptide Protein Res.,* 14, 373, 1979.
42. **Tsubokawa, M., Kawauchi, H., and Li, C. H.,** Isolation and partial characterization of growth hormones from fin whale pituitary glands, *J. Biochem.,* 88, 1407, 1980.
43. **Ashman, K. and Wittmann-Liebold, B.,** A new isocratic HPLC separation for PTH-amino acids, based on 2-propanol, *FEBS Lett.,* 190, 129, 1985.
44. **Roth, M.,** Fluorescence reaction for amino acids, *Anal. Chem.,* 43, 880, 1971.
45. **Palacz, Z., Salnikow, J., Jin, S.-W., and Wittmann-Liebold, B.,** 4-(N-tert-Butyloxycarbonylaminomethyl)-phenylisothiocyanate: its synthesis and use in microsequencing, *FEBS Lett.,* 176, 365, 1984.
46. **Lin, J.-K. and Chang, J. Y.,** Chromophoric labeling of amino acids with 4-dimethylaminoazobenzene-4'sulfonyl chloride, *Anal. Chem.,* 47, 1634, 1975.
47. **Hirano, H. and Wittmann-Liebold, B.,** Protein pico-sequencing with 4-((5-(dimethyl-amino)-1-naphthylsulfonyl)amino)-phenylisothiocyanate *Biol. Chem. Hoppe-Seyler,* 164, 257, 1986.

48. **Salnikow, J., Palacz, Z., and Wittmann-Liebold, B.,** Automated solid-phase sequencing using fluorescent Edman reagents, in *Methods in Protein Sequence Analysis,* Walsh, K. A., Ed., Humana Press, Clifton, N.J., 1987.
49. **Jin, S.-W.,** unpublished results.
50. **Palacz, Z.,** personal communication.

Chapter 7

CHEMICAL METHODS OF AMINO ACID SEQUENCE ANALYSIS FROM CARBOXYTERMINAL END

M. Rangarajan

TABLE OF CONTENTS

I. INTRODUCTION

In contrast to N-terminal group determination and sequential analysis of proteins and peptides from the N-terminal end, only a few chemical methods are available for the determination of C-terminal groups and sequential degradation of proteins and peptides from the C terminus. The most commonly used procedure for the determination of protein and peptide sequences from the C-terminal end is the use of exopeptidases, namely, carboxypeptidases. There are several factors known to influence the rate of release of C-terminal amino acids by carboxypeptidases.[1,2] Despite the many limitations, this has been the most widely used method of all, for C-terminal sequence determination.

A need, therefore, exists for a chemical method which would be generally applicable and which could be used for the controlled stepwise degradation of the peptide chain with unequivocal identification of the amino acids released. This will be invaluable in the case of polypeptides and proteins with blocked N termini. C-terminal sequencing would greatly help in verifying primary structures of proteins predicted from DNA sequences and to detect any posttranslational processing of gene products from known codon sequences. It would be a useful aid in the design of oligonucleotide probes to screen cDNA or gene banks. Automated N-terminal sequencing after CNBr cleavage at methionine residues is currently used, but five or six residues preceding each methionine would be invaluable, because methionine has only a single codon and could be used for the design of minimally redundant oligonucleotide probes to screen gene banks.

II. METHODS

In this review of chemical methods for C-terminal sequencing of peptides and proteins, an attempt has been made to list the various methods published, in chronological order. In most cases, a key reference has been mentioned, except in the case of the isothiocyanate method where several workers have contributed greatly to improving the earlier published procedure.

One of the earliest chemical methods for the sequential degradation of peptides from the C terminus was proposed by Bergmann and Zervas.[3] They converted *N*-benzoyl peptide ester into *N*-benzoyl peptide hydrazide by treatment with hydrazine. The hydrazide was converted to the azide by reacting with nitrous acid, as shown in Figure 1. On treating the azide with benzyl alcohol, it was converted into the benzyl urethane derivative. Catalytic reduction of the N-benzoyl peptide benzyl urethane derivative gave two products, i.e., N-benzoyl peptide$_{-1}$ amide and the original C-terminal amino acid as its aldehyde. The method did not gain popularity because of the extreme complexity of the chemical reactions involved. Several years later, Khorana[4] proposed another very complicated set of reactions for the sequential degradation of peptides. The method involved reacting an acyl peptide such as *N*-benzoyl peptide with a disubstituted carbodi-imide to yield the acyl-peptidyl urea. This set of reactions is set out in Figure 2. Treatment of the acyl-peptidyl urea with ethanolic-sodium hydroxide yielded the shortened peptide and a substituted carbamyl amino acid amide. The amino acid derivative could be identified by hydrolyzing it to the parent amino acid with acid or alkali. Neither of the two methods described so far gained ground in the field of protein sequencing, because of the extreme complexity of the reactions involved and the incomplete removal of the C-terminal amino acid after each cycle.

Kenner et al.[5] published a novel procedure for C-terminal sequencing of peptides a year later. An acylated peptide was treated with diphenylphosphoroisothiocyanatidate in the presence of triethylamine in a nonaqueous medium when it formed a peptidyl-thiohydantoin, as seen in Figure 3. On reacting the peptidyl-thiohydantoin with sodium hydroxide, the C-terminal amino acid was released as its thiohydantoin, leaving behind a shortened peptide.

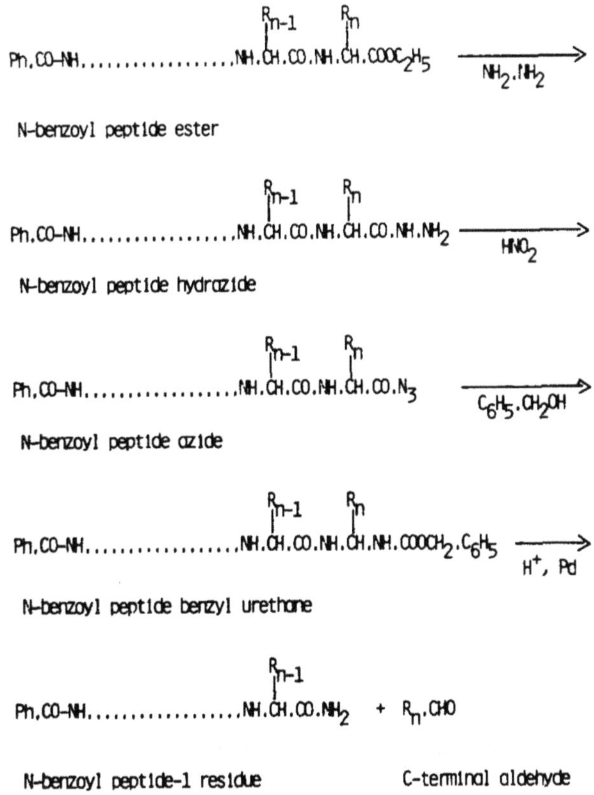

FIGURE 1. Sequential degradation of peptides from the carboxyl end by the method of Bergmann and Zervas.[3]

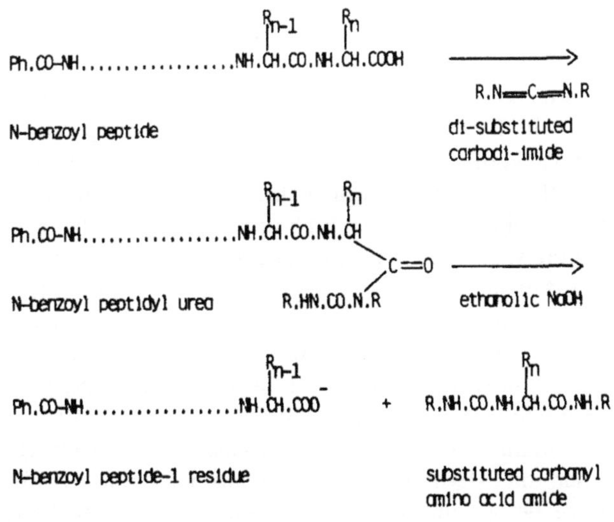

FIGURE 2. Selective degradation of peptides from the carboxyl end by the use of carbodi-imides.

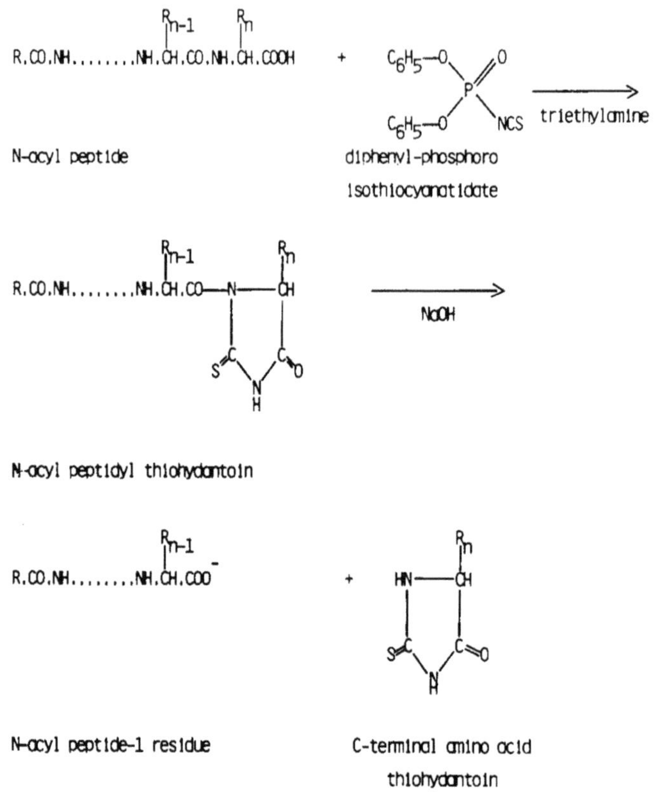

FIGURE 3. Selective removal of the C-terminal residue as a thiohydantoin by the use of diphenyl phosphoroisothiocyanatidate.

The method of Kenner et al.[5] was very similar to the one proposed by Schlack and Kumpf[6] a quarter of a century earlier. An acyl peptide was converted to a peptidyl thiohydantoin by heating with ammonium thiocyanate in acetic anhydride at 100°C for 30 min. The amino acid thiohydantoin corresponding to the C terminus was released from the peptidyl thiohydantoin by treatment with 0.1 *M* sodium hydroxide at room temperature. This is illustrated in Figure 4. The amino acid thiohydantoin was extracted into organic solvents and identified by paper chromatography[7,8] or hydrolyzed with hydrobromic acid[9] at 155°C for 6 hr and identified as the parent amino acid by paper chromatography.[10] The method was successful with some small peptides,[10,11] but was not successful with proteins.[12] Although sodium hydroxide was very efficient in one-cycle degradations of some proteinyl thiohydantoins, the majority of them were insoluble in this reagent, and this led to incomplete cleavage of the C-terminal amino acid thiohydantoin. Stark[13] reinvestigated the method of Schlack and Kumpft[6] and applied the modified procedure to a number of peptides. The peptides were incubated with ammonium thiocyanate in a solution of acetic acid and acetic anhydride to form peptidyl thiohydantoins. The peptidyl thiohydantoin was selectively cleaved with acetohydroxamic acid, and the C-terminal amino acid was determined by difference amino acid analysis of the peptide before and after the degradation cycle. At best, six cycles of degradation were achieved. Cromwell and Stark[14] extended this work to the sequential degradation of proteins. The amino acid thiohydantoin was isolated and identified by thin-layer chromatography on silica gel-coated plates and, also, indirectly by dilute alkali hydrolysis-alkali oxidation-alkali hydrolysis procedure and determination of the parent amino acid. Three cycles of degradation were successfully carried out. These authors reported that aspartic

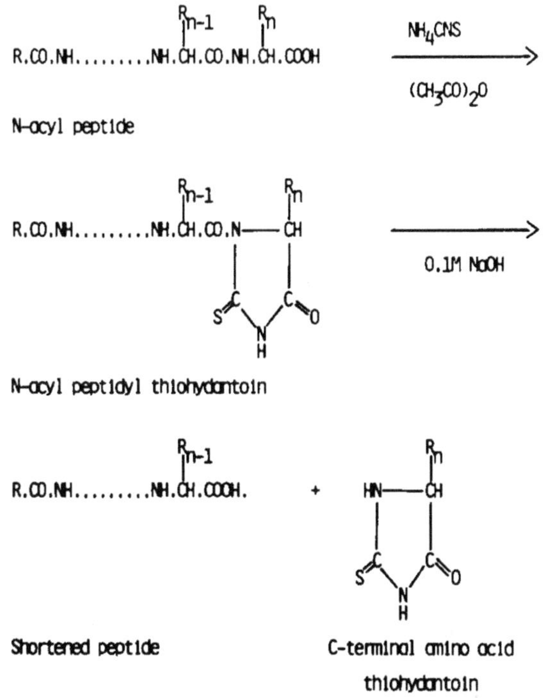

FIGURE 4. The isothiocyanate method for the sequential analysis of peptides from the C-terminal end.

acid and proline did not react to form thiohydantoins. Yamashita[15] improved the procedure for the sequential degradation of proteins by using H^+ form of the cation exchange resin Amberlite IR-120 instead of acetohydroxamic acid or 12 *M* HCl for the cleavage step. He reported successful degradation of aspartic acid and proline in peptides. Yamashita and Ishikawa[16] claimed to have sequenced 10 and 14 amino acid residues from the carboxyl terminus of Ribonuclease A and papain, respectively. Rangarajan and Darbre[17] and Darbre and Rangarajan[18] found that low yields obtained by Stark[13] and Cromwell and Stark[14] were due to two main reasons. The reaction as described was far too slow for general use; it was not possible to exclude atmospheric oxygen during the coupling and cleavage steps, as a result of which several unwanted side reactions occurred, thereby giving lowered yields. Rangarajan and Darbre[17] and Darbre and Rangarajan[18] carried out the coupling and cleavage steps with the peptide or protein attached to a solid phase. A single cycle of degradation could be accomplished in 5 to 6 hr, instead of several days, and the yields were greatly improved. Rangarajan and Darbre[19] originally used a carboxylated polystyrene-based resin to which peptides were attached via their amino groups in the presence of a water-soluble carbodi-imide. Larger polypeptides and proteins were more efficiently coupled to porous glass.[20-22] The amino acid thiohydantoins formed could be identified using thin-layer chromatography,[14,15,23] gas-liquid chromatography,[24,25] and HPLC.[26,27] Kubo et al.[28] reported that substituting thiocyanic acid for thiocyanate salts resulted in improvements in the coupling step. Meuth et al.[27] have described a solid-phase sequential degradation using carbonyl di-imidazole-activated porous glass as the support for the covalent attachment of peptides in high yield, the use of thiocyanic acid in the coupling step which improves efficiency, and the use of HPLC for the identification and quantitation of the amino acid thiohydantoins. They reported a reduction in time of almost 50% for the completion of one cycle of degradation, compared to 5 to 6 hr achieved by Rangarajan and Darbre.[19] The sequence of four residues from the C-terminal end of a 22-residue protein fragment was determined. The thiocyanate method is probably the most thoroughly investigated chemical method to date

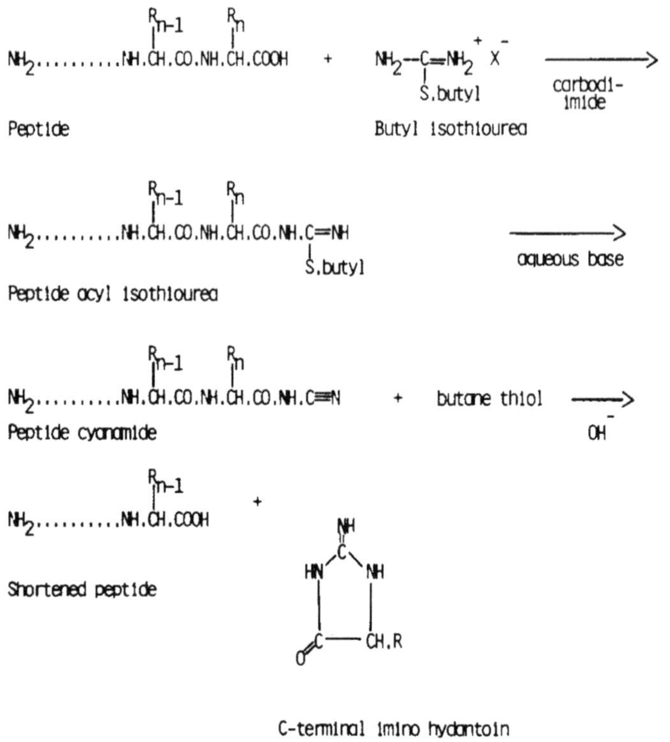

FIGURE 5. Sequential degradation of peptides from the C-terminal end via the cyanamide.

for C-terminal sequencing. In the half century since Schlack and Kumpf[6] published their original procedure, some progress has been made in terms of speed, sensitivity, and yield of the sequential degradation cycle. The method is not without problems, the important one being the very low yields obtained when aspartic acid and proline occur at the C terminus.

While several groups were trying to improve the thiocyanate method, Tarr[29] described some preliminary results at the First International Conference on Solid-Phase Methods in Protein Sequence Analysis of a novel procedure which appeared to possess several advantages over the thiocyanate method. The peptide to be sequenced was first coupled to *S*-butylthiourea using a carbodi-imide at 50°C for 10 to 15 min to yield a peptidyl acyl isothiourea. The cleavage reaction was then performed in aqueous base at pH 10 to 11.5 and was shown to occur in two distinct stages: the alkyl portion of the peptidyl acyl isothiourea was removed first by the base as butane thiol producing a peptidyl cyanamide. The cyano group then suffered nucleophilic attack by the nitrogen of the peptide bond and underwent cyclization, and the C-terminal amino acid was simultaneously released as the iminohydantoin, leaving behind a shortened peptide with a free carboxyl group. These reactions are shown in Figure 5. The author described the gas chromatography of iminohydantoins of a few amino acids. The entire degradation cycle was completed in 40 min. The efficiency of one cycle of degradation was shown to be better than 90%. Good results were obtained with Dansyl di- and tripeptides for one or two cycles. As with the thiocyanate method, problems were encountered when aspartic acid or proline residues were present at the C terminus.

At the Second International Conference on Solid Phase Methods in Protein Sequence Analysis in 1978, another approach to C-terminal sequencing was described by Previero and Coletti-Previero.[30] This involved alcoholysis of an oxazole formed at the C terminus of a peptide. Dipeptides or tripeptides were immobilised by *N*-acylation with polyacrylyl chlo-

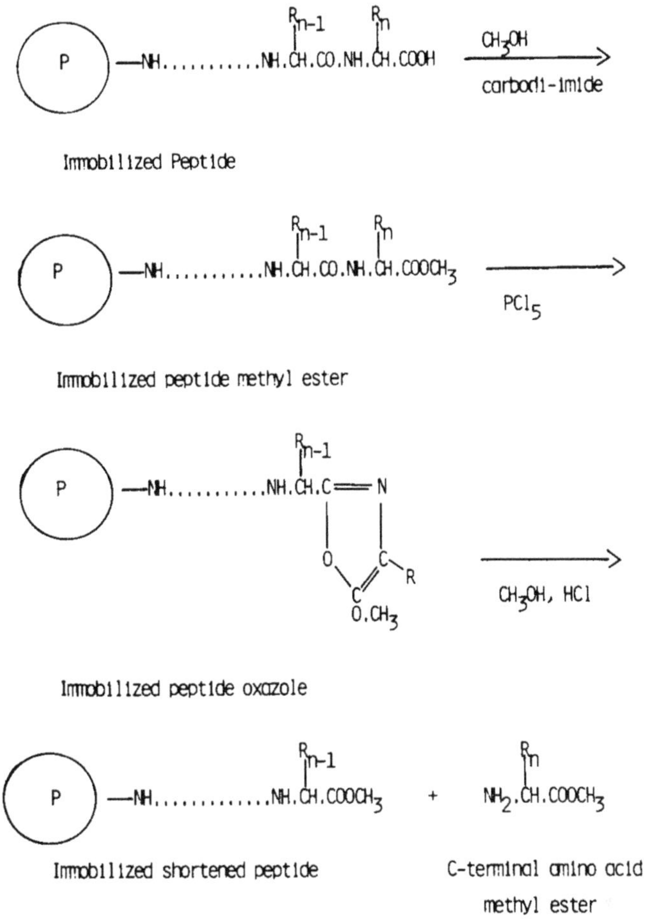

FIGURE 6. C-terminal sequencing of peptides by the method of Previero and Coletti-Previero.[30]

ride and esterified with methanol in the presence of a carbodi-imide. Cyclization to the oxazole was achieved by treating the ester with phosphorous pentachloride. The oxazole was then subjected to acid-catalyzed methanolysis, as shown in Figure 6, which released the C-terminal amino acid as its methyl ester, leaving behind the immobilized shortened peptide. The cleavage of the C-terminal amino acid was incomplete after one cycle, and the cumulative yield of four consecutive cycles was between 80 and 90%. Repetitive degradations were not performed. This procedure does not look very encouraging, judging by results obtained so far.

Parham and Loudon[31,32] have reported a solid-phase method for the removal of C-terminal residues of peptides in good yield. The peptide was first immobilized by coupling its N terminus to activated glass beads. Treatment of the peptide with di-*p*-nitro-phenylphosphoryl azide produced the peptide azide, which on thermolysis was converted to the isocyanate. Decomposition of the isocyanate released the C-terminal amino acid as its aldehyde and the shortened peptide as an amide. Side-chain carboxyl groups of aspartic and glutamic acids were also rearranged in the process to the corresponding amines. Arginine and lysine at the C terminus were converted to the corresponding lactams during the preparation of the acyl azides. The former could be avoided by performing the reaction in *N*-ethyl morpholine half neutralized with acetic acid in dimethyl formamide and the latter could be prevented by acetylation of the peptide prior to exposing it to the various reagents. Further degradation

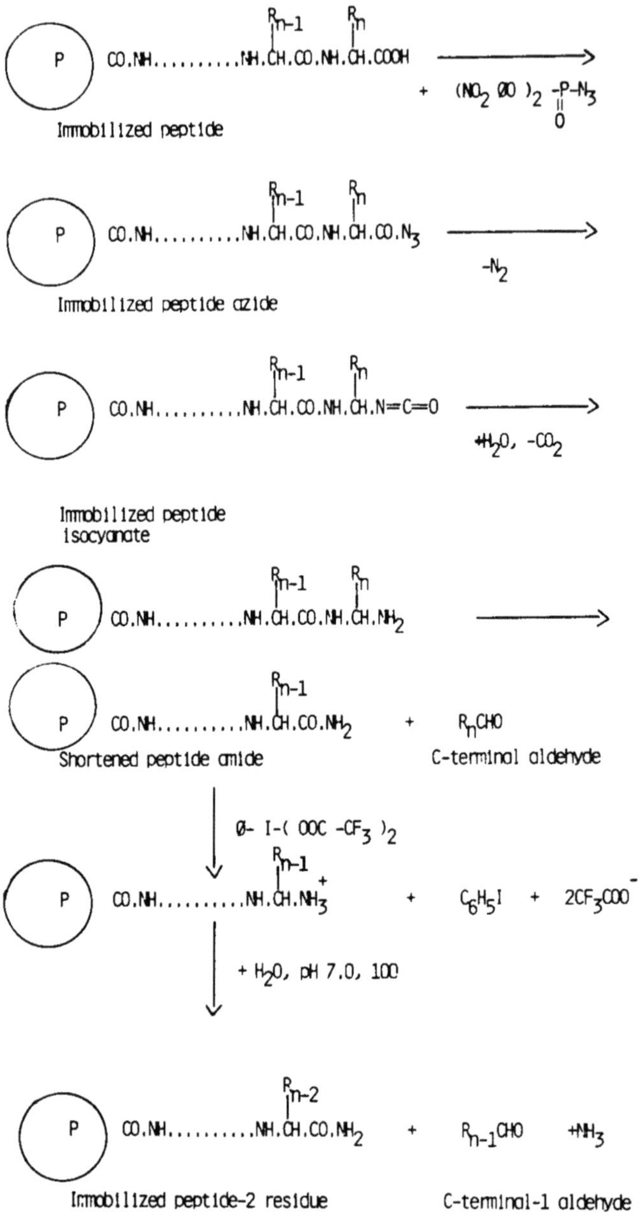

FIGURE 7. Sequential analysis of peptides from the carboxyl terminus of peptides by the method of Parham and Loudon.[31,32]

of the shortened immobilized peptide amide was achieved by treatment with 1,1-bis (trifluoroacetoxy)-iodobenzene under aqueous acidic conditions which resulted in the conversion of the amide to a primary amine via the isocyanate. Hydrolysis of the amine in neutral aqueous buffer cleaved the C-terminal amino acid as an aldehyde, leaving behind the shortened peptide amide attached to the solid support. This is detailed in Figure 7. Repetition of this degradation cycle would yield sequence information from the C-terminal end. Yields of 80% for each step were obtained with a hexapeptide. It was not possible to identify the amino acid aldehyde cleaved off after each cycle. Sequence information was, therefore, obtained by difference amino acid analysis. Under the sequential degradation conditions, side-chain amides of asparagine and glutamine were converted to di-amino-propionic and di-amino-butyric acid, respectively. This procedure has the added advantage that it can be applied to proteins and peptides which have their C-terminal carboxyls blocked by an amide.

III. SUMMARY

Of all the chemical methods outlined in this review, the isothiocyanate method has been the most thoroughly investigated and is probably the one most likely to be applied to a variety of peptides generated from proteins. One or two others show promise and could become the method of choice of many protein chemists if more thoroughly researched.

REFERENCES

1. **Ambler, R. P.,** Enzymatic hydrolysis with carboxypeptidases, in *Methods in Enzymology,* Vol. 11, Hirs, C. H. W., Ed., Academic Press, New York, 1967, 155.
2. **Ambler, R. P.,** Carboxypeptidases A and B, in *Methods in Enzymology,* Vol. 11, Hirs, C. H. W., Ed., Academic Press, New York, 1967, 436.
3. **Bergmann, M. and Zervas, L.,** A method for the stepwise degradation of polypeptides, *J. Biol. Chem.,* 113, 341, 1936.
4. **Khorana, H. G.,** Peptides. III. Selective degradation from the carboxyl end. The use of carbodi-imides, *J. Chem. Soc.,* p. 2081, 1952.
5. **Kenner, G. W., Khorana, H. G., and Stedman, R. J.,** Peptides. IV. Selective removal of the C-terminal residue as a thiohydantoin. The use of diphenyl phosphorisothiocyanatidate, *J. Chem. Soc.,* p. 673, 1953.
6. **Schlack, P. and Kumpf, W.,** Uber eine neue Methode zur Ermittlung der Konstitution von Peptiden, *Hoppe-Seyler's Z. Physiol. Chem.,* 154, 125, 1926.
7. **Dautrevaux, M. and Biserte, G.,** Identification of terminal carboxyl amino acids of peptide chains by paper chromatography of their 2-thiohydantoins, *Comp. R. Acad. Sci.,* 240, 1153, 1955.
8. **Edward, J. T. and Nielsen, S.,** The chromatography of thiohydantoins on paper: determination of the C-terminal amino-acid of bovine plasma albumin, *Chem. Ind. (London),* p. 197, 1953.
9. **Boissonas, R. A.,** Dosage colorimetrique des acides amines separes par chromatographie sur papier, *Helv. Chim. Acta,* 33, 1975, 1950.
10. **Waley, S. G. and Watson, J.,** The stepwise degradation of peptides, *J. Chem. Soc.,* p. 2394, 1951.
11. **Baptist, V. H. and Bull, H. B.,** Determination of the terminal carboxyl residues of peptides and of proteins, *J. Am. Chem. Soc.,* 75, 1727, 1953.
12. **Fox, S. W., Hurst, T. L., Griffith, J. F., and Underwood, O.,** A method for the quantitative determination of C-terminal amino acid residues, *J. Am. Chem. Soc.,* 77, 3119, 1955.
13. **Stark, G. R.,** Sequential degradation of peptides from their carboxyl terminii with ammonium thiocyanate and acetic anhydride, *Biochemistry,* 7, 1796, 1968.
14. **Cromwell, L. D. and Stark, G. R.,** Determination of the carboxyl terminii of proteins with ammonium thiocyanate and acetic anhydride, with direct identification of the thiohydantoins, *Biochemistry,* 8, 4735, 1969.
15. **Yamashita, S.,** Sequential degradation of polypeptides from the carboxyl ends. I. Specific cleavage of the carboxyl-end peptide bonds, *Biochim. Biophys. Acta,* 229, 301, 1971.
16. **Yamashita, S. and Ishikawa, N.,** Sequential degradation of polypeptides from the carboxyl ends. II. Application to polypeptides, in *Proc. 3rd American Peptide Symp.,* Meienhofer, J., Ed., Ann Arbor Science, Ann Arbor, Mich., 1972, 701.
17. **Rangarajan, M. and Darbre, A.,** Peptide sequencing from the carboxyl end using a solid phase method, in *Abstr. Commun. 9th Meeting FEBS,* Hungarian Biochemical Society, 1974, 348.
18. **Darbre, A. and Rangarajan, M.,** Solid-phase sequential degradation of peptides with the thiocyanate method of Stark, in *Solid Phase Methods in Protein Sequence Analysis,* Laursen, R. A., Ed., Pierce Chemical Company, Rockford, Ill., 1975, 131.
19. **Rangarajan, M. and Darbre, A.,** Studies on sequencing of peptides from the carboxyl terminus by using the thiocyanate method, *Biochem. J.,* 157, 307, 1976.
20. **Darbre, A.,** Solid-phase COOH-terminal sequential degradation, in *Methods in Enzymology,* Vol. 47, Hirs, C. H. W. and Timasheff, S. N., Eds., Academic Press, New York, 1977, 357.

21. **Williams, M. J. and Kassel, B.,** A solid phase method for peptide sequencing from the carboxyl terminus, *FEBS Lett.*, 54, 353, 1975.
22. **Kassel, B., Krishnamurti, C., and Friedman, H. L.,** Studies on solid phase carboxyl terminal sequencing, in *Solid Phase Methods in Protein Sequence Analysis*, Previero, A. and Coletti-Previero, M. A., Eds., North-Holland, Amsterdam, 1978, 39.
23. **Rangarajan, M. and Darbre, A.,** Identification of amino acid thiohydantoins directly by thin-layer chromatography and indirectly by gas-liquid chromatography after hydrolysis, *Biochem. J.*, 147, 435, 1975.
24. **Rangarajan, M., Ardrey, R. E., and Darbre, A.,** Gas-liquid chromatography and mass spectrometry of amino acid thiohydantoins and their use in protein sequencing, *J. Chromatogr.*, 87, 499, 1973.
25. **Dwulet, F. E. and Gurd, F. R. N.,** Identification of 2-thiohydantoins by gas chromatography and reconversion to the free amino acids, *Anal. Biochem.*, 82, 385, 1977.
26. **Schlesinger, D. H., Weiss, J., and Audhya, T. K.,** Isocratic resolution of amino acid thiohydantoins by high performance liquid chromatography, *Anal. Biochem.*, 95, 494, 1979.
27. **Meuth, J. L., Harris, D. E., Dwulet, F. E., Crowl-Powers, M. L., and Gurd, F. R. N.,** Stepwise sequence determination from the carboxyl terminus of peptides, *Biochemistry*, 21, 3750, 1982.
28. **Kubo, H., Nakarjima, T., and Tamura, Z.,** Formation of thiohydantoin derivatives of proline from C-terminal of peptides, *Chem. Pharm. Bull.*, 19, 210, 1971.
29. **Tarr, G. E.,** Peptide sequencing from the C-terminus via the cyanamide: preliminary results, in *Solid Phase Methods in Protein Sequence Analysis*, Laursen, R. A., Ed., Pierce Chemical Company, Rockford, Ill., 1975, 139.
30. **Previero, A. and Coletti-Previero, M. A.,** Chemical basis for a new C-terminal sequencing in solid-phase, in *Solid Phase Methods in Protein Sequence Analysis*, Previero, A. and Coletti-Previero, M. A., Eds., North-Holland, Amsterdam, 1978, 49.
31. **Parham, M. E. and Loudon, G. M.,** A new method of determination of the carboxyl-terminal residue of peptides, *Biochem. Biophys. Res. Commun.*, 80, 7, 1978.
32. **Parham, M. E. and Loudon, G. M.,** Carboxyl-terminal sequential degradation of peptides, *Biochem. Biophys. Res. Commun.*, 80, 1, 1978.

Chapter 8

ENZYMATIC METHODS OF PROTEIN/PEPTIDE SEQUENCING FROM CARBOXYTERMINAL END

R. Hayashi

TABLE OF CONTENTS

I. INTRODUCTION

The demand for analysis of the C-terminal sequence of proteins ever increases with new necessity to identify the base sequence corresponding to the proteins concerned. For this purpose, an enzymatic method using carboxypeptidases should be the first choice, because of a lacking of a sufficiently reliable chemical method. This method has successfully been used to deduce more than ten residues of C-terminal sequence in the most favorable conditions.

In the use of carboxypeptidases for end-group analysis of proteins and peptides, the review of Ambler[1,2] is standard and should be perused before performing experiments. It includes preparation of enzyme and substrate, method of digestion, separation and identification of products, and interpretation of result. Although the description principally concerns carboxypeptidases A and B, many of the methods and the comment described there are basically applicable to the use of the other carboxypeptidases.

In addition to metallo-carboxypeptidase (EC 3.4.17), serine carboxypeptidase (EC 3.4.16) has recently been introduced and has provided satisfactory results for C-terminal sequencing of proteins and peptides. In using these enzymes at will, studies are essential and indispensable, not only of substrate specificity of enzymes, but also their general properties, including enzyme chemistry as well as protein chemistry. There are recent reviews for carboxypeptidases A,[3,4] B,[5,6] C,[7-9] P,[10] S,[11] W,[12] and Y.[13-15]

II. PRINCIPLE AND BASIC PRECAUTION

A. Principle

Carboxypeptidase is the enzyme that removes amino acids one residue at a time from the C terminus of proteins and peptides. Thus, C-terminal amino acid sequence is deduced based on the principle that the rate of the sequential release of amino acids during the carboxypeptidase digestion reflects the C-terminal amino acid sequence to be analyzed. In a typical analysis, a sample protein is incubated with carboxypeptidase, and an aliquot of the mixture is withdrawn at appropriate time intervals. The amino acid released in the aliquot is identified and quantitated, and the results are plotted against time as moles of each amino acid released per mole of substrate.

B. Factors Affecting the Rate of Amino Acid Release

Plots of the rate of amino acid releases are sometimes too close to deduce unequivocally the C-terminal sequence, because preceding amino acid residues can be released in an almost same rate with the amino acid residue which is expectable from C terminus. This is due to the fact that both enzyme and substrate contain independent factors affecting the rate as follows.

Carboxypeptidase does not necessarily release all kinds of C-terminal amino acids in an equal rate on account of its own substrate specificity. Depending on the side-chain structure, an amino acid is released faster than the other amino acid when they are placed at the C-terminal position of the peptide. Furthermore, depending on the kind of amino acid residue placed at the penultimate position, the inherent rate of the release of the C-terminal amino acid residue is affected. Even third and fourth amino acid residues from C terminus may affect the rate of the release as well, because the enzyme has characteristic subsites which correspond to several amino acid residues of substrate (see Abramowitz et al.[16] for an example). Thus, an amino acid in the C terminus is released rapidly, but the other amino acid slowly or negligibly, depending upon the side-chain structure and also upon the adjacent amino acid residues. It is impossible at present to appraise the rate of release of amino acid in the relationship between structures of subsites and substrates.

Except small peptides, polypeptides and proteins may have secondary and/or tertiary structures during carboxypeptidase digestion, and such a structure might prevent enzymatic digestion of the C terminus so as to result in slowing down the rate of amino acid releases.

In summary, the rate of appearance of the released amino acids during carboxypeptidase digestion is affected by the characteristics of the enzyme used and, also, by the structure of the substrate. Therefore, in order to avoid the controversy, careful analyses must be performed and careful explanation must be made of the results, taking into consideration that ambiguous factors are inherently involved in enzymatic digestion. Thus, carboxypeptidase digestion should be performed under conditions where proteins and peptides (C-terminal sequences of which are analyzed) could take on a completely unfolded structure during digestion. In sequencing of proteins, reduced and carboxymethylated proteins or previously denatured proteins are preferable for this purpose (see Ambler[1,2]). Figure 1 shows differences in releasing C-terminal amino acid residues from native and denatured ribonucleases. It is also important to confirm the result of a sequence analysis by the use of different carboxypeptidases (see Section IV).

III. PROPERTIES OF CARBOXYPEPTIDASES

Carboxypeptidases are classified into two groups:[13,19] serine carboxypeptidase (EC 3.4.16) and metallo-carboxypeptidase (EC 3.4.17). The former enzyme has been found in molds, yeast, plants, and animals, localized mainly in vacuoles or lysosomes. Properties of carboxypeptidases which have been described as useful enzymes for C-terminal analysis are listed in Table 1. Some of these enzymes are commercially available as an enzyme reagent specially prepared for the analytical purpose.

Until recently, carboxypeptidases A and B, which are obtained from the pancreas, have usually been utilized, but serine carboxypeptidase has also been shown to give satisfactory results for the purpose of C-terminal analysis. One of the serine carboxypeptidases, carboxypeptidase Y (3.4.16.1), has been well characterized and is easily available as a useful reagent for sequencing of the C-terminal end.[13,14]

A. Inhibitors and Active Site

Serine carboxypeptidase is defined as the enzyme which is inhibited by diisopropylfluorophosphate (DFP) and/or phenylmethylsulfonyl fluoride (PMSF). Some of them are sensitive to sulfhydryl reagents such as *p*-chloromercuribenzoate (PCMB) and $HgCl_2$. In contrast to metallo-enzyme, this type of enzyme is not inhibited by metal chelators.

Carboxypeptidase Y (*Saccharomyces cerevisiae*),[13] C_N and C_U (Citrus),[9] S (*Penicillium janthinellum*),[11] Phaseolin (French beans),[17] and a cottonseed (germinating cotyledons)[18] carboxypeptidase have been confirmed as enzymes possessing a serine residue at the active site.

Carboxypeptidase Y has a single histidine[20] and cysteine[21] residue, the modification of which inactivates the activity. The histidine residue is essential for the activity, linked with the serine residue at the active site, but the cysteine residue seems to be located near or at the substrate binding site.[20,22]

Metallo-carboxypeptidase usually has zinc at the active site and is inhibited by the prolonged incubation with metal chelators such as ethylenediamine tetraacetate (EDTA) and *o*-phenanthroline.

B. Specificity

Specificities of two groups of carboxypeptidase are very different. Approximate rates of release of protein amino acids by carboxypeptidases A, B, and Y are briefly summarized in Table 2.

Serine carboxypeptidase is characterized as the enzyme of broad specificity: the enzyme

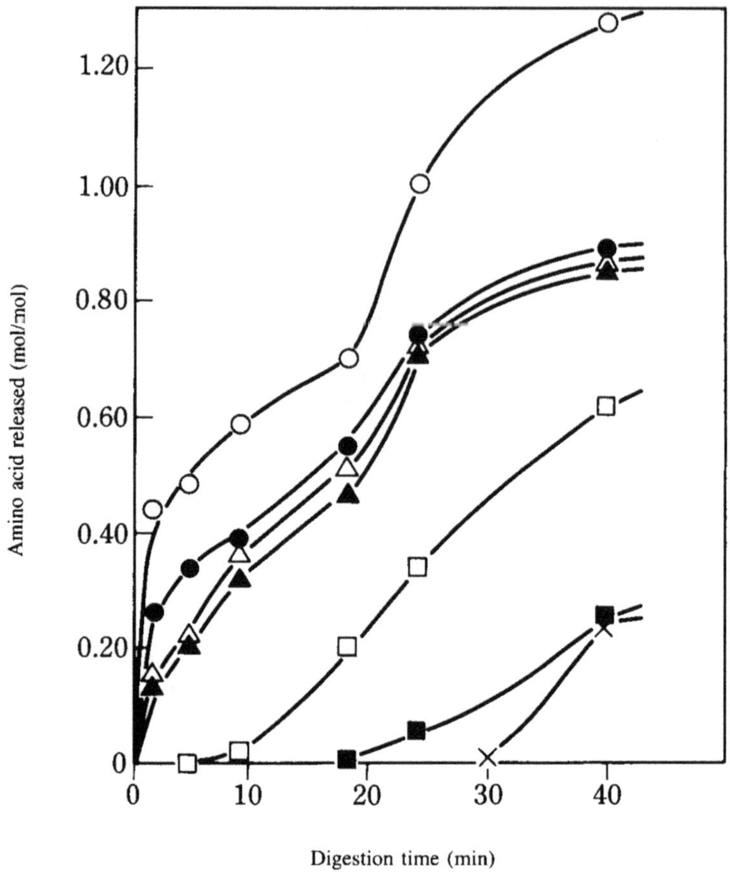

A

FIGURE 1. C-terminal sequencing of bovine pancreatic ribonuclease A by carboxypeptidase Y: comparison of digestions of native and denatured proteins. (A) Digestion of native RNase in 6 *M* urea; (B) digestion of the reduced and carboxymethylated RNase. Experimental conditions and references are described in Table 4. ○, Valine; ●, serine; △, alanine; ▲, aspartic acid; □, phenylalanine; ■, histidine; ×, proline. (From Hayashi, R., Moore, S., and Stein, W. H., *J. Biol. Chem.*, 248, 2296, 1973. With permission.)

could release all the protein amino acids, including proline, in the measurable rate. The ability to release proline is especially useful for C-terminal analyses of proteins and peptides. In contrast, metallo-carboxypeptidase is characterized as the enzyme of narrower specificity: carboxypeptidase A releases most rapidly amino acids with an aromatic or large aliphatic side chain, while B releases the basic amino acids, lysine and arginine, with practically no release of the other protein amino acids.

Serine carboxypeptidase is also unique in esterase activity toward N-blocked amino acid methyl or ethyl ester, in contrast to metallo-carboxypeptidase which hydrolyzes only ester substrates with a free carboxyl group at the C terminus, i.e., HPLA for carboxypeptidase A.

C. Properties of Carboxypeptidases A and B

Carboxypeptidase A is an euglobulin which is brought to solution at high ionic strengths. The enzyme may be stored in crystalline suspension. The solubilized enzyme must be stored

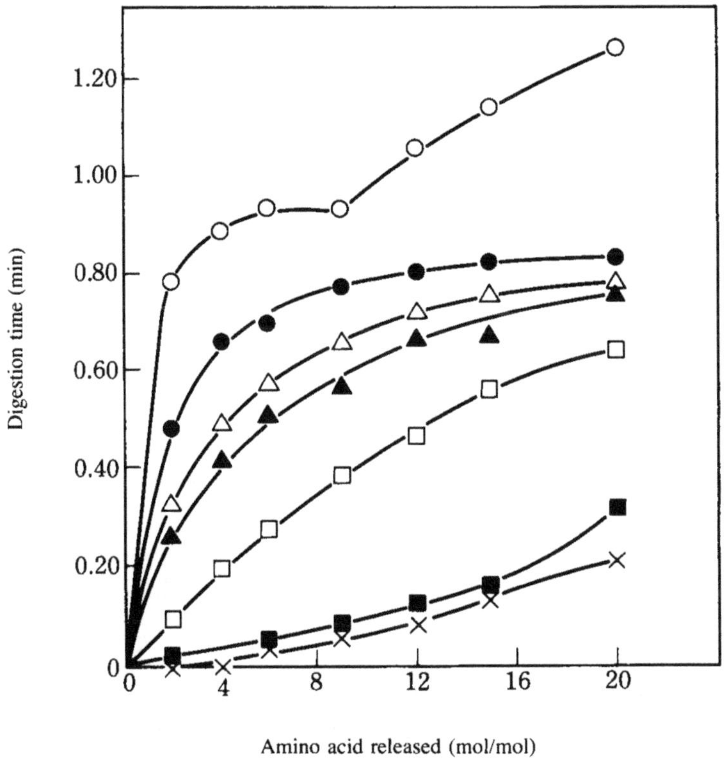

FIGURE 1B.

at 4°C without freezing. Carboxypeptidase B is kept stable in frozen solution at −10°C. The digestion of proteins and peptides with the respective enzyme is performed usually in 0.2 *M N*-ethylmorpholine acetate (pH 7.6 to 8.5) or 0.1 *M* ammonium bicarbonate (pH 8.0). In order to release basic residues in addition to neutral residues, digestion may be performed with an equimolar mixture of carboxypeptidases A and B.

Carboxypeptidase preparations should be free of endopeptidase activity which can lead to erroneous results. The most common endopeptidases found in preparations of carboxypeptidases A and B are pancreatic trypsin and chymotrypsin, both of which are easily inactivated by DFP. Therefore, treatment of preparations of carboxypeptidases A and B with DFP should be performed before use for sequence analysis.[1,2] The DFP-treated enzyme preparations are commercially available. It should be noted here that excess DFP used to inactivate the contaminated endopeptidases should be completely removed before serine carboxypeptidase is used in successive digestion (see Section IV).

Other details in the use of these enzymes for the C-terminal analysis are covered in the comprehensive review by Ambler.[1,2]

D. Properties of Carboxypeptidase Y

Carboxypeptidase Y as a glycoprotein is very soluble in aqueous media. The commercially available lyophilized preparations should be dissolved in water or in 0.01 *M* sodium phosphate buffer (pH 7.0) 1 day before use, since the activity lost during the lyophilization procedure is recovered fully by leaving the solution overnight at 5°C.

The enzyme is relatively stable in the presence of protein denaturants or certain solvents, where substrates of the enzyme are either soluble or are in the denatured form so as to render their C termini available for the enzyme: about 80% of the activity is retained after incubation of the enzyme with 6 *M* urea at 25°C for 1 hr.

Table 1
PROPERTIES OF CARBOXYPEPTIDASES USEFUL FOR C-TERMINAL ANALYSIS

Properties	Carboxypeptidase A	Carboxypeptidase B	Carboxypeptidase C		Carboxypeptidases	
					C_N	C_{Ua}
Origin	Bovine pancreas	Porcine pancreas	Orange peel	Orange leaves	*Citrus natsudaidai*	Mandarin orange
Mol wt[a]	34,489	34,300 (SD)	148,700 (SD) 86,000 (GEL)	175,000 (GE)	93,000 (SD)	96,000 (SD)
$S^0_{20,w}$ ($\times 10^{-13}$ cm/sec)	3.07	3.25	8.14	—	5.5	6.1
$D^0_{20,w}$ (cm^2/sec)	8.67×10^{-7}	8.16×10^{-7}	4.6×10^{-7}	—	5.5×10^{-7}	5.5×10^{-7}
$\bar{v}$	0.732	0.72	—	—	—	—
[η] (mℓ/g)	3.4	—	—	—	—	—
$A^{1\%}_{278\ nm}$	18.6[b]	21.6	—	—	—	—
Isoelectric point	pH 5.95	—	—	pH 4.5	pH 4.3	pH 4.8
Optimum pH	7—9	7—9	5.3—5.7	5	5.5	5.5
Active site	Tyr-248, Glu-270, Arg-145	—	Serine[c]	Unknown	Serine	Serine
Cofactor	Zn (1 g-atom/mol)	Zn (1 g-atom/mol)	None	None	None	None
Inhibitor	Metal chelators (EDTA, *o*-phenanthroline), pyrophosphate, oxalate	Metal chelators, 2,2′-dipyridyl	DFP	?	DFP, PMSF, $HgCl_2$	DFP, $HgCl_2$
Stability	Stable at pH 7—10; stable in crystalline suspension; inactivated by freezing	Stable under frozen state, inactivated by lyophilization	Stable at pH 4—6	Stable at pH 5, unstable below 4 or above pH 6	Stable in pH 4.5—6.0 at 0—45°C; inactivated by freezing and thawing and by lyophilization	
Assay method	Activity for Cbz-Gly-Phe or for HPLA	Activity for Bzl-Gly-Arg	Activity for Cbz-Leu-Phe		Peptidase activity for Cbz-Glu-Phe	
Others	Available as crystalline suspension	Available as crystalline suspension	Inactivated by lyophilization	No effect by DFP, EDTA, and SH-reagents; stored at −20°C	Stored in 0.1 *M* sodium citrate (pH 5.5) at 4°C; weak esterase activity; not inactivated by PCMB; glycoprotein	
Ref.	3, 4	5, 6	7, 8	7, 8	9	

Properties	Carboxypeptidase P	Penicillocarboxypeptidase S₁	Penicillocarboxypeptidase S_2	Carboxypeptidase W	Carboxypeptidase Y
Origin	*Penicillium janthinellum*	*Penicillium janthinellum*		Wheat bran	Bakers' yeast (*Sacchromyces cerevisiae*)
Mol wt[a]	51,000 (GEL) 49,000 (PAGE)	48,000 (PAGE)	128,000—140,000 (SE)	118,000 (GEL)	60,000 (SD) 61,000 (PAGE)
$S^0_{20,w}$ ($\times 10^{-13}$ cm/sec)	—	—	—	6.3	4.23
$D^0_{20,w}$ (cm²/sec)	—	—	—	—	6.1×10^{-7}
$\bar{v}$	—	0.716	0.724	—	0.71
[η] (mℓ/g)	—	—	—	—	4.83×10^{-2}
$A^{1\%}_{278\ nm}$	25.4 (at 280 nm)	26	18.5	19.6 (at 280 nm)	15.0 (at 280 nm)
Isoelectric point	pH 4.1	pH 3.70 and 3.77	pH 4.6—4.7	pH 6.0	pH 3.6
Optimum pH	3.7—4.2	4.3	3.9	4.0	5.5—6.5
Active site	Serine[c]	Serine	Serine	Serine[c]	Serine and histidine
Cofactor	None	None	None	None	None
Inhibitor	DFP, PCMB, monoiodoacetate	DFP, PMSF, PCMB, $HgCl_2$ (S_2 is also inhibited by *o*-phenanthroline and acetate)		DFP, PCMB	DFP, PMSF, ZPCK, PCMB, $HgCl_2$
Stability	Stable in pH 3—7 at 30°C for 20 min; stable as crystalline	Stable at pH 3.5—5.5; inactivated by freeze drying		Stable at pH 4—8; inactivated by freezing and thawing	Stable at pH 5.0—8.0; stored at −20° or in saturated NH_4SO_2 solution; stable in 6 *M* urea for 1 hr
Assay method	Activity for Cbz-Glu-Tyr	Peptidase activity for Cbz-Glu-Tyr		Peptidase activity for Cbz-Glu-Tyr	Peptidase for Cbz-Phe-Leu, esterase for Ac-Tyr-OEt, anilidase for Bz-Tyr-pNA
Others	Optimum pH 5.2 for Cbz-Gly-Lys; available as lyophilized powder containing citrate	S_1 has two isomers; esterase and amidase activities at pH 5.5—6.5; glycoprotein with glucosamine and hexose; stored in frozen state		Inactivated by PCMB; tetrameric protein; available as lyophilized powder	Glycoprotein with glucosamine and 12—22% hexose; carboxamidopeptidase activity at pH 7.5; available as lyophilized powder
Ref.	10	11		12	13—15

[a] Method of determination is shown in parentheses: GEL, gel-filtration; PAGE, polyacrylamide gel-electrophoresis; SD, sedimentation and diffusion coefficients; SE, sedimentation equilibrium.

[b] Calculated from mol wt and molar extinction coefficient.

[c] Parenthesis shows unidentified.

Table 2
RELATIVE RATES OF RELEASE OF PROTEIN AMINO ACIDS BY CARBOXYPEPTIDASES

	Carboxypeptidases		
Side chain of amino acid	**A**	**B**	**Y[a]**
Aliphatic and aromatic	Rapid	Not released	Rapid
Acidic	Slow	Not released	Medium
Basic (lysine and arginine)	Not released	Rapid	Slow
Imino (proline)	Not released	Not released	Medium

[a] Other serine carboxypeptidases[8-13] share most of the properties described here.

Yeast protease A and aminopeptidases, which are inactivated by pepstatin and EDTA, respectively, could contaminate the purified carboxypeptidase Y. The addition of pepstatin and EDTA in digestion mixture may be effective to avoid erroneous results. An addition of EDTA should be done carefully for the following digestion by metallo-carboxypeptidases (see Section IV).

The optimum pH for release of C-terminal neutral and basic amino acids is 6 to 7, while that for the release of C-terminal acidic amino acids is at, or below, pH 5.5. Therefore, the routine digestion for C-terminal analysis is performed in 0.1 *M* pyridine acetate, pH 5.5, or 0.1 *M* ammonium acetate, pH 5.5.

In addition to peptidase activity, carboxypeptidase Y has strong esterase activity toward the chymotrypsin substrate N-Ac-Tyr-OEt and also weak amidase activity toward Cbz-dipeptide amides. Recently, carboxyamidopeptidase activity, which releases amino acid amide from oligo-peptide amides, is also been recognized as a unique activity of this enzyme (see Hayashi[23] and the reference cited therein). Thus, carboxypeptidase Y appears to be quite similar to chymotrypsin both in active site and in action to synthetic substrates, although Y is an exopeptidase, whereas chymotrypsin is an endopeptidase. It should be noted, however, that no endopeptidase activity has been reported during carboxypeptidase Y digestion of proteins and peptides.

Methods described[1,2] for carboxypeptidases A and B can be principally applied in using carboxypeptidase Y for sequence studies of proteins and peptides. The properties of carboxypeptidase Y have also been reviewed in detail.[13-15]

Successful examples provide reliability and stimulate the experimentation in using an enzyme as a reagent. Tables 3 and 4 give examples of carboxypeptidase Y. Readers can see conflicting results in the tables concerning enzyme specificity. Since the history of this enzyme is young, trying the enzyme first might be worthwhile.

IV. SUCCESSIVE DIGESTION BY CARBOXYPEPTIDASES

For sequencing of the C-terminal end, a single carboxypeptidase or a mixture of carboxypeptidases has been widely used to hydrolyze proteins and peptides. However, since each carboxypeptidase has different specificity, the successive digestion by two or three enzymes provides useful information for a large part of the C-terminal sequence. The successive digestion is made in a volatile buffer: 0.1 to 0.2 *M N*- ethylmorpholine acetate or ammonium bicarbonate, pH 7.6 to 8.5, for carboxypeptidases A and B, and 0.1 *M* pyridine acetate, pH 5.5, for carboxypeptidase Y.

In successive digestion, inhibitors used for endopeptidases should be removed so as not to interfere with carboxypeptidases which are used in second and third digestions (see Section III).

Table 3
C-TERMINAL ANALYSIS OF PROTEINS BY CARBOXYPEPTIDASE Y

Protein	Conditions	Result	Ref.
Amyloid protein A (soluble at acidic medium)	0.05 *M* Pyridine-acetate (pH 4.5), 4 hr, protein conc 0.2%, S/E = 71	Deduced as – Gly – Ala – Glu – Asp – Thr – OH	25
Ribonuclease A	0.1 *M* Pyridine-acetate (pH 5.5) containing 6 *M* urea, 40 min, protein conc 1%, S/E = 222	Deduced as –(Val, Pro, His)– Phe – Asp – Ala – Ser – Val – OH (see Figure 1)	26
CM-ribonuclease A	0.1 *M* Pyridine-acetate (pH 5.5), 20 min, S/E = 280	Deduced as –(Val, Pro)– His – Phe – Asp – Ala – Ser – Val – OH (see Figure 1)	26
Cytochrome P-450a	0.2 *M* Pyridine-0.1 *M* sodium acetate (pH 5.5) containing 0.5% SDS, 135 min, S/E = 200	–(Lys, Gly, Ala)– Leu – Phe – Ile – Met – OH 15 min 0.11 0.16 0.16 0.19 0.35 0.43 0.50 45 min 0.19 0.21 0.26 0.31 0.55 0.62 0.95 135 min 0.23 0.29 0.32 0.44 0.76 0.85 1.00	27
CM-brood coagulation factor X	The first carboxypeptidase B digestion for 30 min, followed by Y for 20 min	–Leu – Glu – Arg – OH 2.5 min 0.15 0.42 1.0 5 min 0.30 0.67 1.0 10 min 0.60 0.88 1.0 20 min 0.85 0.98 1.0	28

Note: All experiments were conducted at 25°C except for the carboxypeptidase B digestion in the substrate-to-enzyme ratio (S/E) as indicated, which were conducted at 37°C.

Table 4
C-TERMINAL ANALYSIS OF PEPTIDES BY CARBOXYPEPTIDASE Y

Amino acid released	Time	Example	Comment	Ref.
Proline		Asp - Tyr - Leu - Asn - Gln - Asp - Asp - Pro - Asn - Thr - Tyr	Asp released	29
	60 min	(Asp-Asp) 1.4; Pro 0.70; (Asn-Thr) 1.8; Tyr 1.6		
		Val - Val - Ser - Glu - Pro		29
	10 min	0.54 0.83		
	60 min	0.20 0.91 1.0		
		..... Glu - Arg - Gly - Phe - Phe - Tyr - Thr - Pro - Lys - Ala	Stopped at Gly	26
	2 hr	0.04 0.10 1.4 0.88 1.0 1.0 0.80 1.0		
		..... Gly - Lys - Asp - Pro - Val - Ala - Ala - Gly - Ala - Lys	Pro-Val not split; C-terminal Lys and Gly released	30, 31
	10 min	0 0 0 0.15 0.20 ND		
	20 min	0 0 0 0.21 0.32 ND		
	60 min	0 0 0 0.28 0.60 ND		
	5 hr	0 0 0 1.1 2.0 1.0		
Acidic amino acid		Val - Gly - Ala - Gly - Lys - Leu - Glu	Gly released	30, 31
		0 0.3 0.7 ND 1.2 1.2		
		..... Thr - Pro - Val - Asn - Phe - Thr - Tyr - Ala - Glu	Asn not released (S/E = 30)	32
	10 min	0 0 0.49 0.53 0.66 0.95		
	20 min	0 0.25 0.81 0.84 0.74 1.0		
	40 min	0 0.81 1.0 0.95 1.0 1.1		
	60 min	0 0.94 1.0 0.94 1.0 1.1		
		Thr - Tyr - Gln - Glu - Phe	Dipeptide remained	30, 31
	8 hr	0 0 0.4 0.8 1.0		
		Tyr - Ile -(Asp, Ser)- Ile - Leu - Asp - $(Leu)_3$- Lys Carboxypeptidase A + B	Asp released more easily	33
	8 hr	3.0 1.0		

	14 hr	Carboxypeptidase Y 0.2 0.6 1.6 3.6 1.0		
	2 hr	..... Asp - Phe - Val - Gln - Trp - Leu - Met - Asn - Thr 0.07 0.69 0.85 0.84 0.92 0.81 1.4[a] 0.75	Asp hardly released	26
Basic amino acid	8 hr	Ile - Ala - Glu - Ala - Arg 0 0.4 0.7 0.8	Dipeptide remained	30, 31
	8 hr	Asn - Ala - Val - Glu - Lys 0 0 0.2 0.4 1.0	Dipeptide remained	30, 31
	8 hr	..... Ala - Gly - Lys - Leu - Glu - Ala - Ala - Val - Asn - Lys 0 0.2 0.5 1.2 0.6 1.0 0.8	Gly-Lys not split; Asn released	30, 31
	8 hr	pGx - Pro - Ala - Leu - Lys - Met - Lys 0 0 0 0.6 1.0 1.5		30, 31
	24 hr	.....Glycosyl Ser - Gly - Gly - Pro - Asp - Arg ←	Only Arg released (S/E = 25)	34
		..... Gly - Phe -CySO$_2$H - Lys - Asp - Gly - Tyr - Lys ←	Only Lys released	35
		..... (Ser)- Gly - Pro - Arg ← ← ←	Pro-Arg split	36
		..... Lys - Gly - Tyr - Leu - Gln - His - Leu - Glu - Arg	His released (10 m*M* Na-citrate, pH 5.7, 22°C)	37
	10 min	0.01 0.15 0.43		
	30 min	0.29 0.56 0.99 0.73 0.96		
	60 min	0.94 0.87 1.5 1.0 1.0		
		Arg - Arg - Gln - His, Ala - Arg - Ala - Ser - His - Leu Carboxypeptidase A	His hardly released, but easily by A	38, 39
	20 hr	0.6 0.9 0.15 0.6 0.8		

Table 4 (continued)
C-TERMINAL ANALYSIS OF PEPTIDES BY CARBOXYPEPTIDASE Y

Amino acid released	Example	Comment	Ref.
Homoserine	..... Gln - Lys - Trp - His - Leu - Asn - Asp - Val - Hse 3 hr 0.1 0.1 0.17 1.0 0.70	Asp not released	29
CM-cysteine	Thr - Trp - Thr(P) - Leu -Cys(CM) 2 hr 0.29 1.07	Hydrolyzed at pH 4.9 and 37°C	40
Oxidized amino acid	..... Arg - Gln - Asn - Pro - Met(O_2) - Arg 4 hr 0.7 0.7 0.9 1.0	Met (O_2) released	38, 39
	Phe - Ser - $CySO_3H$ - Gln - Arg 20 hr 0.3 0.9	$CySO_3H$ not released	38, 39
In special sequence	Asn - Thr - Lys - Arg - Asn - Lys - Asn - Asn - Ile - Ala 60 min ← ← ..... Lys - His - Asn - Ile - Thr - Gln 60 min ← ← ←	A sequence (-Lys-X-Asn) stops hydrolysis? (Experiments at pH 6.1, S/E = 500)	41
In complete digestion	Ac - Ser - Asp - Gln - Ala 69 hr ← ← ←	Acetylated peptide	42
	Leu - Val - Gly - Asp - Ile - Gly - Asn - Val - Asn - Hse 24 hr 0.4 1.0 1.0 1.5 2.0 1.8 1.0	N-terminal dipeptide hydrolyzed	43
	Trp - His - Trp - Leu - Gln - Leu - Lys - Pro - Gly - Gln - Pro - Met - Tyr 24 hr 0.46 0.79 1.8 1.0 1.0 1.7 1.8 1.0 1.0	S/E = 500, 4°C	44

Note: All experiments were performed in 0.1—0.2 *M* pyridine-acetate or amonium-acetate buffer (pH 5.5) at 25°C and in the substrate-to-enzyme ratio (S/E) of 50—100, except where otherwise stated. Numbers below the amino acid sequence show moles of amino acid released per mole of peptide.

A. Example 1

The first digestion is performed with carboxypeptidase B in 0.1 *M* ammonium bicarbonate, pH 8.0, and an aliquot of the mixture is subjected to analysis of amino acids released. The remainder of the mixture is successively incubated with carboxypeptidase A and the aliquot is withdrawn at an appropriate time interval for the determination of amino acids released. In this method, it is possible to clearly determine whether any basic residues released are at the C terminus.

B. Example 2

The first digestion by carboxypeptidase B is in 0.1 *M* ammonium bicarbonate, pH 8.5. Then, the reaction mixture is lyophilized and subjected to a second digestion by carboxypeptidase Y in 0.1 *M* pyridine acetate, pH 5.5.[24]

	Ala-	Ser-	His-	Leu-	Gly-	Leu-	Ala-	Arg
First digestion by carboxypeptidase B for 4 hr								0.9
Second digestion by carboxypeptidase Y for 2 min					0.5	0.6	1.0	0.9

C. Example 3

The digestion by a single carboxypeptidase and the successive digestion are effectively combined.

	His-Ile-Ala-Gly-Glu-Ser-Tyr-	Ala-	His-	Gly-	Tyr-	Ile-	Pro-	Val-	Phe
Carboxypeptidase Y digestion									
25°C, 10 min					0.05	0.35	0.56	0.56	0.57
25°C, 60 min					0.06	0.88	0.99	0.82	0.96
25°C, 16 hr		0.77	0.70	0.83	0.82	0.94	1.09	1.02	0.96
Carboxypeptidase A digestion									
37°C, 60 min								0.07	1.05
Successive digestion									
First digestion by carboxypeptidase Y									
25°C, 60 min					0.10	0.42	0.90	0.98	0.99
Second digestion by carboxypeptidase A									
37°C, 60 min		0.03	0.25	0.18	0.94	1.02	1.07	1.00	1.05
37°C, 3 hr		0.25	0.71	0.75	1.07	1.01	0.89	1.05	0.95

V. SEQUENCE ANALYSIS OF PEPTIDE AMIDE BY SERINE CARBOXYPEPTIDASE

Carboxypeptidases P, W, and Y show carboxamidopeptidase action which releases C-terminal amino acid amides at pH 7.5 to 8.0 from various peptide amides.[23] Cathepsin A also shares the property.[45] This activity might be applied to the sequence analysis of peptides having amidated C-terminal groups such as many peptide hormones. However, further demonstration is required to establish whether satisfactory results can be obtained in the C-terminal analysis, because amidase activity (which releases ammonia) is also accompanied by carboxamidopeptidase activity. The following example shows that carboxypeptidase Y releases the C terminus of a peptide amide only as methionine amide at pH 8, but a small amidase activity, followed by the inherent carboxypeptidase activity, is shown at pH 5.5, as seen in the release of small amount of methionine. (Ammonia released was not determined.)

A. Example

Arg-Pro-Lys-Pro-Gln-Gln-Phe-Tyr-Gly-Leu-Met.NH_2 Met

Carboxypeptidase Y digestion for 60 min						
At pH 5.5	0.1	0.9	0.9	1.0	0.9	0.1
At pH 8.0	0	0.5	0.6	0.9	1.2	0

LIST OF ABBREVIATIONS

Ac	Acetyl
Bz	Benzoyl
Bzl	Benzyl
Cbz	Benzyloxycarbonyl
CM	Carboxymethyl
DFP	Diisopropylfluorophosphate
EDTA	Ethylenediamine tetraacetate
HPLA	Hyppuryl-dl-β-phenyllactate
-OEt	Ethylester
-OMe	Methylester
PCMB	*p*-Chloromercuribenzoate
PMSF	Phenylmethylsulfonyl fluoride
-pNA	*p*-Nitroanilide
SDS	Sodium dodecylsulfate
ZPCK	Chloromethylketone derivative of Cbz-Phe

REFERENCES

1. **Ambler, R. P.,** Enzymatic hydrolysis with carboxypeptidases, *Methods Enzymol.*, 11, 155, 1967; 25, 143, 1972.
2. **Ambler, R. P.,** Carboxypeptidases A and B, *Methods Enzymol.*, 11, 436, 1967; 25, 262, 1972.
3. **Pétra, P. H.,** Bovine procarboxypeptidase and carboxypeptidase A, *Methods Enzymol.*, 19, 460, 1970.
4. **Hartsuck, J. A. and Lipscomb, W. N.,** Carboxypeptidase A, in *The Enzymes, Vol. 3*, Boyer, P. D., Lardy, H., and Myrbäck, K., Eds., Academic Press, New York, 1971, 1.
5. **Folk, J. E.,** Carboxypeptidase B (porcine pancreas), *Methods Enzymol.*, 19, 504, 1970.
6. **Folk, J. E.,** Carboxypeptidase B, in *The Enzymes*, Vol. 3, Boyer, P. D., Lardy, H., and Myrbäck, K., Eds., Academic Press, New York, 1971, 57.
7. **Zuber, H.,** Carboxypeptidase C, *Methods Enzymol.*, 45, 561, 1976.
8. **Tschesche, H.,** Carboxypeptidase C, *Methods Enzymol.*, 47, 73, 1977.
9. **Kubota, Y., Funakoshi, T., and Shoji, S.,** Application of citrus carboxypeptidase C to amino acid sequence analysis, *Protein Nucl. Acid Enzyme*, 28, 1407, 1983.
10. **Yokoyama, S.,** The use of Penicillium carboxypeptidase-P in amino acid sequencing, *Protein Nucl. Acid Enzyme*, 28, 1414, 1983.
11. **Hofmann, T.,** Penicillocarboxypeptidases S-1 and S-2, *Methods Enzymol.*, 45, 587, 1976.
12. **Umetsu, H., Abe, M., Sugawara, Y., Nakai, T., Watanabe, S., and Ichishima, E.,** Purification, crystallisation and characterization of carboxypeptidase from wheat bran, *Food Chem.*, 7, 125, 1981.
13. **Hayashi, R.,** Carboxypeptidase Y, *Methods Enzymol.*, 45, 568, 1976.
14. **Hayashi, R.,** Carboxypeptidase Y in sequence determination of peptides, *Methods Enzymol.*, 47, 84, 1977.
15. **Hayashi, R.,** Carboxypeptidase Y as a biochemical reagent, *Protein Nucl. Acid Enzyme*, 28, 1421, 1983.
16. **Abramowitz, N., Schechter, I., and Berger, A.,** On the size of the active site in proteases. II. Carboxypeptidase-A, *Biochem. Biophys. Res. Commun.*, 28, 862, 1967.
17. **Shaw, D. C. and Wells, J. R. E.,** A unique amino acid sequence around a reactive serine residue in peptidases from French beans and yeast, *Biochem. J.*, 128, 229, 1972.
18. **Ihle, J. N. and Dure, L. C., III,** The developmental biochemistry of cottonseed enbryogenesis and germination. II. Catalytic properties of the cotton carboxypeptidase, *J. Biol. Chem.*, 247, 5041, 1972.
19. *Enzyme Nomenclature — 1984*, Academic Press, New York, 1984.

20. **Hayashi, R., Bai, Y., and Hata, T.,** Evidence for an essential histidine in carboxypeptidase Y, *J. Biol. Chem.*, 250, 5221, 1975.
21. **Bai, Y. and Hayashi, R.,** Properties of the single sulfhydryl group of carboxypeptidase Y: effects of alkyl and aromatic mercurials on activities toward various synthetic substrates, *J. Biol. Chem.*, 254, 8473, 1979.
22. **Breddam, K.,** Modification of the single sulfhydryl group of carboxypeptidase Y with mercurials: influence on enzyme specificity, *Carlsberg Res. Commun.*, 48, 9, 1983.
23. **Hayashi, R.,** Carboxamidopeptidase activity of serine carboxypeptidase, in *Prospects for Applied Enzymology*, 3rd Int. Symp. on the Life Sciences, Toyobo Co. Ltd., Shiga Prefecture, Ootsu, 1984, 65.
24. **Hugli, T. E.,** Human anaphylatoxin (C3a) from the third component of complement: primary structure, *J. Biol. Chem.*, 250, 8293, 1975.
25. **Hermodson, M. A., Kuhn, R. W., Walsh, K. A., Neurath, H., Eriksen, N., and Benditt, E. P.,** Amino acid sequence of monkey amyloid protein A, *Biochemistry*, 11, 2934, 1972.
26. **Hayashi, R., Moore, S., and Stein, W. H.,** Carboxypeptidase from yeast: large scale preparation and the application to COOH-terminal analysis of peptides and proteins, *J. Biol. Chem.*, 248, 2296, 1973.
27. **Botelho, L. H., Ryan, D. E., and Levin, W.,** Amino acid-compositions and partial amino acid sequences of three highly purified forms of liver microsomal cytochrome P-450 from rats treated with polychlorinated biphenyls, phenobarbital, or 3-methylcholanthrene, *J. Biol. Chem.*, 254, 5635, 1979.
28. **McMullen, B. A., Fujikawa, K., Kisiel, W., Sasagawa, T., Howald, W. N., Kwa, E. Y., and Weinstein, B.,** Complete amino acid sequence of the light chain of human blood coagulation factor X: evidence for identification of residue 63 as β-hydroxyaspartic acid, *Biochemistry*, 22, 2875, 1983.
29. **Liao, T. -H., Salnikow, J., Moore, S., and Stein, W. H.,** Bovine pancreatic deoxyribonuclease A: isolation of cyanogen bromide peptides; complete covalent structure of the polypeptide chain, *J. Biol. Chem.*, 248, 1489, 1973.
30. **Isobe, T., Black, L. W., and Tsugita, A.,** Primary structure of bacteriophage T4 internal protein II and characterization of the cleavage upon phage maturation, *J. Mol. Biol.*, 102, 349, 1976.
31. **Tsugita, A.,** personal communication.
32. **Joris, B., Van Beeumen, J., Casagrande, F., Gerday, C., Frere, J. M., and Ghuysen, J. -M.,** The complete amino acid sequence of the Zn^{2+}-containing D-alanyl-D-alanine-cleaving carboxypeptidase of *Streptomyces albus G, Eur. J. Biochem.*, 130, 53, 1983.
33. **Austen, B. M., Nyc, J. F., and Smith, E.,** Nicotinamide adenine dinucleotide-specific glutamate dehydrogenase of *Neurospora* VI. Isolation and sequences of eighteen fragments from the cyanogen bromide digest, *J. Biol. Chem.*, 252, 8160, 1977.
34. **Dahr, W., Beyreuther, K., Kordowicz, M., and Krüger, J.,** N-Terminal amino acid sequence of sialoglycoprotein D (Glycophorin C) from human erythrocyte membranes, *Eur. J. Biochem.*, 125, 57, 1982.
35. **Reddy, M. N., Keim, P. S., Heinrikson, R. L., and Kézdy, F.,** Primary structural analysis of sulfhydryl protease inhibitors from pineapple stem, *J. Biol. Chem.*, 250, 1741, 1975.
36. **Schuppan, D., Glanville, R. W., and Timple, R.,** Covalent structure of mouse type-IV collagen: isolation, order and partial amino-acid sequence of cyanogen-bromide and tryptic peptides of pepsin fragment P1 from the α1 (IV) chain, *Eur. J. Biochem.*, 123, 505, 1982.
37. **Howard, J. B., Lorsbach, T. W., Ghosh, D., Melis, K., and Stout, C. D.,** Structure of *Azotobacter vinelandii* 7Fe ferredoxin: amino acid sequence and electron density maps of residues, *J. Biol. Chem.*, 258, 508, 1983.
38. **Hugli, T. F., Morgan, W. T., and Müller-Eberhard, H. J.,** Circular dichroism of C3a anaphylatoxin: effects of pH, heat, guanidinium chloride, and mercaptoethanol on conformation and function, *J. Biol. Chem.*, 250, 1479, 1975.
39. **Hugli, T. F.,** personal communication.
40. **Shoji, S., Titani, K., Demaille, J. G., and Fischer, E. H.,** Sequence of two phosphorylated sites in the catalytic subunits of bovine cardiac muscle adenosine 3′:5′-monophosphate-dependent protein kinase, *J. Biol. Chem.*, 254, 6211, 1979.
41. **Jacobsen, H., Demandt, A., Mondy, A. J., and Sunby, F.,** Sequence analysis of porcine GLI-1, *Biochim. Biophys. Acta*, 493, 452, 1977.
42. **Aquila, H., Misra, D., Eulitz, M., and Klingenberg, M.,** Complete amino acid sequence of the ADP/ATP carrier from beef heart mitochondria, *Z. Physiol. Chem.*, 363, 345, 1982.
43. **Wang, C. -M., Nguyen, N. Y., Yonaha, K., Robey, F., and Liu, T. Y.,** Primary structure of rabbit C-reactive protein, *J. Biol. Chem.*, 257, 13610, 1982.
44. **Tanaka, T., Kita, H., Murakami, T., and Narita, K.,** Purification and amino acid sequence of mating factor from *Saccharomyces cerevisiae, J. Biochem.*, 82, 1681, 1977.
45. **Hayashi, R. and Kawamura, Y.,** unpublished data.

Chapter 9

PROTEIN/PEPTIDE SEQUENCE ANALYSIS BY MASS SPECTROMETRY

H. C. Krutzsch

TABLE OF CONTENTS

I. INTRODUCTION

The determination of protein and polypeptide primary structure has always been an important area of research in biochemistry. However, structure analysis has received even stronger emphasis in recent years, because of the prominent role it has assumed in areas ranging from applications in pharmacology to understanding the mechanisms of neoplastic cellular transformation. There are a number of methodologies currently available to the protein biochemist for polypeptide sequence analysis, with the most widely used being the Edman degradation procedure. The use of mass spectrometry for elucidation of polypeptide primary structure has also found broad application, and can be used in situations where the Edman technique fails, as, for example, when the polypeptide has a blocked N-terminus. The use of mass spectrometric techniques also allows other determinations to be made on the polypeptide, such as identifying posttranslational modifications or the presence of unusual amino acids.

Within the broad area of mass spectrometry exists a variety of methodologies that can be applied to the molecular analysis of polypeptides and proteins. This chapter will present an overview of the procedures available to the mass spectroscopist for these analyses. For organizational purposes, the subject will be divided into topics related to the means of introducing the sample into the mass spectrometer. During these discussions, references to detailed descriptions of each technique will be cited so the reader can gain additional information about the subject at hand if so desired. First to be covered will be direct probe introduction of samples to be analyzed. Fast atom bombardment (FAB) and other desorbtion techniques will be discussed here. Second to be described will be applications using liquid chromatography-mass spectrometry (LC-MS) techniques. Third, the use of gas chromatography-mass spectrometry (GC-MS) for polypeptide sequence analysis will be presented and the oligiopeptide reduction-trimethylsilylation and dipeptidyl peptidase methodologies will be described.

II. DIRECT PROBE ANALYSIS OF POLYPEPTIDES

A. Fast Atom Bombardment Mass Spectrometry of Polypeptides

Of all the mass spectrometric techniques in use today for polypeptide structure determinations, fast atom bombardment, or FAB analysis, is employed most often. The introduction of FAB mass spectrometry has made this instrumentation a convenient approach to sequence analysis of polypeptides, with sensitivities comparable to Edman sequencing techniques. This methodology, a variant of secondary ion mass spectrometry (SIMS), has two attractive advantages. First, the polypeptide needs no prior derivitization to render it suitable for analysis, and second, molecules containing up to about 100 residues[1] are accessible to this methodology. The ability to reach such high masses is the result of the recent development of high field magnets that allow analysis at high accelerating voltages, thus preserving ion beam stability and sensitivity.[2] In the future, application to larger molecules may, hopefully, succeed. However, at present, most use for this technique has been with peptides derived from larger protein molecules by enzymatic or chemical means or with intact isolates. Several reviews of this subject have been recently published that are useful for further reading.[3-5]

The preparation and purity of the peptide to be analyzed by FAB are critical for obtaining the best possible spectral data. In principle, it is possible to analyze mixtures of peptides, and some mapping studies have been done by detecting the constituent peptide protonated molecular ions.[6] However, variations in ionizing efficiency, fragmentation, etc. make sequencing mixtures with FAB rather difficult.[7] The peptide sample, as in other sequencing methodologies, should be pure and preferentially isolated by means such as HPLC. It also should be as free of salts as possible, because salt decreases the efficiency of ionization and because the molecular weight ions, as well as other fragment ions, appear not only at the value expected for their protonated form, but also at the values of their natriated and kaliated forms.[7] This causes a drop in sensitivity, which can be considerable if a number of Asp and Glu residues appear in the peptide. Other contaminants simply cause extraneous peaks, or chemical noise, in the mass spectrum, making interpretation more difficult. Peptides derived from HPLC systems employing all volatile buffer systems are the best candidates, because they have minimum salt loads and because the buffers can be easily removed from the peptide to be analyzed.

Once the sample is purified and prepared, it is introduced into a low vapor pressure liquid matrix, such as glycerol, or a 5:1 mixture of dithiothreotol and dithioerythretol.[8] A small volume (1 to 2 $\mu\ell$) of this mixture is then placed on the FAB probe tip and introduced into the mass spectrometer. The material on the probe tip should be gold plated where it comes into contact with the sample.[7] Two inert gases have been most used as the bombarding species, argon and xenon,[7,9] with xenon being the preferred species, as less fragmentation of the peptide occurs,[10] usually resulting in greater sensitivity, especially in detection of the protonated molecular ion. The choice of inert gas will depend on what is required in a particular experiment, for example, whether more or less fragment ion formation is desired. The sensitivity depends on the peptide under analysis and can range from several picomoles to several nanomoles of material being required.[7,10] Factors that influence the sensitivity include sample purity, peptide size, the presence of prosthetic groups, the array of amino acids, and other variables.[2,4,10,12,13] For example, only 50 pmol of angiotensin I is required to give a good spectrum,[7] because it is only ten residues in length and contains a positively charged Arg residue which allows facile production of the protonated molecular ion. To increase the signal resulting from the protonated molecular ion, a variety of acidic materials, such as acetic acid, oxalic acid, or thioglycolic acid,[13] has been added to the sample matrix.

During analysis, a variety of reactions may occur with the sample on the probe tip that can cause confusion in data interpretation. One of the reactions that can occur during analysis is reduction of disulfide bonds by the glycerol matrix.[4,10,13] Another reaction that can occur between the peptide and glycerol is adduct formation, causing the molecular weight to be higher than that expected from other data.[9] Dehydrogenation and dehydration[10] reactions can also produce alterations in mass values, as can the cationation reactions described previously.

Once the mass spectrum has been obtained for a particular polypeptide, the first characteristic to be ascertained is the value for the mass of the protonated molecular ion. Generally, this ion is readily observed in the higher mass area of the spectrum as an envelope of unit mass signals, due to the polyisotopic weights of the atomic constituents, and is usually the most intense signal in this region.[4]

However, not all peptides exhibit a readily discernable protonated molecular ion, and the mass value of this species can be altered by reactions occurring in the matrix or during the ionization/desorbtion process as described above. An excellent piece of ancillary data that can be used at this point, and later in interpretation of the rest of the spectrum, is an amino acid composition determination. With the techniques available at the present time, such as the PITC amino acid analysis methodologies,[14] very little sample is needed for such a determination, and amounts usually in the range of 20 to 50 pmol are sufficient to obtain satisfactory data. Once an amino acid composition is in hand, the expected molecular weight of the peptide can be calculated and compared to the one observed in the mass spectrum. If the MS signal is low, the amino acid data can be used to locate or confirm the protonated molecular ion in the mass spectrum. If the two values are in agreement, then the results are straightforward, and the presence of amide groups, unusual or undetected amino acids, or posttranslational modifications can be ruled out. If the molecular weight determined by the mass spectrum is lower than that calculated by amino acid analysis, then the presence of carboxamide groups can be inferred. Other occurrences that can make this value lower include loss of water or ammonia, loss of a posttranslational modification that removes part of the amino acid side chain, or loss of some other neutral group such as the side chain of a particular residue. When the protonated molecular ion has a mass value greater than that calculated from composition data, the presence of undetected amino acids, such as tryptophan or posttranslational modifications, can be inferred. Other more artifactual events, such as those occurring on the probe, can also cause an increase in molecular weight; these include addition of a molecule of glycerol to the peptide or the reduction of a disulfide bond. The latter possibilities can sometimes be ruled out by examining sequentially accumulated mass spectra to observe if a change in mass is occurring in a time-dependent manner.[10] Additional causes for a higher-than-expected molecular weight can be modifications such as the presence of the oxidized form of an amino acid, for example, oxidized methionine or tryptophan. In some instances, the signal from molecular ions modified by addition of an extraneous moiety, such as a sodium ion in place of a hydrogen ion, can help to confirm the mass value of the molecular ion observed in the mass spectrum. In this case, in addition to the signal resulting from the protonated molecular ion, a satellite ion signal will be observed that is 22 mass units higher than the molecular ion. Other data present in the spectrum, such as signals resulting from doubly protonated molecular ions and molecular ion clusters, also help in this regard.

After the molecular weight of the polypeptide has been determined, the next task is to determine the amino acid sequence of the polypeptide from other ions present in the mass spectrum, ions that result from fragmentation of the protonated molecular ion. In this regard, the amino acid composition data for the peptide are also useful. If the composition shows the presence of both leucine and isoleucine, the placement of these residues will have to be determined by other means, such as by inference from nucleotide sequence data or from Edman degradation data, since these two residues are of identical mass. Also indistinguishable on the basis of mass are lysine and glutamine, but chemical modifications, such as acetylation, can be carried out that allow one to determine the correct identity of these two residues.[3] There are two pathways for decomposition of the protonated molecular ion into fragment ions that give useful sequence information, both of which result from peptide bond fission.[10,15] The first is peptide bond cleavage with charge retention on the amino terminal fragment and with loss of the carboxy fragment as a neutral species, forming a series of acylium and aldiminum ions. The second is the reverse of the first, with charge retention on the carboxyl fragment, producing a series of ammonium ions. These two pathways are complimentary, and data from one can be used to support and confirm data from the other. These fragmentations are often accompanied by hydrogen transfer processes[10] and relatively abundant satellite ions accompany these fragment ions. Because fragment ions arise from the sample matrix and from trace low molecular weight impurities, ions appearing below about mass 300 are not useful for sequence assignments. A schematic presentation of these mass spectral splitting patterns is shown in Figure 1. A variety of publications, a few of which are cited here,[12,16] can be consulted for reproductions of actual spectra and for further discussion of their interpretation.

The intensities of the fragment ions that yield sequence information are normally lower than the intensity of the protonated molecular ion, usually in the range of 10% of that intensity.[3] Fragmentations of peptide bonds will vary in efficiency, depending on their individual bond strengths, which depend somewhat on their immediate microenvironment, and this can also contribute to losses in sensitivity.[3,17] Fragmentation efficiency appears to roughly correlate with the susceptibility of a peptide bond to acid hydrolysis, and peptide bonds connecting two hydrophobic residues are likely to fragment less efficiently than peptide bonds connecting two hydrophilic residues. In some instances, fission at a particular peptide bond occurs only with a low efficiency, making a sequence assignment at this point rather

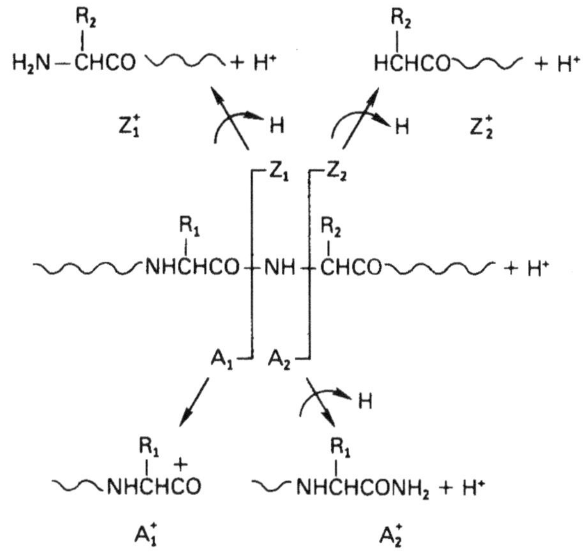

FIGURE 1. Schematic illustration of significant mass spectral fragmentations used for FAB polypeptide sequence analysis.

difficult.[17] At times, these possible difficulties can be overcome by merely applying more sample and rerunning the mass spectrum. In other cases, sequence information leading to a complete structure determination will come from other methodologies such as an Edman degradation or from a sequence deduced from the polynucleotide that codes for it.

One technique that can facilitate interpretation of complex spectra is acetylation of the peptide prior to analysis with a 1:1 mixture of H_3 and D_3 acetic anhydride.[18] This labels all sequence-determining ions that result from the cleavage of the peptide bond with charge retention on the amino-terminal fragment, and the signals from the amino-terminal fragments appear as doublets separated by 3 amu.

Another technique that can assist in the interpretation of complex spectra is the use of carboxypeptidase.[19] The polypeptide of interest is incubated with this enzyme, and aliquots are taken at various intervals and analyzed in the FAB mass spectrometer. In this instance, a series of protonated molecular ions is obtained that differs from one another by the mass equivalent to the removed amino acids. Since the protonated molecular ions are the ones observed, higher sensitivity is gained. The same operation has been carried out with aminopeptidases.[20] Of course, if both termini are blocked, this method will not be of help. The procedure can also be curtailed by the limited ability of these enzymes to cleave peptide bonds involving certain amino acids, such as proline.

The negative ionization mode of FAB mass spectrometry has also been successfully applied to the study of the mass spectra of polypeptides.[21] In some instances, better results are obtained in this mode, in terms of sensitivity and sequence information. As with other aspects of peptide chemistry, favorable results with one or another technique depend on the particular peptide under study.

To illustrate the scope of application of this technique for polypeptide structural analysis, some previously reported structural determinations are shown in Table 1. In some of the examples, areas where the amino acid sequence could not be ascertained are indicated by asterisks.

Another technique that may become useful in the area of FAB mass spectrometry for polypeptide structure determination is the one known as tandem mass spectrometry,[22,23] where two mass spectrometers are coupled together. In this methodology, the first mass

Table 1
SELECTED EXAMPLES OF POLYPEPTIDES SEQUENCED BY FAB METHODOLOGY

No.	Polypeptide[a,b]	Ref.
1	D-R-V-Y-I-H-P-F-H-L	7
2	I-N-L-K-A-L-A-A-L-A-K-K-I-L-NH2	10
3	A-V-S-E-I-Q-F-M-H-N*L*G-K-H-L-S-S-M-E-R-V-E-W-L*R-K-K-L-Q-D-V-H-N-F	10
4	G*I-V-E*Q*C*C*A*S*V-C-S-L*Y-Q-L-E-Y-C-N	10
5	F-V-N*Q*H*L*C-G-S-H-L-E-A-L-Y*L*V-C-G-E-K-G-F-Y*T-P*K-A	10
6	Y-G-G-F*M-T-S-E-K-S-Q-T-P*V*L*T	16
7	F-G-G-F	18
8	Y-G-G-F-M	18
9	M-K-R-P-P-G-F-S-P-F*R	38
10	Fl-G-S-E-S*G-N-D-V*R[c]	38

[a] Single letter code used to denote amino acids.
[b] * Indicates peptide bond for which no sequence-determining ion was observed in the mass spectrum.
[c] Fl = formyl group.

spectrometer is used to generate and separate the protonated polypeptide molecular ion, and the second mass spectrometer is used to analyze the fragment ions derived from collision of this species with some gas molecules just before it enters the second mass spectrometer. By this means, all of the extraneous chemical noise arising from the sample, matrix, etc. in the first mass spectrometer is filtered out, yielding much cleaner mass spectra and allowing useful fragmentation patterns to be observed more readily. While some peptide mixtures have been analyzed in the usual FAB mass spectrometer, either to gain sequence information or to do a type of peptide mapping, for example, in conjunction with DNA sequencing,[24] the use of tandem mass spectrometry should make this type of procedure much more amenable to producing more consistently interpretable data.

Mass spectral methodologies other than FAB for desorbing the underivatized polypeptide from the sample probe to gain structural information have also been reported, though to a lesser extent. These will be briefly described in the following sections, beginning with secondary ion mass spectrometry (SIMS), followed by plasma desorbtion, laser desorbtion, and field desorbtion methodologies.

B. Secondary Ion Mass Spectrometry of Polypeptides

At the present time, SIMS has been used only infrequently to gain structural information on polypeptides. Most of the literature published at the current time appears to have only dealt with small peptides of known structure. When SIMS is used with no liquid matrix for sample dispersion, desorbtion of sample is usually achieved by bombardment with cesium ions,[25] but Ar and Xe ions have also been used.[26] However, the observed intensity of protonated molecular ions is lower than for the other two methods mentioned previously, due to the lower efficiency of ionization produced by SIMS, which arises from sample decomposition considerations.[27] To increase the ionization efficiency of this method, the polypeptide can be modified to form a quaternary ammonium derivative prior to analysis.[27] This derivatization also augments the production of fragment ions useful for structure analysis.

SIMS desorbtion has also been performed on polypeptides in the presence of a liquid matrix, usually glycerol, using cesium ions for bombardment.[28] Again, the liquid matrix approach has not as yet been applied to enough samples to assess its capability, but, judging from comparable results achieved with FAB methodology, it should have the same scope of utility as already described for the FAB technique. In fact, in a comparison of the two, liquid SIMS with cesium ions as the bombarding species was found to be about three times as sensitive as FAB using xenon atoms as the bombarding species.[28]

For best results, the sample preparation procedure used for SIMS should be as rigorous as previously described for FAB analysis. Ideally, samples should be salt free and as pure as possible. Again, HPLC purification using volatile buffers would be the best route to obtain the cleanest samples. As in other methodologies for structural analysis, an amino acid analysis is a useful ancillary piece of data that will help in the interpretation of the mass spectrum.

The mass spectrum obtained for polypeptides from the liquid SIMS methodology is quite similar to that obtained by FAB. Usually the most abundant ion in the spectrum is the protonated molecular ion, with fragmentations useful for structural elucidation occurring at the peptide bonds. The highest molecular weight species investigated to date are in the 3000-dalton range,[28] but, as already done with FAB, larger polypeptides should also be amenable to analysis.

The scope of utility for this method of desorbtion for polypeptide sequence determinations will await future applications.

C. Plasma Desorbtion Mass Spectrometry of Polypeptides

Another mass spectral methodology that has recently been applied to polypeptide sequence analysis is known as plasma desorbtion mass spectrometry. As in FAB mass spectrometry, the underivatized protein or polypeptide is used for analysis, but ionization is achieved, in this case, by bombardment from a ^{252}Cf source, and a liquid sample matrix is not used. Several reviews[29,30] on this methodology have recently appeared, and a comparison of FAB and plasma desorbtion MS has been published.[31] While this methodology has not received the wider application of FAB mass spectrometry and is usually used to establish polypeptide molecular weights,[29] it has been used in a variety of situations to confirm proposed structures,[32] to help elucidate structures,[33] and to examine the behavior of known polypeptides[34,35] under the analytical conditions used. As with FAB mass spectrometry, polypeptides and proteins with molecular weights up to 10 kdaltons have been successfully ionized,[29] in one instance, a protein of molecular weight 25 kdaltons has been observed to give a molecular ion,[36] so the upper limit of the molecular weight of proteins amenable to desorbtion by this method may not yet have been reached.

As with the other methodologies for sequence analysis, the polypeptide should be as pure as possible. It should be free of salts and preferably isolated by HPLC using volatile buffers, for much the same reasons as were described for FAB analysis. Polypeptides amenable to sequence analysis are in the same molecular weight range as with the FAB method, and larger proteins will have to be fragmented into smaller pieces to gain any significant quantity of sequence information. At present, it appears that no work has been done on peptide mixtures.

Samples usually are dissolved at a concentration of 1 $\mu g/\mu\ell$ in a solvent such as trifluoroacetic acid, and then approximately 3 to 25 $\mu\ell$ of sample solution is electrosprayed onto a thin aluminum target.[33,37] At the end of an analytical run, most of the sample can be recovered from the target if so desired. The amount of sample analyzed has usually been reported to be in the 1- to 10-nmol range.[33,37] At the present time, no data have been published on what the lower limit of sensitivity might be, or on what factors have an influence on this aspect. The mass spectrometer used in this methodology has a lower resolution than does the typical FAB mass spectrometer, and unit mass resolution is not obtained above about m/z 600, but, by running the analysis long enough to collect many ions, mass measurements of good accuracy can be made.[33]

The mass spectrum obtained for a polypeptide usually exhibits an abundant protonated molecular ion, as well as an abundant natriated molecular ion.[33,37] This is different from a

FAB spectrum, where the protonated molecular ion predominates in salt-free samples. The rest of the mass spectrum obtained from the two methods are usually quite similar,[31,37] with the same fragmentation series being used for sequence information as was previously described for the FAB technique. However, the intensities of these ions are not normally as intense as those observed in FAB mass spectra.[31] As with FAB spectra, ions below about m/z 300 are not useful for sequence information due to interference from trace impurities.[37]

D. Field Desorbtion Mass Spectrometry of Polypeptides

In field desorbtion mass spectrometry, an underivatized polypeptide sample is desorbed by means of a heated and electrically charged filament in order to obtain its mass spectrum. This methodology has only been applied to polypeptide structure analysis in a rather small number of cases.

A recent publication has presented a review of field desorbtion (FD) mass spectrometry of peptides, compares it to FAB methods for peptide analysis, and also includes illustrations of mass spectra.[38] The mass spectrum of the polypeptide obtained from field desorbtion usually contains the protonated molecular ion in reasonable abundance, but not much fragmentation information,[38] which is a drawback if sequence information is desired. However, if the molecular ion is made to collide with some gas molecules, such as those introduced into the mass spectrometer as a slight air leak, then fragmentation of the molecular ion can occur to yield useful sequence information.[39,40] As with the other methods of desorbtion, these fragmentations arise from splitting of the peptide bonds in the polypeptide. This lack of appreciable fragmentation in the FD mass spectra of polypeptides without the aid of collision activation has been put to good use to do peptide mapping. In one case, for example, a study of hemoglobin variants was undertaken by enzymatic digestion of these proteins followed by FD mass spectrometry maps on the resulting mixtures.[41] In other instances, the use of Edman degradations[42] or exopeptidase digestion[43] followed by FD analysis of the shortened polypeptide has also been used to gain sequence information.

At the present time, polypeptides with molecular weights up to about 2000 daltons have been analyzed by field desorbtion mass spectrometry.[38] The ability of this technique to desorb larger molecules successfully will await further applications.

The preparation of sample for the FD method should follow the same procedures outlined for FAB mass spectrometry; that is, the sample should be as pure and as free of salts as possible.[38] This is best accomplished, as previously stated, through the use of HPLC methods. The requirement for the sample to be free of salts is crucial and is even more important than in FAB determinations.[38] As with other techniques for structure analysis, good amino acid analysis data are useful. Generally, the amount of material used for determinations appears to be in the range of about 10 nmol,[38] but no study defining the lower attainable limits seems to be available.

Because the amount of presently available data on FD applications to polypeptide structure analysis is not very large, no conclusions can be made as to the scope of this mass spectral technique.

E. Laser Desorbtion Mass Spectrometry of Polypeptides

The use of a laser beam to desorb underivatized polypeptides so they can be ionized and analyzed in the mass spectrometer is a fairly recent development. The desorbtion procedure itself appears to operate through a rapid heating process.[44] At the present time, the methodology has only been applied to a small number of oligiopeptides of known structure,[45,46] so the scope and sensitivity of the technique remains to be determined. The methodology looks promising for the examples with which it has been tried, and both strong quasi-molecular ions and fragment ions useful in structure determination have been observed.[45] This fragmentation pattern is the same as the patterns previously described for FAB and plasma desorbtion mass spectrometry. While both time of flight[46] and Fourier transformation[45] mass spectrometry have been used, the higher resolving power of the latter instrument should prove more useful, especially with polypeptides of higher molecular weight. This methodology has an advantage over FAB techniques in that it does not require a liquid matrix for the sample, an advantage also previously cited for the plasma desorbtion method. Further work with laser desorbtion techniques for polypeptide structure analysis will allow a more complete evaluation of this method.

F. Mass Spectrometry of Methylated-Acetylated Polypeptides

The use of methylated-acetylated derivatives of polypeptides, which renders them volatile enough to be desorbed by simple heating of the direct inlet probe, has been used for the structural analysis of a number of peptides. Several reviews have been written on this area of mass spectrometry, some of which are cited here.[7,16,48]

While techniques such as FAB have supplanted this methodology, it still has useful applications for solving structural analysis problems. In one instance, the entire primary structure of an enzyme was determined using this method of sequence analysis alone.[49] This methodology is only suitable for structural analysis of peptides; therefore, larger protein molecules must be further treated to produce smaller fragments that are amenable to analysis. Because this methodology relies on desorbtion by heating, molecules longer than about 10 to 12 residues are not usually amenable to analysis. This size limitation restricts the utility of the method and makes it less widely applicable than, for example, the FAB method.

In general, for best results, the peptide should be a single species, purified by HPLC techniques, for the same reasons that have been stressed in the section on FAB. Mixtures have been successfully analyzed, though, by slowly heating the probe so that there is successive distillation of the various components of the mixture.[50] Ions in the spectra are grouped together by their similar rates of appearance and disappearance in a series of mass spectra.

A valuable ancillary piece of data, as in the other techniques, is a good amino acid analysis. This allows the calculation of an expected molecular weight and can indicate the presence of problematic residues, such as methionine or histidine, which can easily be quaternized[51] if care is not taken during the methylation procedure. If the amino acid analysis shows the presence of arginine, the peptide must be treated with aqueous hydrazine prior to derivatization to convert it to ornithine by removing the guanido group.

The methylation-acetylation derivatization procedure involves two general steps.[50] In the first step, the polypeptide is acetylated by treating it with a mixture of acetic anhydride in methanol (3:1 v/v) which also contains a trace of triethylamine. After 3 hr, the mixture is evaporated to dryness *in vacuo*. In the second step, the acetylated peptide is dissolved in dimethyl sulfoxide and an aliquot of the methyl sulfinyl carbanion in dimethyl sulfoxide is added. After about 15 min, methyl iodide is added. After about 1 to 2 min of reaction time, the methylation reaction is quenched by addition of distilled water. This is an important step, because too long of an exposure to methyl iodide can cause quaternization of susceptible amino acids, as mentioned above. The methylation reaction occurs on the NH group of peptide bonds and acetylamido NH groups and also occurs on hydroxyl, carboyxl, primary amido (twice), and imidozyl and indolyl NH groups.[50] Finally, the acetylated and methylated peptide is extracted with chloroform several times. The combined washes were dried down, and the residue is redissolved in a small volume of chloroform to facilitate application to the direct inlet probe of the mass spectrometer. Generally about 10 to 20 nmol of peptide is required for successful application of this technique.[48] The reaction scheme for this procedure is shown in Figure 2.

The mass spectrometric analysis of the derivatized sample has been done in both the EI mode[16] and in the CI mode.[52] In the EI mode, fragmentation occurs with charge retention

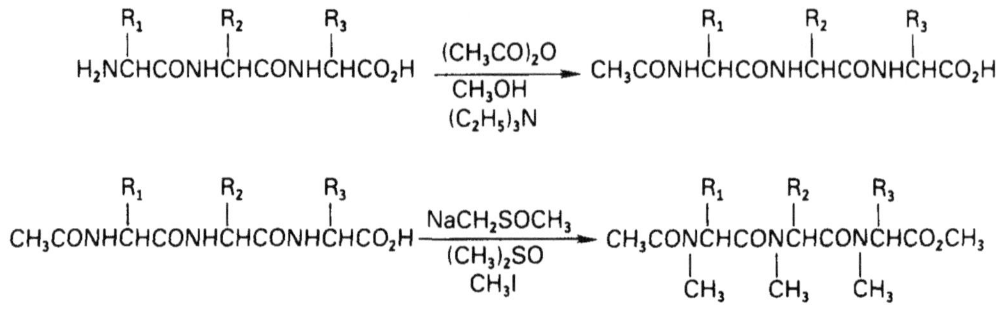

FIGURE 2. Outline of reactions used to prepare acetylated-methylated peptides.

on the amino terminal fragment, generating a series of signals useful for sequence determination that differ from one another by the mass value of the next amino acid from the amino terminus in the peptide chain.[50] These ions generally decrease in intensity as they go down the series from the peptide N terminus, and the molecular ion is usually not very intense. A number of publications contain illustrations of actual mass spectra, some of which are cited here for further reference.[48,50] When the mass spectrum is obtained in the CI mode, both the fragmentation series arising from retention of charge on the N-terminal fragment as well as those arising from retention of charge on the C-terminal fragment are observed.[52] This generally allows more sequence information to be obtained, and the protonated molecular ion is more visable. Two gases are used in obtaining CI spectra, methane and isobutane, with isobutane usually being the preferred reagent. A number of publications containing pictures of CI mass spectra of acetylated-methylated peptides have appeared, one of which is cited here for further reference.[52] The fragmentation patterns useful for sequence analysis are shown in schematic fashion in Figure 3.

As with FAB spectra, the amino acid analysis is a useful piece of data to have on hand when interpreting the mass spectrum for a methylated-acetylated peptide. If both leucine and isoleucine are present, other data sources must be consulted to correctly place these residues. If glycine is present, and the time for methylation was allowed to go too long, partial conversion of this residue to alanine can occur, yielding possibly confusing fragmentation data.[52] All of the other residues have unique mass values, so no confusion will arise in their placement in the peptide when it is deduced from the fragmentation pattern in the mass spectrum.

As stated earlier, a large number of peptides have been successfully sequenced by this methodology. To illustrate the scope of this technique, some applications are listed in Table 2.

III. LIQUID CHROMATOGRAPHY-MASS SPECTROMETRY OF POLYPEPTIDES

This methodology is attractive because it couples an instrument that can separate component mixtures to a mass spectrometer and, thereby, eliminates the intermediate sample handling steps that would be necessary with conventional HPLC preparative techniques.

In the first reports of liquid chromatography-mass spectrometry (LC-MS) analysis of peptides, the peptides were derivatized by acetylation-methylation to convert them into a volatile form compatible with the instruments being used.[53] However, recently, use of thermospray liquid chromatography-mass spectrometry has allowed underivatized peptides to be analyzed.[54] This new technique has been shown to give mass spectra which usually contain the protonated molecular ion along with small signals due to cationated molecular

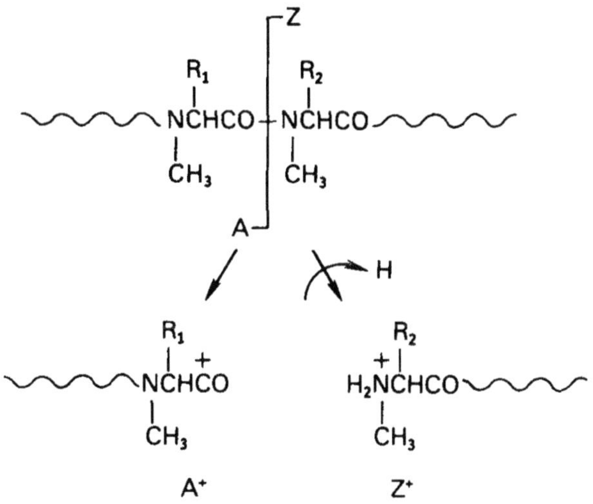

FIGURE 3. Schematic illustration of significant mass spectral fragmentations used for acetylated-methylated peptide sequence analysis.

Table 2
SELECTED EXAMPLES OF POLYPEPTIDES SEQUENCED BY AC-ME METHODOLOGY

No.	Polypeptide[a]	Ref.
1	Y-G-G-F-M	16
2	L-E-Q-V-P-Y	50
3	V-G-L-A-P-V-A	50
4	A-N-N-K	50
5	Q-Y-Y-T-V-F	50
6	V-L-A-H-T-K	51
7	F-S-V	52
8	V-L-S-E-G	52
9	F-D-A-S-V	52
10	W-M-D-F	52

[a] Single letter code used to denote amino acids.

ions. The spectra contain very few fragmentaiton ions, however, so sequence analysis of the polypeptide may not be feasible by this method.

In one application of LC-MS, determinations were done on about 4 nmol each of several peptides with chain lengths in the range of 5 to 13 residues.[55] In another application, renin substrate was analyzed.[56] In another publication, thermospray LC-MS was used to analyze tryptic and carboxypeptidase Y digests of polypeptides[57] in which about 2 nmol of sample were sufficient for studies of peptides up to 15 residues in length. Other studies indicate that sensitivities in the picomolar range may be attainable. At this point, the literature that has appeared on this technique indicates that it has useful applications in peptide structure work, but that the scope and sensitivity have yet to be determined in any detail.

IV. POLYPEPTIDE SEQUENCE ANALYSIS BY GAS CHROMATOGRAPHY-MASS SPECTROMETRY

Another technique for polypeptide sequence analysis that has been studied in detail involves introduction of samples by use of a gas chromatograph. This technique, known as gas chromatography-mass spectrometry (GC-MS), will be presented in the following two sections. The first will describe the procedure as applied to small oligiopeptide mixtures. The second will present the use of GC-MS for analysis of dipeptide mixtures derived from dipeptidyl peptidase digestion of larger polypeptides and proteins.

A. Polypeptide Sequence Analysis with Oligiopeptide Reduction/Trimethylsilylation and Gas Chromatography-Mass Spectrometry

This approach to polypeptide structure determination employs limited acidic or enzymatic digestion of the polypeptide to create mixtures of oligiopeptides of two to six residues in length. The resulting mixture is then converted into volatile derivatives suitable for analysis by gas chromatography-mass spectrometry (GC-MS). When the identities of these derivatized fragments are deduced from GC-MS, they are aligned into a family of overlapping short peptides to give the sequence of the starting polypeptide.

Although the oligiopeptide aspect of the GC-MS methodology does not have the sensitivity of the dipeptidyl peptidase aspect, it has the ability to do sequence analysis on fragments derived from a variety of positions in a protein molecule, thus making it a good partner to DNA sequence analysis. For further reference, several reviews of the oligiopeptide/GC-MS technique are cited here.[58-61] Usually between 50 and 400 nmol of polypeptide are used in the procedure described below.

As with all other methodologies for polypeptide sequence determination, the material under analysis should be a single species, preferably having been isolated by HPLC. Again, an amino acid determination on the molecule is a useful piece of data.

For sample preparation, the polypeptide is first converted into a mixture of two- to six-residue oligiopeptides, either by a combination of enzymes or by mild acid treatment. Six enzymes have been employed for digestion as single species, in combination, or in series.[59-63] The six enzymes are trypsin, chymotrypsin, elastase, thermolysin, proteinase K, and pepsin. The buffer employed for these enzymatic digestions is usually ammonium acetate, except for digestions with pepsin in which the buffer is dilute formic acid. When mild acid treatment is used to generate the oligiopeptide mixture, the polypeptide is treated with 6 *N* HC1 at 110°C for 12 to 60 min.[59,64] While mild acid treatment yields oligiopeptides that may not be obtainable by enzymatic means, hydrophilic areas of the polypeptide are hydrolyzed faster than hydrophobic areas, and partial primary amide and tryptophan destruction can occur.[63]

Secondly, the oligiopeptide mixutre is made volatile by conversion to a mixture of trimethylsilylated aminoalcohols. If arginine is present, it is converted into ornithine with hydrazine hydrate[59] prior to the derivatization procedure. In the first derivatization step, the oligiopeptides are converted to their methyl esters.[59,63] Next, these compounds are trifluoroacetylated,[59,63] then reduced to the corresponding polyaminoalcohols with hexadeuterodiborane.[59,63] This reaction scheme is outlined in Figure 4. In the final step, the polyaminoalcohol mixture is trimethylsilylated using trimethylsilyldiethylamine in pyridine.[60] Under these conditions, only the active hydrogens on hydroxyl groups are reactive. Aliquots of this solution are then subjected to GC-MS analysis.

When GC-MS analysis of a trimethylsilylated polyaminoalcohol mixture is carried out, a 0.32-mm I.D. × 15- to 30-m fused silica SE 30 capillary GC column, fitted with an on-column injector, is used,[65] and column temperature is programmed from 80 to 330°C. The mass spectrometer is operated in the electron impact mode, and continuous scanning is done

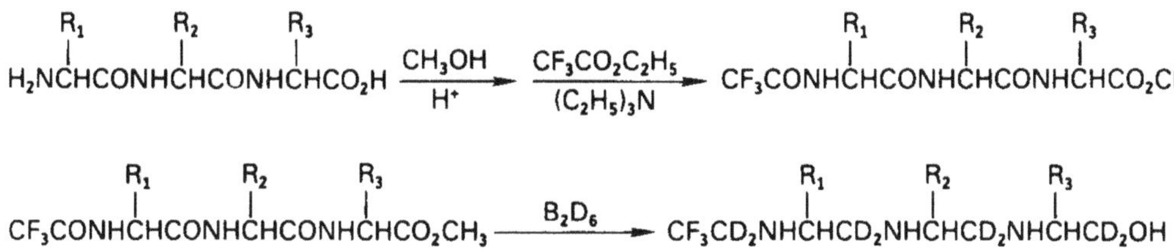

FIGURE 4. Outline of reactions used to prepare peptide trifluoromethyl polyaminoalcohol derivative.

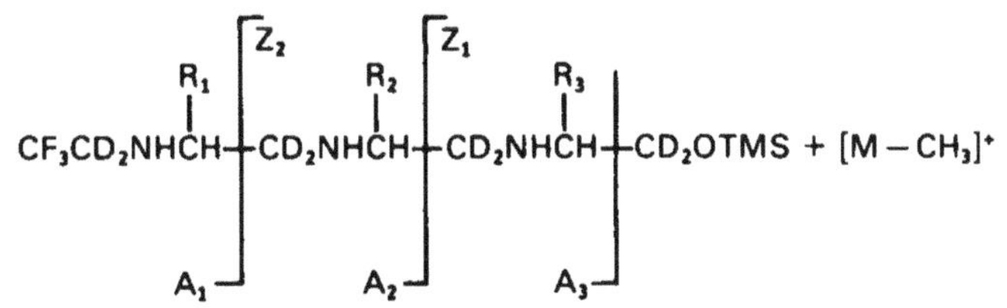

FIGURE 5. Schematic illustration of significant mass spectral fragmentations used for trimethylsilylated polyaminoalcohol sequence analysis.

during a run, with scan data stored on a computer interfaced to the mass spectrometer. The mass spectrum of trimethylsilyl polyaminoalcohols is relatively simple, and sequence ions result from cleavage of the carbon-carbon bonds in the polyethylenediamine backbone, with charge retention on either fragment, giving rise to an A and Z ion series.[66] The loss of a trimethylsilyl methyl group from the parent radical ion gives the signal in the mass spectrum that allows assignment of the molecular weight of the trimethylsilyl aminoalcohol. This mass spectral pattern is shown in diagram form in Figure 5. Actual examples of mass spectra can be found in a number of publications, some of which are cited here.[60,66,67]

The derivatization procedure was designed, in part, to aid in the identification of the trimethylsilyl aminoalcohol, with the use of deuterated reducing agent allowing discrimination between reduced aspartic acid and threonine, and the use of trifluoroacetyl instead of acetyl groups prior to reduction providing a more balanced array of diagnostic ions from the final derivatized product. In this procedure, Leu and Ile residues cannot be distinguished from one another, unless they appear at the C-terminus of the oligiopeptide.[65] If the polypeptide is derived from tissue culture experiments, dideutereoleucine can be used to effect a distinction.[68] If the work is done in conjunction with DNA sequence analysis, Ile and Leu identities can be deduced from the DNA-derived sequence.

The oligiopeptide/GC-MS methodology has been used to solve a number of structural analysis problems. Both whole polypeptide sequences,[67,69,70] as well as partial peptide sequences done in conjunction with DNA sequencing,[71,72] have been elucidated. To illustrate the scope of this method, a small list of representative polypeptides that have been sequenced is shown in Table 3. As this table shows, polypeptides can have a variety of chain lengths and amino acid residues. Partial sequence analyses, of which no examples are listed here, have also been performed on much larger polypeptides than the ones illustrated above and have allowed assignment of DNA reading frames and confirmation of sequence data.[71]

B. Polypeptide Sequence Analysis with Dipeptidyl Peptidases and Gas Chromatography-Mass Spectrometry

The other approach to polypeptide sequence analysis with GC-MS utilizes enzymatic dipeptidyl peptidase (DP) digestion of the polypeptide to produce dipeptide fragments and trimethylsilylation of the resulting mixture to render it suitable for GC-MS analysis. The

Table 3
SELECTED EXAMPLES OF POLYPEPTIDES SEQUENCED WITH OLIGIOPEPTIDES/GC-MS

No.	Polypeptide[a]	Selected oligiopeptides	Ref.
1	G-L-G-T-L-Y-F-L-V-K-G-Hse	G-L-G-T	60
		T-L-Y	
		L-Y-F	
		Y-F-L	
		F-L-V-K	
		K-G-Hse	
2	L-L-G-Y-G-L-T-Hse	L-L-G-Y-G	60
		G-L-T-Hse	
3	I-T-A-Q-L-K	I-T-A-Q	64
		T-A-Q-L	
		A-Q-L-K	
4	L-Y-I-L-Y-V-L-F-F-G-F-T-S-K-H-Z-S-Hse	L-Y-I-L	68
		I-L-Y	
		L-Y-V-L	
		V-L-F-F	
		F-F-G-F	
		T-S-K	
		S-K-A-Z	
		S-Hse	

[a] Single letter code used to denote amino acids.

sequence of the polypeptide is deduced by aligning these dipeptides from other information, such as the identities of the "one-residue frame-shift" dipeptides,[73] homology, etc. The "frame shift" dipeptides are generated from DP digestion of the polypeptide after it has been modified by addition or subtraction of one amino acid residue.[74] Several detailed publications on the dipeptidyl peptidase/GC-MS methodology that have previously appeared are cited here.[61,73,74] With the DP technique, 1 nmol of polypeptide is normally used, but lesser amounts (to about 5 pmol) can also be readily analyzed because of the high sensitivities available with GC-MS. Because of rapid GC-MS run times (approximately 30 min per run), about 12 digestions a day can be analyzed.

As is true of other MS sequencing techniques, the polypeptide under study should be a homogeneous species and free of salts, preferably having been isolated by HPLC. As with other methods, the amino acid composition of the polypeptide is a valuable piece of ancillary data.

For sample preparation, the polypeptide is digested completely into dipeptide fragments using one of the dipeptidyl peptidase enzymes. There are two types of DP enzymes, both of which are exopeptidases. Dipeptidyl aminopeptidase (DAP) has a specificity for the amino terminus of the peptide,[73] and dipeptidyl carboxypeptidase (DCP) has a specificity for the carboxyl terminus of the peptide.[74,75] Therefore, digestions can be done commencing either from the N- or C-terminus of the sample polypeptide. Thus, a polypeptide with a blocked N-terminus, for example, which cannot be analyzed by the Edman method, is still accessible to the DP methodology, as it is with all other protein sequencing techniques using mass spectrometry.

Digestions with DAP and DCP are carried out in a small Teflon®-capped vial using a volatile buffer, and are terminated by freeze-drying after 4 hr. The two enzymes (DAP and DCP) used to effect digestion have a broad scope of activity, and digestion yields are lowered only when a peptide bond involving the imino group of proline is encountered. Polypeptides

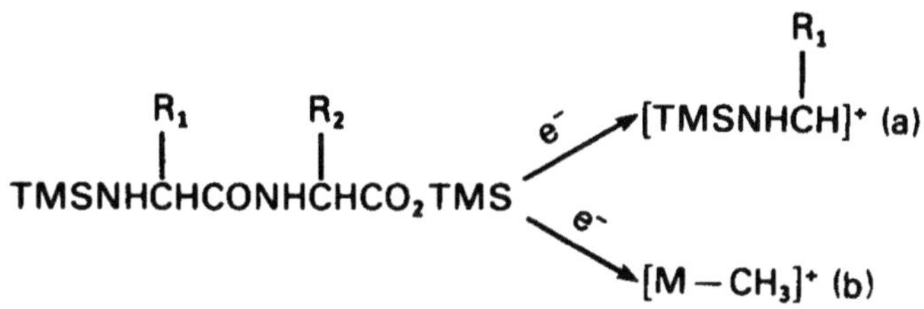

FIGURE 6. Schematic illustration of significant mass spectral fragmentations used for trimethylsilylated dipeptide identification.

up to about 250 residues, in some cases, have been found to be degraded by these enzymes. The major use, however, has been with peptides of about 30 residues or less.

The dipeptides are then converted to their trimethylsilyl derivatives in the same vial using bis(trimethylsilyl)trifluoroacetamide;[73,74] no transfer of material is necessary. When trimethylsilylation is carried out, the active hydrogens of amino, carboxyl, imidazoyl, indolyl, primary amide, and guanido groups react.[76] One proton on the α-amino, primary amido, and aminoethyl-Cys is exchanged, both protons on the α-amino Gly and ε-amino Lys groups are replaced, and three of the four protons of the guanido group of Arg are displaced. Protons of the secondary amide group of dipeptides generally do not react, but protons of the ring NH groups in diketopiperazines react. During trimethylsilylation, dipeptides containing N-terminal Asp usually form cyclic imides, and some Gly-X (X = Gly, Ala, Met, Gln, Glu) dipeptides form diketopiperazines.

For GC-MS analysis of the trimethylsilylated dipeptide mixture, a 0.30-mm × 15- to 25-m fused silica GC capillary column coated with bonded OV-1 or SE-54 can be used,[61,74] and the temperature is programmed from 110 to 270°C to effect elution. The identification of trimethylsilylated dipeptides from their mass spectrum is quite straightforward. Except when cyclization reactions occur, identification is made from two ion masses.[76] One is the ion resulting from cleavage of the central CH-CO bond, with charge retention on the amino terminal portion, which is the sequence-determining ion of the dipeptide. The second ion mass is the one resulting from loss of a trimethylsilyl methyl group from the parent ion radical, which is the molecular weight-determining ion of the dipeptide. When cyclization occurs, the sequence-determining ion is not present, but the molecular weight-determining ion is sufficient for identification in these cases. This splitting pattern is shown schematically in Figure 6.

Dipeptides containing Leu are distinguished from those containing Ile by the ratio of intensities of the ion masses at 41/43 amu, where the intensity is higher at 43 than at 41 for Leu. In addition, when Ile is at the dipeptide N-terminus, an ion mass that arises from the loss of the Ile side chain is observed in the mass spectrum. Examples of trimethylsilylated dipeptide mass spectra can be found in an earlier publication.[76]

The dipeptidyl peptidase GC-MS approach to polypeptide sequencing analysis has been applied to a broad spectrum of sequencing situations, ranging from partial to complete sequence determinations, to confirmation of the structures of synthetic peptides, determination of blocking groups, confirmation and/or completion of Edman-derived sequences, and sequence determinations on N-terminal blocked peptides. To illustrate the scope of the DP method, Table 4 shows examples selected from the many polypeptides that have been analyzed. The first four examples show complete sequence analyses that were accomplished with the DP method. The next three show putative polypeptide sequences that were confirmed by this approach. Partial sequences solved with the DP method are shown in examples 8, 9, and 10. Finally, in the last two examples, the DP method was used for difficult determinations, ones that escaped definition by other methods. In example 11, an unusual blocking

Table 4
SELECTED EXAMPLES OF POLYPEPTIDES SEQUENCED WITH DP/GC-MS

No.	Polypeptide[a]	Treatment	Dipeptides	Ref.
1	F-S-W-G-A-E-G-Q-K	DAP	F-S,W-G,A-E,G-Q	77
		DCP	S-W,G-A,E-G,Q-K	
2	Ac-S-S-S-G-T-P-D-L-P[b]	DCP	Ac-S,S-S,G-T,P-D,L-P	78
	Ac-S-S-S-G-T-P-D-L	DCP	Ac-S-S,S-G,T-P,D-L	
3	G-A-H-D-A-Q-G-T-L-S-K	DCP	A-H,D-A,Q-G,T-L,S-K	79
		CPB + DCP	G-A,H-D,A-Q,G-T,L-S	
4	F-S-G-S-G-S-G-T-F-L-Q-I-S-R	DAP	F-S,G-S,G-S,G-T,F-L,Q-I,S-R	73
		Abu + DAP[c]	U-F,S-G,S-G,S-G,T-F,L-Q,I-S	
5	A-P-D-Y-K-P-A-H-R	DAP	A-P,D-Y,K-P,A-H	77
6	G-R-P-Q-D-E-N-P-R	DAP	G-R,P-Q,D-E,N-P	77
7	Ac-A-N-E-R-A-D-L-I-A-Y-L-K-Q-A-T-K	DCP	Ac-A-N,E-R,A-D,L-I,A-Y,L-K,Q-A,T-K	—
8	—S-L-Y-L-Q-Hse	DCP	S-L,Y-L,Q-Hse + others	80
9	—Y-S-L-T-I-S-S-Hse	DCP	Y-S,L-T,I-S,S-Hse + others	81
10	P-I-A-P-T-V-L-L-F-P-P—	DAP	P-I,A-P,T-V,L-L	82
11	My-G-Q-T[d]	DCP	My-G,Q-T	83
12	Fl-M-L-F[e]	DCP	Fl-M,L-F	84

[a] Single letter code used to denote amino acids.
[b] Ac = acetyl group.
[c] Abu and U denote α-amino butyric acid.
[d] My = myristyl group.
[e] Fl = formyl group.

group, myristic acid, was identified. In example 12, a biologically active peptide with a blocked N-terminus was successfully sequenced, including identification of the formyl blocking group. As the table also shows overall, the dipeptidyl peptidase/GC-MS methodology has a broad spectrum of application for solving polypeptide structure determination problems.

V. CONCLUDING REMARKS

In summary, mass spectrometry is a valuable tool for polypeptide sequence analyses. As work progresses in this area and as more applications are made, the usefulness of this technique should increase. As is evident from the preceeding discussions, each type of mass spectral methodology has its strengths and weaknesses, and much like other analytical techniques, a particular MS method may be employed to solve a particular sequencing problem, or several MS methods may be combined to solve various aspects of a structure determination.

While the majority of protein and peptide structure determinations will continue to be carried out either by the Edman method or inferred from the DNA sequence, mass spectrometry should play a useful role in this process, and, as stated above, with continued improvement, this role should increase in importance. In some applications, such as carrying out structure determinations on N-blocked peptides, determining posttranslational modifications, or the identity of unknown amino acids, and in some areas of peptide mapping and DNA sequence analysis, it should be the method of choice. Overall, these technologies can be combined to yield a potent means for elucidation of the total structures of proteins and peptides, further augmenting the protein biochemist's ability to carry out these determinations with greater facility and accuracy.

REFERENCES

1. **Barber, M., Bordoli, R. S., Elliot, G. J., Horoch, N. J., and Green, B. N.,** Fast atom bombardment mass spectrometry of human proinsulin, *Biochem. Biophys. Res. Commun.*, 110, 753, 1983.
2. **Dell, A. and Taylor, G. W.,** High-field-magnet mass spectrometry of biological molecules, *Mass Spectrom. Rev.*, 3, 357, 1984.
3. **Fraser, B. A.,** Fast atom bombardment mass spectrometry: applications to peptide structural analysis, in *Modern Methods in Protein Chemistry,* L'Italien, J., Ed., Plenum Press, New York, 1986.
4. **Barber, M., Bordoli, R. S., Segwick, R. D., Taylor, R. N., and Green, B. N.,** Fast atom bombardment mass spectrometry, in *Biochemistry, Medicine, and Environmental Research,* Vol. 8, Frigerio, A., Ed., Elsevier, Amsterdam, 1983.
5. **Rinehart, K. L., Jr.,** Fast atom bombardment mass spectrometry, *Science,* 218, 254, 1982.
6. **Morris, H. R., Panico, M., and Taylor, G. W.,** FAB-mapping of recombinant-DNA protein products, *Biochem. Biophys. Res. Commun.*, 117, 299, 1983.
7. **Buko, A. M., Phillips, L. R., and Fraser, B. A.,** Peptide studies using a fast atom bombardment high field mass spectrometer and data system. I. Sample introduction, data acquisition, and mass calibration, *Biomed. Mass Spectrom.*, 10, 324, 1983.
8. **Gower, J. L.,** Matrix compounds for fast atom bombardment mass spectrometry, *Biomed. Mass Spectrum.*, 12, 191, 1985.
9. **Martin, S. A., Costello, C. E., and Biemann, K.,** Optimization of experimental procedures for fast atom bombardment mass spectrometry, *Anal. Chem.*, 54, 2362, 1982.
10. **Buko, A. M., Phillips, L. R., and Fraser, B. A.,** Peptide studies using a fast atom bombardment high field mass spectrometer and data system. II. Characteristics of positive ionization spectra of peptides, M/Z 858 to M/Z 5729, *Biomed. Mass Spectrom.*, 10, 408, 1983.
11. **Morris, H. R.,** Biomolecular mass spectrometry, *Int. J. Mass Spectrom. Ion Phys.*, 45, 331, 1982.
12. **Buko, A. M. and Fraser, B. A.,** Peptide studies using a fast atom bombardment high field mass spectrometer and data system. IV. Disulfide-containing peptides, *Biomed. Mass Spectrom.*, 12, 577, 1985.
13. **Barber, M., Bordoli, R. S., Elliot, G. J., Tyler, A. N., Bill, J. C., and Green, B. N.,** Fast atom bombardment (FAB) mass spectrometry: a mass spectral investigation of some of the insulins, *Biomed. Mass Spectrom.*, 11, 182, 1984.
14. **Heinrikson, R. L. and Meredith, S. C.,** Amino acid analysis by reverse phase high performance liquid chromatography: precolumn derivatization with phenylisothiocyanate, *Anal. Biochem.*, 136, 65, 1984.
15. **Williams, D. H., Bradley, C. V., Santikarn, S., and Bojsen, G.,** Fast-atom-bombardment mass spectrometry. A new technique for the determination of molecular weights and amino acid sequences of peptides, *Biochem. J.*, 201, 105, 1982.
16. **Morris, H. R., Dell, A., Etienne, A. T., Judkins, M., McDowell, R. A., Panico, M., and Taylor, G. W.,** New methods for the characterization of biopolymers, *Pure Appl. Chem.*, 54, 267, 1982.
17. **Roepstorff, P., Hojrup, P., and Moller, J.,** Evaluation of fast atom bombardment mass spectrometry for sequence determination of peptides, *Biomed. Mass Spectrom.*, 12, 181, 1985.
18. **Morris, H. R., Panico, M., Barber, M., Bordoli, R. S., Sedgewick, R. D., and Tyler, A.,** Fast atom bombardment: a new mass spectrometric method for polypeptide sequence analysis, *Biochem. Biophys. Res. Commun.*, 101, 623, 1981.
19. **Bradley, C. V., Williams, D. H., and Hanly, M. R.,** Peptide sequencing using the combination of Edman degradation, carboxypeptidase digestion, and fast atom bombardment mass spectrometry, *Biochem. Biophys. Res. Commun.*, 104, 1223, 1982.
20. **Self, R. and Parente, A.,** The combined use of enzymatic hydrolysis and fast atom bombardment mass spectrometry for peptide sequencing, *Biomed. Mass Spectrom.*, 10, 78, 1983.
21. **Buko, A. M., Phillips, L. R., and Fraser, B. A.,** Peptide studies using fast atom bombardment high field mass spectrometer and data system. III. Negative ionization: mass calibration, data acquisition and structural characterization, *Biomed. Mass Spectrom.*, 10, 387, 1983.
22. **McLafferty, F. W., Ed.,** *Tandem Mass Spectrometry,* Wiley-Interscience, New York, 1983.
23. **Hunt, D. F., Bone, W. M., Shabanowitz, J., Rhodes, J., and Ballard, J. M.,** Sequence analysis of oligiopeptides by secondary ion/collision activated dissociation mass spectrometry, *Anal. Chem.*, 53, 1704, 1981.
24. **Gibson, B. W. and Biemann, K.,** Strategy for the mass spectrometric verification and correction of the primary structures of proteins deduced from their DNA sequences, *Proc. Natl. Acad. Sci. U.S.A.*, 81, 1965, 1984.
25. **Westmore, J. B., Ens, W., and Standing, K. G.,** Secondary ion mass spectra of the neuropeptides leucine-enkephalin and methionine-enkephalin, *Biomed. Mass Spectrom.*, 9, 119, 1982.
26. **Kambara, H.,** Comparative study of [Xe] + and [Ar] + bombardment in secondary ion mass spectrometry for bioorganic compounds, *Org. Mass Spectrom.*, 17, 29, 1982.

27. **Kidwell, D. A., Ross, M. M., and Colton, R. J.,** Sequencing of peptides by secondary ion mass spectrometry, *J. Am. Chem. Soc.*, 106, 2219, 1982.
28. **Aberth, W., Straub, K. M., and Burlingame, A. L.,** Secondary ion mass spectrometry with cesium ion primary beam and liquid target matrix for analysis of bioorganic compounds, *Anal. Chem.*, 54, 2029, 1982.
29. **Busch, K. L. and Cooks, R. G.,** Mass spectrometry of large, fragile, and involatile molecules, *Science*, 218, 247, 1982.
30. **Macfarlane, R. D.,** Californium-252 plasma desorbtion mass spectrometry — large molecules, software and the essence of time, *Anal. Chem.*, 55, 1247A, 1983.
31. **Fohlman, J., Peterson, P. A., Roepstorff, P., Hojrup, P., Kamensky, I., Sawe, G., Hakansson, P., and Sundqvist, B.,** Comparison of Californium-252 plasma desorption and fast atom bombardment mass spectrometry for analysis of small peptides, *Biomed. Mass Spectrom.*, 12, 380, 1985.
32. **Chait, B. T., Gisin, B. F., and Field, F. H.,** Fission fragment ionization mass spectrometry of Alamethicin I, *J. Am. Chem. Soc.*, 104, 5157, 1982.
33. **Fales, H. M., McNeal, C. J., Macfarlane, R. D., and Shimhigashi, Y.,** Californium-252 plasma-desorption mass spectrometry of polyethylene-diamine linked enkephalin peptides, *Anal. Chem.*, 57, 1616, 1985.
34. **Chait, B. T., Agosta, W. C., and Field, F. H.,** Fission fragment ionization (Cf-252) mass spectrometry. Positive and negative spectra and decomposition mechanisms for seven compounds, *Int. J. Mass Spectrom. Ion Phys.*, 39, 339, 1981.
35. **Macfarlane, R. D.,** *Biochemical Applications of Mass Spectrometry, First Supplementary Volume*, Waller, G. R. and Dermer, O. C., Eds., Wiley-Interscience, New York, 1980, 1209.
36. **Sundqvist, B., Roepstorff, P., Hedin, A., Hakansson, P., Salehpour, M., Sawe, G., Fohlman, J., Kamensky, I., and Lindberg, M.,** Molecular weight determinations of proteins by Californium plasma desorption mass spectrometry, *Science*, 226, 696, 1984.
37. **Sundqvist, B., Kamensky, I., Hakansson, P., Kjellberg, J., Salehpour, M., Widdiyaskera, S., Fohlman, J., Peterson, P. A., and Roepstorff, P.,** Californium-252 plasma desorption time of flight mass spectroscopy of proteins, *Biomed. Mass Spectrom.*, 11, 242, 1984.
38. **Przybylski, M.,** Fast atom bombardment and field desorption mass spectrometry: comparative aspects of analytical development and bioanalytical application, *Fres. Z. Anal. Chem.*, 315, 402, 1983.
39. **Desiderio, D. M. and Sabbitini, J. Z.,** Field desorption collision activation linked scanning mass spectrometry of underivatized oligiopeptides, *Biomed. Mass Spectrom.*, 8, 565, 1981.
40. **Weber, R. and Levsen, K.,** Collision induced dissociation of field desorbed di- and tripeptides, *Biomed. Mass Spectrom.*, 7, 314, 1980.
41. **Wada, Y., Hayashi, A., Fujita, T., Matsuo, T., Katakuse, I., and Matsuda, H.,** Structural analysis of hemoglobin variants by mass spectrometry, *Int. J. Mass Spectrom. Ion Phys.*, 48, 209, 1983.
42. **Shimanishi, Y., Hong, Y. -M., Kitagishi, T., Matsuo, T., Matsuda, H., and Katakuse, I.,** Sequencing of peptide mixtures by Edman degradation and field-desorption mass spectrometry, *Eur. J. Biochem.*, 112, 251, 1980.
43. **Tsugita, A., Van Den Broek, R., and Przybylski, M.,** Exopeptidase digestion in combination with field desorption mass spectrometry for amino acid sequence determination, *FEBS Lett.*, 137, 19, 1982.
44. **Posthumus, M. A., Kistemaker, P. G., Meuzelaar, H. L. C., and Ten Noever de Braun, M. C.,** Laser desorption-mass spectrometry of polar nonvolatile bio-organic molecules, *Anal. Chem.*, 50, 985, 1978.
45. **Wilkins, C. L., Weil, D. A., Yang, C. L. C., and Ijames, C. F.,** High mass analysis by laser desorption fourier transform mass spectrometry, *Anal. Chem.*, 57, 520, 1985.
46. **Tabet, J. -C. and Cotter, R. J.,** Laser desorption time-of-flight mass spectrometry of high mass molecules, *Anal. Chem.*, 56, 1662, 1984.
47. **Morris, H. R.,** Biomolecular structure determination by mass spectrometry, *Nature*, 286, 447, 1980.
48. **Morris, H. R., Dell, A., and McDowell, R. A.,** Extended performance using a high field magnet mass spectrometer, *Biomed. Mass Spectrom.*, 8, 463, 1981.
49. **Morris, H. R.,** Research on peptides and glycopeptides, *Philos. Trans. R. Soc. London Ser. A*, 293, 39, 1979.
50. **Morris, H. R., Williams, D. H., and Ambler, R. P.,** Determination of the sequences of protein-derived peptides and peptide mixtures by mass spectrometry, *Biochem. J.*, 125, 189, 1971.
51. **Morris, H. R.,** Studies towards the complete sequence determination of proteins by mass spectrometry: a rapid procedure for the successful permethylation of histidine containing peptides, *FEBS Lett.*, 22, 257, 1972.
52. **Mudgett, M., Bowen, D. V., Field, F. H., and Kindt, T. J.,** Peptide sequencing: the utility of chemical ionization mass spectrometry, *Biomed. Mass Spectrom.*, 4, 159, 1977.
53. **Roepstroff, P., McDowall, M. A., Games, M. P. L., and Games, D. E.,** Peptides sequencing by combined liquid chromatography/mass spectrometry, *Int. J. Mass Spectrom. Ion Phys.*, 48, 197, 1983.

54. **Blakley, C. R., Carmody, J. C., and Vestal, M. L.,** Combined liquid chromatograph/mass spectrometer for involatile biological samples, *Clin. Chem.*, 26, 1467, 1980.
55. **Pilosof, D., Kim, H. Y., Dyckes, D. F., and Vestal, M. L.,** Determination of nonderivatized peptides by thermospray liquid chromatography/mass spectrometry, *Anal. Chem.*, 56, 1236, 1984.
56. **Blakley, C. R. and Vestal, M. L.,** Thermospray interface for liquid chromatography/mass spectrometry, *Anal. Chem.*, 55, 750, 1983.
57. **Kim, H. Y., Pilosof, D., Dyckes, D. F., and Vestal, M. L.,** On-line peptide sequencing by enzymatic hydrolysis, high performance liquid chromatography, and thermospray mass spectrometry, *J. Am. Chem. Soc.*, 106, 7304, 1984.
58. **Bieman, K.,** *Biochemical Applications of Mass Spectrometry, First Supplemental Volume*, Waller, G. R. and Dermer, O. C., Eds., Wiley-Interscience, New York, 1980, 469.
59. **Carr, S. A., Herlihy, W. C., and Biemann, K.,** Advances in gas chromatographic mass spectrometric protein sequencing. I. Optimization of the derivatization chemistry, *Biomed. Mass Spectrom.*, 8, 51, 1981.
60. **Herlihy, W. C., Anderegg, R. J., and Biemann, K.,** Advances in gas chromatographic mass spectrometric protein sequencing. II. Application to membrane proteins, *Biomed. Mass Spectrom.*, 8, 62, 1981.
61. **Krutzsch, H. C.,** Polypeptide sequence analysis using gas chromatography-mass spectrometry, in *Microcharacterization of Polypeptides: A Practical Manual*, Shively, J. E., Ed., Humana Press, Clifton, N.J., 1986, chap 15.
62. **Nau, H. and Biemann, K.,** Amino acid sequencing by gas chromatography-mass spectrometry using trifluoro-dideuteroalkylated peptide derivatives. III. The primary structure of the carboxypeptidase inhibitor, *Anal. Biochem.*, 73, 175, 1976.
63. **Carr, S. A., Hauschka, P. V., and Biemann, K.,** Gas chromatographic-mass spectrometric sequence determination of osteocalcin, a gamma-carboxyglutamic acid-containing protein from chicken bone, *J. Biol. Chem.*, 256, 9944, 1981.
64. **Gerber, G. E., Anderegg, R. J., Herlihy, W. C., Gray, C. P., Biemann, K., and Khorana, H. G.,** Partial primary structure of bacteriorhodopsin: sequencing methods for membrane proteins, *Proc. Natl. Acad. Sci. U.S.A.*, 76, 227, 1979.
65. **Samy, T. S. A., Hahm, K. -S., Modest, E. L., Lampman, G. W., Keutmann, H. T., Umezawa, H., Herlihy, W. C., Gibson, B. W., Carr, S. A., and Biemann, K.,** Primary structure of macromycin, an antitumor antibiotic protein, *J. Biol. Chem.*, 258, 183, 1983.
66. **Nau, H. and Biemann, K.,** Amino acid sequencing by gas chromatography-mass spectrometry using perfluoro-dideuteroalkylated peptide derivatives. II. Interpretation of the mass spectra, *Anal. Biochem.*, 73, 154, 1976.
67. **Carr, S. A. and Biemann, K.,** Gas chromatographic mass spectrometric sequencing of peptides and proteins containing gamma-carboxyglutamic acid, *Biomed. Mass Spectrom.*, 7, 172, 1980.
68. **Herlihy, W. C., Kidwell, D., Meeusen, B., and Biemann, K.,** Mass spectrometric differentiation of leucine and isoleucine in proteins derived from bacteria or cell culture, *Biochem. Biophys. Res. Commun.*, 102, 335, 1981.
69. **Hudson, G. and Biemann, K.,** Mass spectrometric sequencing of proteins. The structure of subunit I of monellin, *Biochem. Biophys. Res. Commun.*, 71, 212, 1976.
70. **Khorana, H. G., Gerber, G. E., Herlihy, W. C., Gray, C. P., Anderegg, R. J., Nihel, K., and Biemann, K.,** Amino acid sequence of bacteriorhodopsin, *Proc. Natl. Acad. Sci. U.S.A.*, 76, 5046, 1979.
71. **Herlihy, W. C., Royal, N. J., Biemann, K., Putney, S. D., and Schimmel, P. R.,** Mass spectra of partial protein hydrolysates as a multiple phase check for long polypeptides deduced from DNA sequences: NH_2-terminal segment of alanine tRNA synthetase, *Proc. Natl. Acad. Sci. U.S.A.*, 77, 6531, 1980.
72. **Hoben, P., Royal, N., Cheung, A., Yamao, F., Biemann, K., and Soll, D.,** *Escherichia coli* glutaminyl-tRNA synthetase. II. Characterization of the glnS gene product, *J. Biol. Chem.*, 257, 11644, 1982.
73. **Krutzsch, H. C. and Pisano, J. J.,** Analysis of dipeptides by gas chromatography-mass spectrometry and application to sequencing with dipeptidyl aminopeptidases, in *Methods in Enzymology*, Vol. 47, Hirs, C. H. W. and Timasheff, S. N., Eds., Academic Press, New York, 1977, 391.
74. **Krutzsch, H. C.,** Polypeptide sequencing with dipeptidylpeptidases, in *Methods in Enzymology*, Vol. 91, Hirs, C. H. W. and Timasheff, S. N., Eds., Academic Press, New York, 1983, 511.
75. **Krutzsch, H. C.,** Determination of polypeptide amino acid sequences from the carboxyl terminus using angiotensin I converting enzyme, *Biochemistry*, 19, 5290, 1980.
76. **Krutzsch, H. C. and Pisano, J. J.,** Separation and sequence of dipeptides using gas chromatography and mass spectrometry of their trimethylsilylated derivatives, *Biochemistry*, 17, 2791, 1978.
77. **Kira, J. -I., Deibler, G. E., Krutzsch, H. C., and Martenson, R. E.,** Amino acid sequence of porcine myelin basic protein, *J. Neurochem.*, 44, 134, 1985.
78. **Fairwell, T., Krutzsch, H., Hempel, J., Jeffrey, J., and Jornvall, H.,** Acetyl-blocked N-terminal structures of sorbitol and aldehyde dehydrogenases, *FEBS Lett.*, 170, 281, 1984.
79. **Deibler, G. E., Martenson, R. E., Krutzsch, H. C., and Kies, M. W.,** Sequence of guinea pig myelin basic protein, *J. Neurochem.*, 43, 100, 1984.

80. **Rao, D. N., Rudikoff, S., Krutzsch, H., and Potter, M.,** Structural evidence for independent joining region gene in immunoglobulin heavy chains from anti-galactan myeloma proteins and its potential role in generating diversity in complementary-determining regions, *Proc. Natl. Acad. Sci. U.S.A.*, 76, 2890, 1979.
81. **Pawlita, M., Potter, M., and Rudikoff, S.,** Kappa-chain restriction in anti-galactan antibodies, *J. Immunol.*, 129, 615, 1982.
82. **Yarmush, M. L., Krutzsch, H. C., and Kindt, T. J.,** Amino acid sequence analysis of immunoglobulin light chains by gas chromatographic-mass spectrometric techniques: structural identity of nominal and latent b9 molecules, *Mol. Immunol.*, 17, 319, 1980.
83. **Henderson, L. E., Krutzsch, H. C., and Oroszlan, S.,** Myristyl aminoterminal acylation of murine retrovirus proteins: an unusual posttranslational modification, *Proc. Natl. Acad. Sci. U.S.A.*, 80, 339, 1980.
84. **Marasco, W. A., Phan, S. H., Krutzsch, H., Showell, H. J., Feltner, D. E., Nairn, R., Becker, E. L., and Ward, P. A.,** Purification and identification of formyl-methionyl-leucyl-phenylalanine as the major peptide neutrophil chemotactic factor produced by *Escherichia coli*, *J. Biol. Chem.*, 259, 5430, 1984.

Chapter 10

A NOVEL APPROACH TO AMINO ACID SEQUENCING

R. A. Jue and R. F. Doolittle

TABLE OF CONTENTS

I. INTRODUCTION

Traditional protein chemists cannot help but be impressed by the imaginative procedures developed by molecular biologists in their assault on DNA. It was not altogether surprising, then, when two different laboratories independently reported a novel approach to protein structure patterned on DNA sequencing procedures.[1,2] DNA sequencing was pioneered by Sanger, who used polyacrylamide gels to separate populations of radioactive DNA fragments that were distinguished by the relative occurrences of the four bases; amazingly, the sequence could be read directly off the gel.[3] In its present form,[4] the method uses an enzymatic copying operation to label the DNA and a set of nucleotide analogues to interrupt the polymerization preferentially at one or the other of the four bases. Maxam and Gilbert[5] used DNA labeled exclusively at one end and chemically fragmented it under appropriate conditions for each of the bases; they, too, used polyacrylamide gel electrophoresis to separate the nested fragments. Both the Sanger and Maxam-Gilbert procedures depend on the generation of a random set of fragments covering the full length of the DNA fragment being studied. Of the two, however, the chemical fragmentation of end-labeled material offers the more reasonable prospect for a protein sequencing procedure.

There are a number of obstacles in studying proteins this way that did not occur in the development of DNA sequencing. For one, even if there were ways to selectively cleave next to each of the 20 amino acids, it would mean running 20 lanes in the gel electrophoresis instead of the four that are needed for DNA. As it happens, it isn't yet possible to cleave preferentially next to each kind of amino acid, so that is not as big a problem as it might be. On the other hand, it also means that a complete amino acid sequence is not yet attainable by these methods. On another level, the resolution of polyacrylamide gels in the separation of polypeptides is not nearly as good as it is for polynucleotides. DNA polymers are uniformly charged, each nucleotide contributing the same amount of negative charge. Proteins must be unfolded with dodecyl sulfate (SDS), the charge that determines the electrophoretic mobility being mainly contributed by the bound detergent. In spite of these problems, it is possible to determine the arrangement of certain amino acids in a polypeptide chain, and this information can be put to great use in a number of important situations.

The first hurdle to be overcome was to find a method for labeling the polypeptide chain exclusively at one end. Next, conditions had to be found for the random fragmentation of the chain at specific residues. Finally, electrophoretic systems had to be developed that could resolve polypeptide chains that differ in length by a single amino acid residue. At present, none of these problems has been solved to the point where they are universally applicable. Nonetheless, considerable success has been achieved in some realms, and several proteins have already yielded to characterization by this general strategy. We have applied the method to the three nonidentical chains of fibrinogen with considerable success,[1] and Jay has characterized β-galactosidase and an immunity repressor protein from a λ phage.[2] The information obtained in some of these studies, while far from exact sequence determination, has already proved useful, especially in conjunction with DNA sequencing where it has been used for confirming inferred amino acid sequences.[1]

II. END-LABELING PROCEDURES

As noted, the first problem we faced was how to label a polypeptide chain exclusively at one end. The difficulty stems from the terminal α-amino group not being readily distinguished from the amino groups of lysine residues, and the terminal α-carboxyl group not being significantly different from the side-chain carboxyls of aspartic and glutamic acids. There are a number of ways of circumventing the problem, however, the most straightforward of which involves the judicious use of a stepwise degradation reagent. The polypeptide is

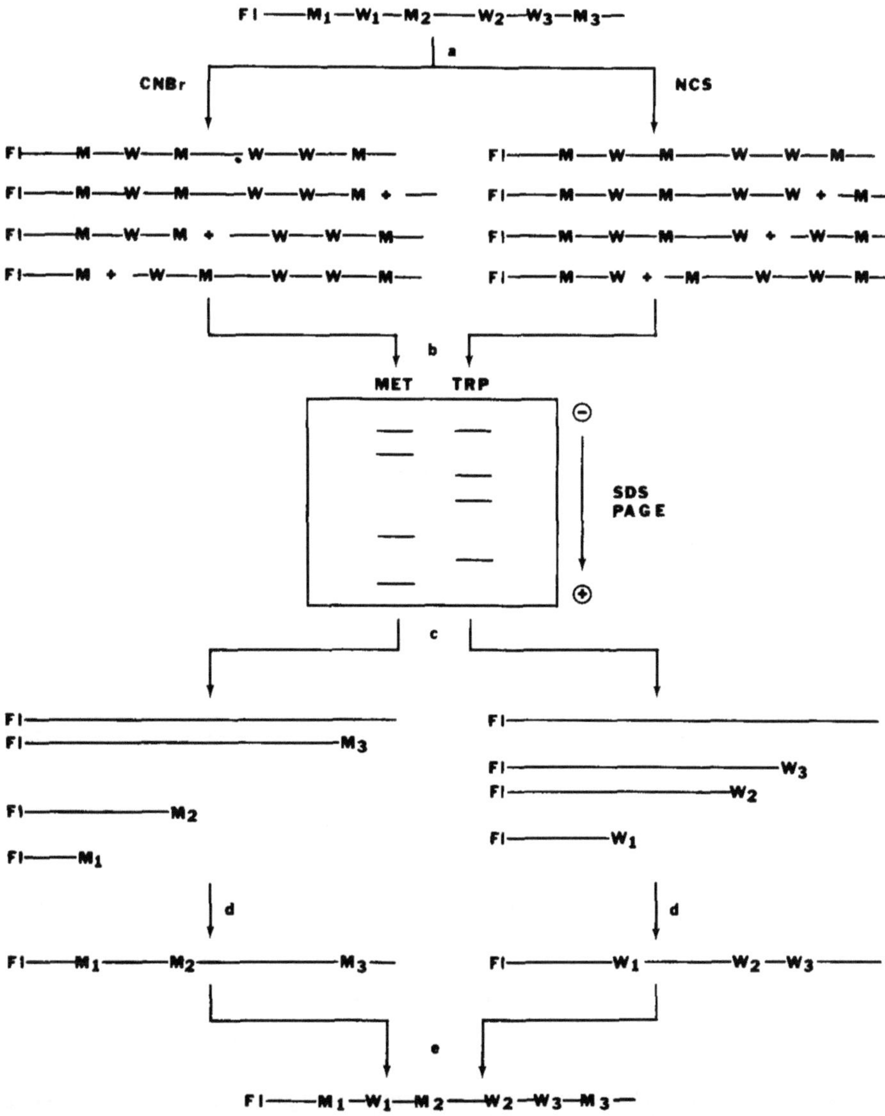

FIGURE 1. Outline of steps involved in the exclusive labeling of a terminal-amino group by fluorescein isothiocyanate (FITC). (Reprinted with permission from *Biochemistry*, 24, 162. Copyright 1985 American Chemical Society.)

coupled with a reagent like phenylisothiocyanate (PITC), which reacts with all amino groups;[6] care must be taken to insure that every group is completely labeled. Then, the preparation is run through a single cleavage cycle with an appropriate acid, the result of which is cyclization and release of the terminal amino acid. The derivatized lysine side chains are not affected by the acid step and remain firmly masked throughout the rest of the procedure. The result is a derivatized polypeptide chain with a single free amino group at its amino terminus. The preparation is then exposed to some fluorescent or radioactive reagent that reacts with amino groups, and a labeled product emerges that is ready for fragmentation and electrophoresis (Figures 1 and 2).

In our experience, PITC was not a very satisfactory reagent in this first step of the procedure, because the derivatized protein turned out to be an intractable mass that was difficult to handle thereafter. Instead, we have used thioacetylthioglycolic acid (TATG), a water-soluble thioacylating agent that works on a similar cyclization principle in releasing

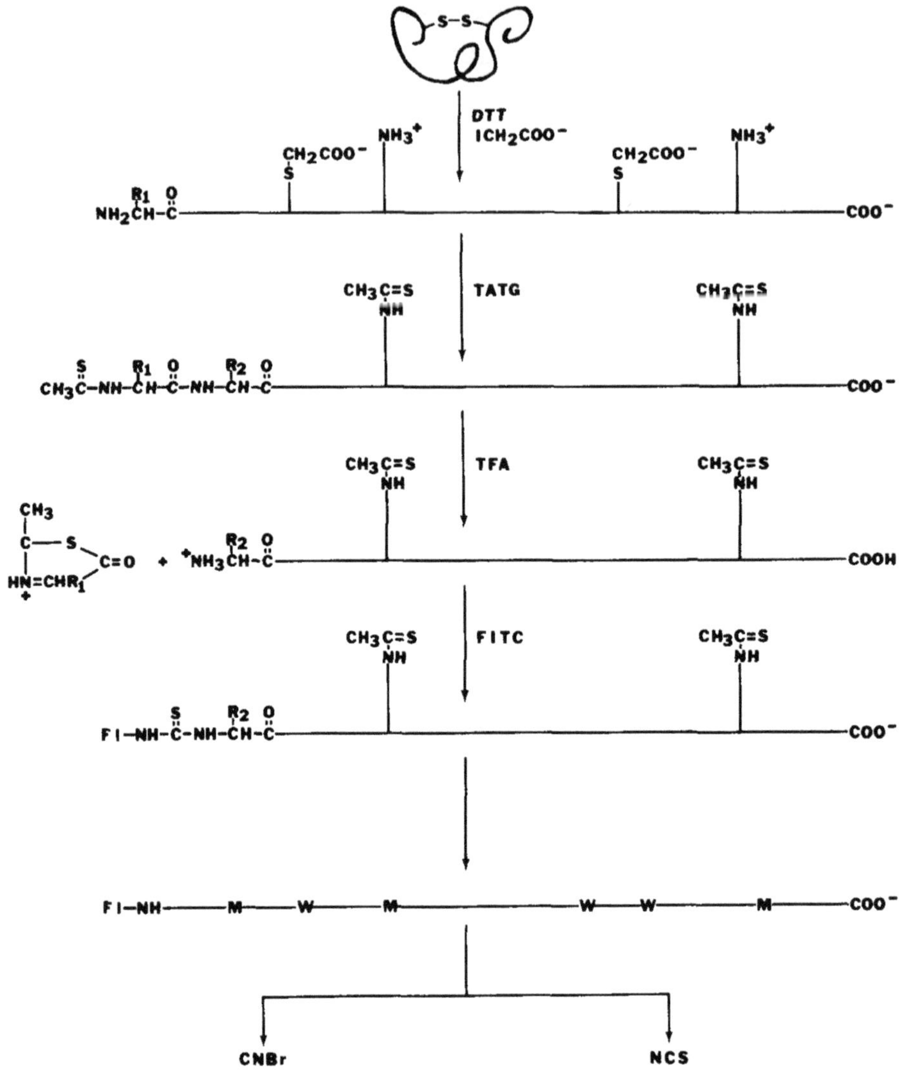

FIGURE 2. Schematic outline of steps involved in partial cleavages of methionines and tryptophans by cyanogen bromide and *N*-chlorosuccinimide, respectively. The "ladder" on the gel is read from the bottom upwards in this case. (Reprinted with permission from *Biochemistry,* 24, 162. Copyright 1985 American Chemical Society.)

the terminal amino acid.[7] The solubility properties of the resulting thioacylated polypeptide chains are not altogether ideal, but preparations are generally much easier to handle than PITC-treated material.

In either case, a number of different labeling reagents are available for attachment to the newly exposed amino group of the penultimate residues. The Bolton-Hunter reagent,[8] prelabeled with radioactive iodine, can be used, especially if very high sensitivity is required. In our developmental work, however, where we were still working out suitable conditions for all the reactions, we found it very useful to have a fluorescent label that could be directly monitored during electrophoresis, and we used fluorescein isothiocyanate (FITC). Although small amounts of this label are lost during subsequent steps involving acidic solvents, it has the virtues of safety and convenience (Figure 3). More recently we have employed an ^{35}S-labeling reagent which has the advantages of sensitivity, long half-life (87 days), and relative safety (compared with iodine radioisotopes). In this regard, we purchased t-butoxycarbonyl 35-*S*-methionyl hydroxysuccinimide (Amersham); it was necessary to convert the

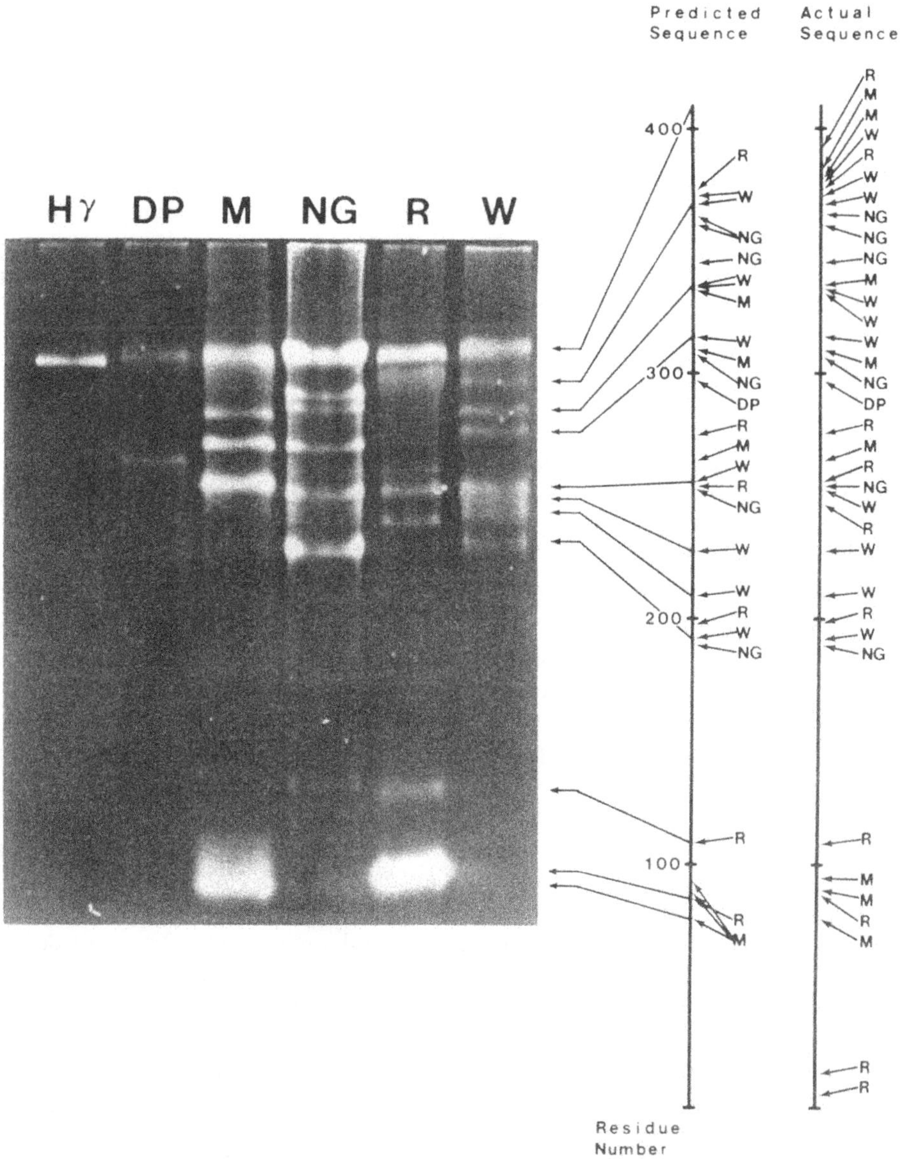

FIGURE 3. Photograph of a fluorescein-labeled human fibrinogen γ-chain sequencing gel. The lanes include intact human γ chain (Hγ) (1.5 μg) and human γ chains (5 to 60 μg) subjected to partial specific cleavage methods. Partial cleavages were selective for aspartylprolines (DP), methionines (M), asparaginylglycines (NG), arginines (R), and tryptophans (W). To determine the sequence, begin at the bottom of the gel and read upward. Since the chain was labeled at the amino terminus, reading upward gives a sequence from the amino terminus to the carboxy terminus. The predicted sequence is shown to the right of the gel. Each letter represents a band on the gel. An arrow extending from a particular amino acid on the predicted sequence indicates which band on the gel that particular amino acid corresponds to. The actual sequence is shown to the right of the predicted sequence. (Reprinted with permission from *Biochemistry*, 24, 162. Copyright 1985 American Chemical Society.)

methionyl residue to its sulfone by mild oxidation before use in order to prevent loss of the label in subsequent cyanogen bromide (CNBr) treatments.

A. Other Labeling Approaches

The strategy outlined above is straightforward, but it can be thwarted if the starting polypeptide chain has a blocked amino terminus.[9] If the block involves a cyclized glutamine (pyrrolidone carboxylic acid), then it is sometimes possible to remove the block enzymatically.[10] Blocks due to acylation are more troublesome and have so far been unyielding to chemical or enzymatic removal.

Blocked amino termini, whatever their nature, can be gotten around in various ways. For example, lysine side chains can be fully derivatized — in this case without regard to the need for subsequent removal of the terminal residue.[11] The protein is then subjected to a specific cleavage at some infrequent site, preferably near its amino terminus. The labeling reagent, whether fluorescent or radioactive, is added, and, after a suitable incubation period, the preparation is examined on a gel. If there is only a single band, partial fragmentation is performed as usual. If there are two or more bands, then it is necessary to purify each of them by electroelution before continuing.

The specific cleavage used at this step may be either enzymatic or chemical. In the case of the α and β chains of fibrinogen, for example, thrombin removes short peptides (the fibrinopeptides) from the amino termini, leaving virtually full-length chains for labeling and ready characterization.[1] It should be possible to use this same approach on other pre- or proproteins, including various zymogens and hormone precursors, as long as an activation enzyme comparable to thrombin is available.

In the case of most blocked proteins, however, it will be necessary to use one of the chemical cleavage methods. The general frequency of Asp-Pro and Asn-Gly bonds is such that one can expect them to occur once in every 350 and 320 residues, respectively, which means about once each in an average polypeptide chain. As such, they are ideal candidates. Conditions for the effective cleavage of Asp-Pro and Asn-Gly bonds in high yields have been described.[12,13]

In some special cases, one can specifically attach a label to a single active site residue in a polypeptide chain, and if that site is near either end of the polypeptide chain, the fragmentation procedures can be executed directly. Any suitably situated active site can be used if it is close enough to either end of the chain so that cleavages do not occur on the "wrong" side. In the case of fibrinogen, for example, a cross-linking site exists very near the carboxy terminus of the γ- chain,[14] and we were able to incorporate a fluorescent substitute donor[15] into a position 14 residues from the end. Happily, none of the 13 wrong-side residues were targets for fragmentation. A fringe benefit of this special situation was that no chemical treatment of the protein was required at the first stage, and, as a result, the handling of the preparation in all successive operations was greatly facilitated. Even better, the final electrophoresis resolution was greatly improved (Figure 4), suggesting that a certain amount of microheterogeneity is ordinarily introduced during the amino-group blocking steps.

III. FRAGMENTATION SCHEMES

In the limit, it would be desirable if procedures were available for specifically cleaving bonds adjacent to each one of the 20 amino acids. So far, procedures that cleave next to methionine, cysteine, tryptophan, arginine, and glutamic acid have been successfully employed (Table 1). In addition, the locations of Asn-Gly and Asp-Pro bonds have been determined by virtue of the special sensitivities of those bonds. In general, chemical methods work better than enzymatic ones, primarily because they yield more random patterns. Thus, even though staphylococcal protease breaks preferentially at glutamic acid residues,[16] some of these sites are usually much more vulnerable to the enzyme than others, making it difficult to get a genuinely random population of labeled fragments.

A. Chemical Cleavages

Random cleavage next to methionine residues by CNBr in 70% formic acid has proved very effective,[17] although it must be kept in mind that CNBr does not break methionyl-

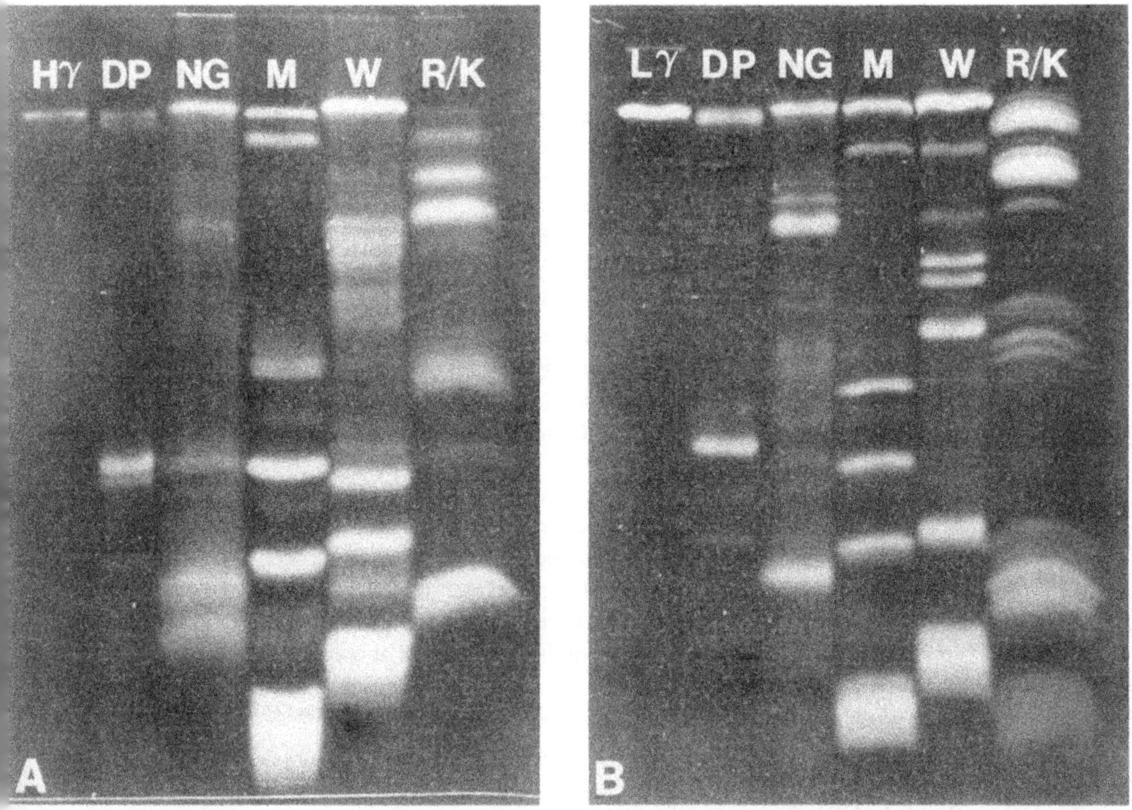

GURE 4. Sequencing gels of dansyl-labeled human and lamprey fibrinogen γ chains. (A) Intact human γ chains (Hγ) (5 μg) d human γ chains (10 to 60 μg) that were subjected to selective cleavages; (B) intact lamprey γ chains (Lγ) (4 μg) and selectively agmented lamprey γ chains (10 to 60 μg). Selective cleavages were specific for aspartylprolines (DP), asparaginylglycines (NG), ethionines (M), tryptophasn (W), and, finally, both arginines and lysines (R/K). Because the chain was labeled at the carboxy minus, the sequence reads *down* from the amino terminus to the carboxy terminus. (Reprinted with permission from *Biochemistry*, , 162. Copyright 1985 American Chemical Society.)

Table 1
DEMONSTRATED CLEAVAGE SITES[a]

Chemical	
Met	Cyanogen bromide
Cys	2-Nitro-5-thiocyanobenzoic acid
Trp	*N*-Chlorosuccinimide
Asp-Pro	50% Formic acid
Asn-Gly	2 *M* Hydroxylamine
Enzymatic	
Arg	Trypsin, 4 *M* urea
Glu	Staph, protease, SDS

[a] Exact recipes are provided by Jue and Doolittle[1] and Jay.[2]

threonine bonds.[18] The conditions for partial fragmentation are similar to those used for complete breakage, except the times are considerably reduced. Tryptophan cleavages with *N*-chlorosuccinimide (NCS) also yield excellent results.[19] These reactions are conducted in 80% acetic acid at room temperature for 30 to 60 min. Detailed recipes for these cleavages, as well as those for Asp-Pro and Asn-Gly breakage, have been described.[1] An effective method for cleaving at cysteines with 2-nitro-5-thiocyanobenzoic acid (NTCB) has been described,[20,21] and Jay[2] has used the reagent successfully in his characterization of β-galactosidase and the 434 immunity repressor protein.

B. Enzyme Cleavages

When enzymes are used to cleave labeled proteins, care must be taken to unfold the protein so that each of the targeted sites has an equal opportunity for encounter. In the case of trypsin, digestions are carried out in 4 *M* urea, under which conditions the enzyme is still sufficiently active. If, as is usually the case, the lysine amino groups on the labeled protein have been blocked, then the only bonds cleaved are those adjacent to arginines. Similarly, staphylococcal protease is known to be active in the presence of SDS,[22] a detergent that effectively disrupts protein folding. As such, digestions with this enzyme may be conducted in SDS environments for suitably chosen times.[2]

C. Other Possible Cleavages

It should be possible to work out conditions whereby preferential cleavage is obtained for bonds involving other amino acids than those cited above. For example, specific cleavage next to proline residues has been obtained by the use of sodium borohydride in the presence of 0.1 *M* NaOH,[23] and it is well known that preferential cleavage can be effected at aspartic acids by heating in dilute acetic acid.[24] In the latter case, an extension of the cleavages to asparagines may be achieved by first treating the labeled protein with base under conditions that promote deamidation. Also, tyrosine and histidine sites, as well as tryptophan, are susceptible to oxidation by *N*-bromosuccinimide.[25] As was the case of the cleavages demonstrated listed in Table 1, much effort will likely have to be expended to find exact conditions for optimum fragmentation.

Other enzymatic cleavages are also possible (Table 2). Chymotrypsin can be used under conditions similar to trypsin, and although the rates of cleavage for tryptophan, tyrosine, and phenylalanine are not the same — they occur in roughly that order — it should still be possible to sort them out. Tryptophan sites will be obvious because of the simultaneous appearance of bands in the NCS lane, and tyrosine and phenylalanine can be differentiated

Table 2
SOME POTENTIAL CLEAVAGE SITES

Chemical	
Asp	1% Acetic acid
Asn	1% Acetic acid after mild alkali pretreatment
Pro	1 *M* $NaBH_4$, 0.1 *M* NaOH
His, Tyr	*N*-Bromosuccinimide[a]
His	*N*-Bromosuccinimide after tetronitromethane
Enzymatic	
Phe, Trp, Tyr	Chymotrypsin, 4 *M* urea
Phe, Trp	Chymotrypsin after tetranitromethane
Gln	Staph protease after mild alkai pretreatment

[a] Trp also breaks, but will be distinguished by NCS (Table 1).

by prior treatment of a companion preparation with tetronitromethane to modify the tyrosines.[26] Finally, it should be possible to identify glutamines by subjecting a preparation to partial deamidation under mildly basic conditions, followed by neutralization and digestion with staphylococcal protease.

IV. POLYACRYLAMIDE ELECTROPHORESIS

It is clearly more difficult to separate polypeptide chains differing by a single residue on gels than it is to separate polynucleotides differing by one nucleotide; but it is not impossible. In general, the choice of conditions for SDS gel electrophoresis depends on the size of the target protein and the number of sites that are subjected to attack. In our studies on fibrinogen, the molecular weights of the constituent chains ranged from 47,000 to 90,000, and we used a modified Laemmli[27] system with 1 *M* Tris buffer and 22 to 25% acrylamide gels. In his characterization of β-galactosidase (molecular weight = 120,000), Jay used 12% acrylamide gels in a conventional Laemmli system; for the 25,000 molecular weight immunity repressor, he used 16% acrylamide. As in the case of DNA sequencing, it will likely be necessary to use gels of different acrylamide concentrations in most cases in order to obtain full-length characterizations. Alternatively, it may be necessary to run gels for varying lengths of time. There is still much to be learned about whether short or long gels can be made more highly resolving. For small fragments, we have found that urea-containing systems[28] greatly improve the resolution.

V. SENSITIVITY

The original goals of this new approach to protein characterization were to make it simple and sensitive. Although sensitivity is primarily limited by the detection of the labeled fragments, there are also handling considerations, especially in the modification and labeling reactions where the protein must be precipitated and washed several times. Losses can be minimized by the inclusion of an inert or derivatized carrier protein that itself cannot be labeled. Succinylated or amidinated albumin molecules, for example, make good carriers. It is also possible to carry out the initial modification and labeling reactions on an unpurified mixture of unknown proteins and then purify the target materials by electroelution from gels.

It is informative to note the amounts of labeled fragments that are required for a reasonably thorough characterization of an unknown protein. If fluorescein isothiocyanate is used as a labeling agent, then for a polypeptide chain of the order of 50,000 mol wt, it is necessary to apply somewhere between 5 and 50 μg of material to each lane, depending on the approximate number of expected cleavages. Usually, one would apply 5 μg in the cases of the Asp-Pro and Asn-Gly cleavage, and 30 μg for the CNBr lanes, inasmuch as methionyl bonds are about 6× as frequent as either of the first mentioned. All told, about 300 μg of labeled protein is the minimum amount necessary if five or six sets of reactions are to be run in duplicate. These estimates are based on our observation that as little as 30 pmol of an unfragmented fluroescein-labeled polypeptide can be visualized on a gel and readily photographed.

Radioactive labels are obviously more sensitive. Jay,[2] who used an ^{125}I Bolton-Hunter reagent, conducted his characterization of β-galactosidase on only 50 μg of protein. We ourselves are currently exploring the use of ^{35}S as a labeling agent with the hope that a comparable sensitivity can be obtained with a material that is safer and more convenient to handle than ^{125}I.

VI. COMPUTER SEARCHING

It is possible to identify unknown proteins on the basis of the arrangement of a few key amino acids. Thus, if only the order of methionines, tryptophans, and the dipeptides Asn-Gly and Asp-Pro is known for a protein, it is still a relatively simple matter to find if the sequence of the protein or anything similar has already been reported. In this regard, a candidate list of similar proteins can be retrieved by a suitable computer search of a protein sequence collection.

The search regimen is quite simple. First, the computer copies out each sequence for examination and deletes all residues except the ones of interest (Met, Trp and Asn-Gly, and Asp-Pro). The abbreviated sequence is then progressively tested for a minumum of three matches in every set of four. A cutoff is set at some number of times the condition must be met; the cutoff is proportional to the length of the (abbreviated) sequence. The progressive evaluation allows for both species differences and occasional errors or omissions in the experimental determination of the arrangement.

VII. CONCLUDING REMARKS

In summary, we have developed a procedure for determining the arrangement of several different amino acids in a polypeptide chain. The method is quick, simple, reproducible, and inexpensive, and it yields a large amount of valuable information. As additional selective cleavages are undertaken or become available, more primary protein structure information ought to be obtainable by this method. The procedure has usefulness in assessing the homology between two or more proteins, and it can be used to ensure that DNA-inferred sequences are in the proper frame and otherwise accurate. It also provides important information for the undertaking of an amino acid sequence by more traditional routes. Finally, when used in conjunction with computer searching, it may also prove useful in the identification of unknown proteins.

REFERENCES

1. **Jue, R. A. and Doolittle, R. F.,** Determination of the relative positions of amino acids by partial specific cleavages of end-labeled proteins, *Biochemistry*, 24, 162, 1985.
2. **Jay, D. G.,** A general procedure for the end labeling of proteins and positioning of amino acids in the sequence, *J. Biol. Chem.*, 259, 15572, 1984.

3. **Sanger, F. and Coulson, A. R.,** A rapid method for determining sequences in DNA by primed synthesis with DNA polymerase, *J. Mol. Biol.*, 94, 441, 1975.
4. **Sanger, F., Nicklen, S., and Coulson, A. R.,** DNA sequencing with chain-terminating inhibitors, *Proc. Natl. Acad. Sci. U.S.A.*, 74, 5463, 1977.
5. **Maxam, A. L. and Gilbert, W.,** A new method for sequencing DNA, *Proc. Natl. Acad. Sci. U.S.A.*, 74, 560, 1977.
6. **Edman, P.,** Method for determination of the amino acid sequence in proteins, *Acta Chem. Scand.*, 4, 283, 1950.
7. **Mross, G. A. and Doolittle, R. F.,** A new stepwise degradation procedure for determining the amino acid sequences of peptides attached to inert supports, *Fed. Proc. Fed. Am. Soc. Exp. Biol.*, 30, 1241, 1971.
8. **Bolton, A. E. and Hunter, W. M.,** The labelling of proteins to high specific radioactivities by conjugation to a 125-I-containing acylating agent, *Biochem. J.*, 133, 529, 1973.
9. **Doolittle, R. F.,** Sequencing peptides and proteins lacking free α-amino groups, *Mol. Biol. Biochem. Biophys.*, 25, 38, 1977.
10. **Doolittle, R. F. and Armentrout, R. A.,** Pyrrolidonyl peptidase. An enzyme for selective removal of pyrrolidone carboxylic acid residues from peptides, *Biochemistry*, 7, 516, 1968.
11. **Hunter, M. J. and Ludwig, M. L.,** Amidination, *Methods Enzymol.*, 25, 585, 1972.
12. **Piszkiewicz, D., Landon, M., and Smith, E. L.,** Anomalous cleavage of aspartyl-proline peptide bonds during amino acid sequence determination, *Biochem. Biophys. Res. Commun.*, 40, 1173, 1970.
13. **Bornstein, P. and Balian, G.,** Cleavage at Asn-Gly bonds with hydroxylamine, *Methods Enzymol.*, 47, 132, 1977.
14. **Chen, R. and Doolittle, R. F.,** γ-γ Cross-linking site in human and bovine fibrin, *Biochemistry*, 10, 4486, 1971.
15. **Lorand, L. and Ong, H. H.,** Labeling of amine-acceptor cross-linking sites of fibrin by transpeptidation, *Biochemistry*, 5, 1747, 1966.
16. **Homard, J. and Drapeau, G. R.,** Staphylococcal protease: a proteolytic enzyme specific for glutamoyl bonds, *Proc. Natl. Acad. Sci. U.S.A.*, 69, 3506, 1972.
17. **Gross, E. and Witkop, B.,** Nonenzymatic cleavage of peptide bonds: the methionine residues in bovine pancreatic ribonuclease, *J. Biol. Chem.*, 237, 1856, 1962.
18. **Carpenter, F. H. and Shiigi, S. M.,** Cyanogen bromide treatment of N-acetylmethionyl residues without cleavage, *Biochemistry*, 13, 5159, 1974.
19. **Schechter, Y., Patchornik, A., and Burstein, Y.,** Selective chemical cleavage of tryptophanyl peptide bonds by oxidative chlorination with N-chlorosuccinimide, *Biochemistry*, 15, 5071, 1976.
20. **Degani, Y. and Patchornik, A.,** Cyanylation of sulfhydryl groups by 2-nitro-5-thiocyanobenzoic acid. High-yield modification and cleavage of peptides at cysteine residues, *Biochemistry*, 13, 1, 1974.
21. **Jacobsen, G. A., Schaffer, M. H., Stark, G. R., and Vanaman, T. C.,** Specific chemical cleavage in high yield at the amino peptide bonds of cysteine and cystine residues, *J. Biol. Chem.*, 248, 6583, 1973.
22. **Cleveland, D. W., Fischer, S. G., Kirschner, M. W., and Laemmli, U. K.,** Peptide mapping by limited proteolysis in sodium dodecyl sulfate and analysis by gel electrophoresis, *J. Biol. Chem.*, 252, 1102, 1977.
23. **Shimamura, M., Inoue, Y., and Inoue, S.,** Reductive cleavage of Xaa-Proline peptide bonds by mild alkaline borohydride treatment employed to release O-glycosidally linked carbohydrate units of glycoproteins, *Arch. Biochem. Biophys.*, 232, 699, 1984.
24. **Shultz, J., Allison, H., and Grice, M.,** Specificity of the cleavage of proteins by dilute acid. I. Release of aspartic acid for insulin, ribonuclease and glucagon, *Biochemistry*, 1, 694, 1962.
25. **Hermodson, M. A.,** Chemical cleavage of proteins, in *Methods in Protein Sequence Analysis* Elzinga, M., Ed., Humana Press, Clifton, N.J., 1982, 313.
26. **Riordan, J. F. and Vallee, B. L.,** Nitration with tetranitromethane, *Methods Enzymol.*, 25, 515, 1972.
27. **Laemmli, U. K.,** Cleavage of structural proteins during the assembly of the head of bacteriophage T4, *Nature*, 227, 680, 1970.
28. **Swank, R. T. and Munkres, K. D.,** Molecular weight analysis of oligopeptides by electrophoresis in polyacrylamide gel with sodium dodecyl sulfate, *Anal. Biochem.*, 39, 462, 1971.

Chapter 11

ANALYSIS OF MODIFIED AMINO ACIDS

W. K. Paik, H. W. Lee, and S. Kim

TABLE OF CONTENTS

I. INTRODUCTION

More than 140 posttranslationally modified amino acid derivatives in protein have thus far been reported,[1-3] and several comprehensive review articles on the analysis of these amino acids have recently been published, particularly in *Methods of Enzymology*.[4-6] Therefore, with the limited space available, only those which have been our major interest during the past two decades, namely, posttranslationally methylated amino acid derivatives,[7-9] will be reviewed here. Additional materials for the synthesis and analysis of methylated amino acids are available in articles written by Benoiton[10] and Huszar.[11]

II. NATURAL OCCURRENCE OF VARIOUS METHYLATED AMINO ACID DERIVATIVES

Since the last comprehensive compilation in 1980,[9] the number of proteins which have been reported to contain methylated amino acid residues has steadily grown. We, therefore, updated the original list. Table 1 lists 20 additional new references; they include proteins such as α-amylase, HMG-1 and HMG-2 proteins, nuclear acidic phosphoprotein C-23, citrate synthase, elongation factors, initiation factor, and heat shock proteins. Also included are newly observed amino acids such as *S*-methylmethionine, *S*-methylcysteine, and δ-*N*-methylarginine.

III. ANALYSIS OF *N*-METHYLATED AMINO ACIDS

Except carboxyl methylesterified glutamic and aspartic acids, most of the *N*- and *S*-methylated amino acids are quite stable during treatment with 6 *N* HC1 at 110°C, under which hydrolysis of protein is carried out. This property of the methylated amino acid residues in protein makes their identification relatively easily amenable to most of the common analytical methods.

Most of the analytical techniques currently available have been applied to the analysis of the methylated amino acids: paper chromatography and electrophoresis, thin-layer chromatography and electrophoresis, ion-exchange chromatography, gas chromatography, and high performance liquid chromatography (HPLC).

Since the methylated amino acids, in general, are present in minor quantities in protein, it is advisable to carry out an initial purification of protein hydrolyzates prior to the analysis. This process not only concentrates the methylated amino acids, but also avoids possible overloading of the system with other amino acids.

In the following section, we will focus our major attention only on the methylated lysines, arginines, and histidines, because major research efforts have been directed towards these amino acids and, also, because the presence of some of the naturally occurring methylated amino acids listed in Table 1 have been observed only by a single research group. The readers are, therefore, referred to the individual reference sources for these analytical methods.

A. Partial Purification of N-Methylated Amino Acids from Acid-Hydrolyzates of Proteins and Other Biological Samples

Prior to the analysis of the methylated amino acids in the acid-hydrolyzates of proteins or other biological materials, it is highly desired to partly purify the amino acids, particularly when paper or thin-layer chromatography is to be employed. This partial purification can be achieved by the use of an ion-exchange resin in H^+ form, thus, obtaining basic amino acids. In the following *Illustrative Procedure,* the partial purification of basic methylated amino acids from brain and other organs carried out by Kakimoto et al.[83] is described. One reason to choose this example is that the authors also removed lysine and arginine by

Table 1
NATURAL OCCURRENCE OF VARIOUS METHYLATED AMINO ACID DERIVATIVES

Protein	Source	Authors and year of finding	Ref.
		ε-N-Monomethyllysine	
Flagella protein	*Salmonella typhimurium*	Ambler and Rees (1959)	12
	Spirillium serpens	Glazer et al. (1969)	13
Histone	Thymus, wheat germ, kidney	Murray (1964)	14
	Ascites carcinoma	Comb et al. (1966)	15
	Pea	Fambrough and Bonner (1968)	16
Myosin	Skeletal muscle	Hardy and Perry (1969)	17
		Huszar and Elzinga (1969)	18
	Physarum polycephalum	Venkatesan et al. (1969)	19
Actin	Amoeba	Weihing and Korn (1970)	20
Ribosomal protein	*Blastocladiella emersonii*	Comb et al. (1966)	15
Opsin	Bovine retina	Reporter and Reed (1972)	21
Tooth matrix protein	Human	Kalasz et al. (1978)	22
Ferredoxin	*Sulfolobus acidocaldarius*	Minami et al. (1985)	23
Elongation factor Tu	*Salmonella typhimurium*	Ames and Niakido (1979)	24
	Escherichia coli		
Elongation factor 1α	*Mucor* (fungus)	Hiatt et al. (1982)	25
Free amino acid	Brain (bovine)	Matsuoka et al. (1969)	26
	Plasma (human)	Perry et al. (1969)	27
	Plasma (sheep)	Weatherall and Haden (1969)	28
	Urine (human)	Asatoor (1969)	29
		ε-*N*-Dimethyllysine	
Nuclear protein	Thymus (calf)	Kim and Paik (1965)	30
Histone	Thymus (calf)	Paik and Kim (1967)	31
Flagella protein	*Salmonella typhimurium*	Glazer et al. (1969)	13
Myosin	Soleus muscle (cat)	Kuehl and Adelstein (1970)	32
	Physarum polycephalum	Venkatesan et al. (1975)	19
Actin	Amoeba	Weihing and Korn (1970)	20
Myofibrilla protein	Skeletal muscle (rabbit)	Hardy et al. (1970)	33
Opsin	Bovine retina	Reporter and Reed (1972)	21
Ribosomal protein	HeLa cells	Chang et al. (1976)	34
Elongation factor 1α	*Mucor* (fungus)	Hiatt et al. (1982)	25
Elongation factor Tu	*Salmonella typhimurium*	Ames and Niakido (1979)	24
	Escherichia coli		
Free amino acid	Urine (human)	Kakimoto and Akazawa (1970)	35
	Plasma (human)	Kakimoto and Akazawa (1970)	35
		ε-N-Trimethyllysine	
Histone	Thymus	Hempel et al. (1968)	36
Cytochrome *c*	Wheat germ	DeLange et al. (1969)	37
	Neurospora crassa		
	Yeast	DeLange et al. (1970)	38
	Protozoa	Hill et al. (1971)	39
	Spinacea oleracea (spinach)	Brown and Boutler (1973)	40
Myosin	Skeletal muscle (rabbit)	Hardy et al. (1970)	33
	Physarum polycephalum	Venkatesan et al. (1975)	19
Actin	Amoeba	Weihing and Korm (1970)	20

Table 1 (continued)
NATURAL OCCURRENCE OF VARIOUS METHYLATED AMINO ACID DERIVATIVES

Protein	Source	Authors and year of finding	Ref.
Ribosomal protein	*Escherichia coli*	Alix and Hayes (1974)	41
		Chang et al. (1974)	42
	HeLa cells	Chang et al. (1976)	34
Calmodulin	Testis (rat)	Jackson et al. (1977)	43
α-Amylase	Wheat	Motojima and Sakaguchi (1982)	44
Citrate synthase	Porcine heart	Bloxham et al. (1981)	45
Heat shock protein	Chicken fibroblast	Wang et al. (1982)	46
Elongation factor Tu		L'Italien and Laursen (1979)	47
		Ohba et al. (1979)	48
Elongation factor	*Mucor* (fungus)	Hiatt et al. (1982)	25
Free amino acid	*Laminariaceae* (kelp)	Takemoto et al. (1964)	49
	Seed of *Reseda luteola L.*	Larsen (1968)	50
	Urine (human)	Kakimoto and Akazawa (1970)	35
	Plasma (human)	Tomita and Nakamura (1970)	35
	Placenta (human)	Tomita and Nakamura (1977)	51
	ε-N-Trimethyl-δ-Hydroxylysine		
Cell wall	Diatom	Nakajima and Volcani (1970)	52
	N^G-Monomethylarginine		
Histone	Thymus (calf)	Paik and Kim (1967)	53
Acidic protein	Liver nuclei (rat)	Friedman et al. (1969)	54
Myelin basic protein	Myelin (human)	Baldwin and Carnegie (1971)	55
Heat shock protein	Chicken fibroblast	Wang et al. (1982)	46
Histon HI[a]	*Euglena gracilis*	Tuck et al. (1985)	56
Free amino acid	Seed of *vicia faba L.* (bean)	Kasai et al. (1976)	57
	Placenta (human)	Tomita and Nakamura (1977)	51
	N^G,N^G-Dimethylarginine (Asymmetric)		
Histone	Thymus (calf)	Paik and Kim (1967)	53
Myelin basic protein	Myelin (human)	Baldwin and Carnegie (1971)	55
Myosin	Leg muscle (rat, chicken)	Reporter and Corbin (1971)	58
HnRNP protein	Liver nuclei (rat, chicken)	Boffa et al. (1977)	59
	Nuclei (*Physarum polycephalum*)	Christensen et al. (1977)	60
HMG-2	Nuclei (rat)	Boffa et al.	61
	Nucleolus (Novikoff hepatoma)		
C23 acidic phosphoprotein		Lischwe et al. (1982)	62
Ribosomal protein	HeLa cells	Chang et al. (1976)	34
Tooth matrix protein	Human	Kalasz et al. (1978)	22
Scleroderma antigen	Nucleolus (Novikoff hepatoma)	Lischwe et al. (1986)	63
Free amino acid	Urine (human)	Kakimoto and Akazawa (1970)	35
	Serum (human)	Kakimoto and Akazawa (1970)	35
	Placenta (human)	Tomita and Nakamura (1977)	51
	Seed of *Vicia faba L.* (bean)	Kasai et al. (1976)	57

Table 1 (continued)
NATURAL OCCURRENCE OF VARIOUS METHYLATED AMINO ACID DERIVATIVES

Protein	Source	Authors and year of finding	Ref.
	N^G,N'^G-Dimethlyarginine (Symmetric)		
Histone	Thymus (calf)	Paik and Kim (1967)	53
Myelin basic protein	Myelin (human)	Baldwin and Carnegie (1971)	55
Tooth matrix protein	Human	Kalasz et al. (1978)	22
Free amino acid	Urine (human)	Kakimoto and Akazawa (1970)	35
	Serum (human)	Kakimoto and Akazawa (1970)	35
	Seed of *Vicia faba L.* (bean)	Kasai et al. (1976)	57
	δ-N-Monomethylarginine		
	Root tuber (*Trichosanthes cucumeroides*)	Inukai et al. (1968)	64
	δ-N-Methylornithine		
Free amino acid	Brain (bovine)	Matsuoka et al. (1969)	26
	3-*N*-Methylhistidine		
Myosin	Skeletal muscle (rabbit)	Johnson et al. (1967)	65
	Physarum polycephalum	Venkatesan et al. (1975)	19
Actin	Skeletal muscle (rabbit)	Johnson et al. (1967)	65
	Muscle (bovine)	Asatoor and Armstrong (1967)	66
Histone	Erythrocyte (avian)	Gershey et al. (1969)	67
Opsin	Retina (bovine)	Reporter and Reed (1972)	21
Free amino acid	Urine (human)	Tallan et al. (1954)	68
	δ-N-Methylglutamine		
Ribosomal protein	*Escherichia coli*	Lhoest and Colson (1977)	69
	***N*-Dimethylproline**		
Cytochrome *c*-557	*Crithidia oncopelti*	Pettigrew and Smith (1977)	70
Histone (H2B)	*Asteria rubens* (starfish)	Martinage et al. (1985)	71
	***N*-Monomethylalanine**		
Ribosomal protein	*Escherichia coli*	Whittman-Liebold and Pannenbecker	72
	***N*-Trimethylalanine**		
Ribosomal protein	*Escherichia coli*	Lederer et al. (1977)	73
Myosin	Skeletal muscle (rabbit)	Henry et al. (1982)	74
	***N*-Methylmethionine**		
Ribosomal protein	*Escherichia coli*	Brosius and Chen (1976)	75
IF-3ℓ	*Escherichia coli*	Brauer and Wittman-Liebold (1977)	76

Table 1 (continued)
NATURAL OCCURRENCE OF VARIOUS METHYLATED AMINO ACID DERIVATIVES

Protein	Source	Authors and year of finding	Ref.
		***S*-Methylcysteine**[a]	
O^6-Methylguanine methyltransferase	*Escherichia coli*	Olsson and Lindahl (1980)	77
		***S*-Methylmethionine**[a]	
Cytochrome *c*[a]	*Euglena gracilis*	Farooqui et al. (1985)	78
		Glutamic Acid 5-Methyl Ester or (γ-Glutamyl Methyl Ester)	
Chemotactic membrane protein	*Escherichia coli*	Kleene et al. (1977)	79
	Salmonella typhimurium	Van DerWerf and Koshland (1977)	80
		D-Aspartic Acid β-Methyl Ester	
Membrane protein	Human erythrocytes	Janson and Clarke (1980)	81
		L-Isoaspartic Acid α-Carboxy Methyl Ester	
Adrenocorticotropin		Aswad (1984)	82

[a] Presence is demonstrated in enzymatic incubation in vitro.

individual specific enzymes after isolating the basic amino acids and prior to the analysis. When ion-exchange chromatography was to be employed subsequently, the removal of lysine, arginine, or histidine by the enzymes might not be necessarily required.

1. Illustrative Procedures[83]

One to two hundred milligrams of the protein were hydrolyzed in 15 volumes of 6 *N* HCl at 110°C for 24 to 48 hr in a sealed tube. The hydrolyzate was filtered, and the insoluble materials were washed with 30 mℓ of water. The combined filtrates were passed through a 1 × 4-cm column of Amberlite IR-120 (100 to 200 mesh, H^+ form) and the resin was washed with 20 mℓ of water and then with 60 mℓ of 20% pyridine. This treatment elutes acidic, neutral, and imidazoleamino acids. This fraction was used for the determination of 3-*N*-methylhistidine.

The retained amino acids, lysine, arginine, and their methylated derivatives, were eluted with 20 mℓ of 3 *N* NH_4OH. The eluate was evaporated to dryness. This dried sample could be used for analysis by paper, thin-layer, or ion-exchange chromatography. The sample could also be treated to remove lysine and arginine as follows. The sample was dissolved in 3 mℓ of 0.2 *M* phosphate buffer, pH 6.0, and 5 mg of lysine decarboxylase (specific activity, 15 μmol of lysine decarboxylated per 20 min per milligram enzyme protein), and 50 μg of pyridoxal phosphate in 0.25 mℓ of water were added. The mixture was incubated at 37°C for 2 hr. The reaction was terminated by immersing the tube in boiling water for 3 min. After cooling, 10 mg of the arginine deiminase (EC 3.5.3.6) (specific activity, 5 μmol of arginine deiminated per 20 min per milligram enzyme protein) was added and incubated at 37°C for 2 hr. Then 10 mℓ of 10% trichloroacetic acid was added, and the supernatant was passed through a 1 × 3-cm column of Amberlite IR-120 (100 to 200 mesh, H^+ form).

methods using an automatic amino acid analyzer.[96-100] However, they are all derived from slight modifications of various methods described herein. A comprehensive review on the chromatographic identification of various methylated amino acids has been published by Horakova and Deyl.[4]

2. Manual Column Chromatography

While investigating the identities of various methylated amino acids in human urine, Kakimoto and Akazawa[35] analyzed the aliphatic amino acids on an Amberlite IR-120 column with batchwise increase of ammonium hydroxide concentration from 0.2 to 4.0 *N*. The aliphatic amino acids were obtained by the method essentially similar to that described in Section III.A.

A more comprehensive isolation and identification method of various methylated amino acids from physiological fluids and protein hydrolyzates was carried out by Markiw.[86] Again, the aliphatic amino acids, which were partly purified from the acid-hydrolyzates, were applied on a 1.1 × 70-cm column of Dowex-50W × 8 (NH_4^+) and eluted with 2000 mℓ of a linear gradient from 0.0 to 0.2 *M* NH_4OH in 10-mℓ fractions. After collecting about 150 fractions, the reservoir containing 0.2 *N* NH_4OH was substituted with 1.5 *N* NH_4OH, and elution was continued until the emergence of arginine.

This method offers a means to prepare various methylated amino acid derivatives in comparatively large quantities and relatively pure qualities. However, it suffers from poor reproducibility of the elution time as well as the elution sequence; for example, in contrast to the elution sequence reported by Markiw,[86] ε-*N*-trimethyllysine was first eluted followed by ε-*N*-dimethyllysine, ε-*N*-monomethyllysine, and lysine under the identical condition employed by us.[101] We did not pursue the problem further, however, this discrepancy suggests that every investigator should establish the elution condition of any particular amino acid on this column system.

E. Gas Chromatography

Kalasz et al.[22] attempted the identification of various methylated amino acids by gas chromatography after preparing their corresponding ethyl ester *N*-trifluroacetyl derivatives. However, the resolution was not satisfactory, since the separation of N^G,N^G-dimethylarginine (asymmetric) and N^G,N'^G-dimethylarginine (symmetric) was not achieved.

F. High Performance Liquid Chromatography (HPLC)

In spite of its extremely high sensitivity, HPLC has not been commonly utilized for qualitative or quantitative analysis of methylated amino acids. To the best of our knowledge, only one report is available so far in the literature. Davis et al.[102] determined the amount of free ε-*N*-trimethyllysine in plasma on HPLC after removing acidic and neutral amino acids by Dowex resin prior to its application on HPLC. According to our preliminary observation, if acidic and neutral amino acids were not removed, the resolution of methylated amino acids was not proper. Further improvement in this procedure is under study before this otherwise highly useful technique can be applied in identification of methylated amino acids.

IV. ANALYSIS OF CARBOXYL *O*-METHYLATED AMINO ACIDS

In contrast to *N*-methylated amino acid residues in protein, *O*-methylated residues, namely, carboxyl methylesters of glutamyl or aspartyl residues, are labile in both weak alkali and strong acid.[9] Therefore, proteolytic digestion of the methylated protein is the only method available to hydrolyze peptide bonds while preserving such ester bonds. However, the recovery of the ester after proteolysis is not complete and is very much dependent on the kind of protein-methyl esters. This variability in the ester stability is most likely due to the side-chain amino acids neighboring the ester linkage; free γ-glutamyl and β-aspartyl methyl esters are relatively more stable in alkaline than their protein-methyl esters.[103] Owing to this lability, when the protein is radiolabeled as carboxyl methylesters, the amount of alkali-hydrolyzable radiomethyl methanol can be measured to quantifying the amount of protein-methyl esters.[104]

A. Glutamyl γ-Methyl Ester

Glutamyl γ-methyl ester is formed posttranslationally on specific glutamyl residues of chemotactic bacterial membrane proteins[79,80,105] by protein-carboxyl methyltransferase, a *che*R gene product.[106] This carboxyl-methylation is known to be a biochemical signal by which bacterial flagella motility is controlled. In order to identify the presence of glutamyl methyl ester residues, methyl-accepting chemotactic protein (MCP) is radiomethyl labeled, subjected to proteolysis, and the hydrolyzate is analyzed either on a standard automatic amino acid analyzer, paper chromatography, or on electrophoresis.[79,80,105] Carboxyl methylesterified MCP can be prepared from chemotactic bacteria either in vivo or in vitro according to the methods described.[80,107]

1. Enzymatic Digestion

In order to achieve complete hydrolysis of peptide bonds, combination of several proteases are used.[79,80,105] Typically, the radiomethyl-labeled MCP is suspended in 50 m*M* sodium acetate buffer, pH 5.2, containing 0.2 m*M* EDTA and 1 m*M* 2-mercaptoethanol. Papain (0.5 mg/mℓ) is added and the mixture is incubated at 37°C for 16 to 20 hr. The solution is then acidified to pH 2 with 1 *M* HCl and freeze dried. The residue is resuspended in 5 m*M* Tris-HCl buffer, pH 8.0 to 8.5, containing 0.625 m*M* $MnCl_2$. Leucine aminopeptides (100 μg/mℓ) and prolidase (200 μg/mℓ) are added, and this mixture is incubated at 37°C for 4 hr. The sample is then acidified to pH 2.0 and freeze dried. The amount of radiolabel recovered throughout the entire digestion was reported to be 60[79] to 90%.[105]

2. Analysis

a. Ion-Exchange Chromatography

A standard automatic amino acid analyzer is suited for the analysis of radiomethyl-labeled carboxyl methylesters. A protein digest obtained from enzymatic proteolysis is mixed with standard glutamyl γ-methyl and aspartyl β-methyl esters (0.2 μmol each) and chromatographed on the 50-cm column using 0.2 *N* sodium citrate buffer, pH 5.23, at 55.5°C, and a flow rate of 66 mℓ/hr. The fractions are collected and aliquots are analyzed for ninhydrin color reaction and radioactivity. Under these conditions, aspartyl β-methyl and glutamyl γ-methyl esters elute at 57 and 93 min, respectively.[80] Here, comigration of authentic ester and radioactive methyl ester should be observed. It should be cautioned, however, that when *methyl*-^{3}H)-labeling is used, ^{3}H-methyl ester does not exactly coincide with authentic ester due to an isotope effect. Kleene et al.[79] reported that glutamyl γ-(*methyl*-^{3}H)ester elutes about 35 sec earlier than the nonradioactive standard on their chromatogram.

b. Paper Chromatography and Electrophoresis[105]

The protein digest is suspended in a minimal volume of water or 50% methanol (v/v) to minimize the salt content. Cloudy samples are filtered through a 0.45-μm Millipore filter. Descending paper chromatography has been carried out in Whatman 3MM paper using two different solvent systems: (1) *n*-butanol/acetic acid/water (12:3:5, v/v) and (2) *n*-butanol/pyridine/acetic acid/water (20:10:5:2, v/v). A sample and standard (30 to 50 μg) are run in parallel and located by spraying the chromatogram with ninhydrin in acetone. The radioactivity profile is determined by cutting the sample lane into 1-cm sections, eluting the paper

Table 4
COMPARISON OF RELATIVE MOBILITIES OF METHYLATED AMINO ACID STANDARDS

	R_{GluOMe} on paper, chromatography — solvent[a]		R_{GluOMe} on high voltage paper electrophoresis at pH 1.9 (origin at anode)
	i	ii	
N^G-Methylarginine	0.62	0.74	2.08
Aspartate β-methyl ester	0.38	0.52	1.20
1-*N*-Methylhistidine	0.39	0.53	2.27
3-*N*-Methylhistidine	0.49	0.65	2.27
ε-*N*-Methyllysine	0.81	0.88	2.46

[a] Composition of the solvent: (i) *n*-butanol/acetic acid/water (12:3:5, v/v); (ii) *n*-butanol/pyridine/acetic acid/water (20:10:5:2, v/v).

into 0.5 mℓ of water for 30 min, and counting for radioactivity in a liquid scintillation spectrometer.

Electrophoresis is performed on Whatman 3MM paper in a Savant flat-bed system with continuous cooling. Voltage is initially set at 100 V for 10 min, increased to 1000V for 5 min, and then increased to 2000 V for 200 min. Electrode buffer is 1.5 *M* acetic acid/0.58 *M* formic acid, pH 1.9.

The mobilities of authentic amino acids relative to the migration of glutamyl γ-methyl ester (R_{GluOMe}) are listed in Table 4.[105]

B. Aspartyl β-Methyl Ester

Aspartyl β-methyl ester has been found as one of the products formed by mammalian protein carboxyl *O*-methyltransferase (EC 2.1.1.24). This ester is also extremely labile in weak alkali and forms substoichiometric quantity in most cases. Interestingly, the aspartyl β-methyl esters from human erythrocyte membrane proteins were shown to be in the D configuration[108] and were implicated to be an intermediate for the repair of racemized D-aspartyl residues occurring in aged protein.

Because of the low extent of the ester formation in the membrane proteins, radiomethyl-labeled proteins have to be prepared. As in the case of bacterial MCP, both in vivo and in vitro methylesterification can be carried out.[109,110]

1. Enzymatic Digestion

Digestion of (*methyl*-^{3}H)erythrocyte membrane proteins was carried out with an excess amount of protease. Although the recovery of the ester after the digestion was low,[81] aspartyl β-methyl ester was the only product found (not glutamyl γ-methyl ester).

In a 400-μℓ plastic microfuge tube, 20 μℓ of membranes (0.1 mg protein, 3.2 pmol of *methyl*-^{3}H group) and 80 μℓ (0.16 mg) of carboxypeptidase Y (Sigma Chemical Company) in citrate buffer (pH 5.0) are incubated at 37°C for 16 to 25 hr. The reaction is terminated by the addition of 20 μℓ of 1 *M* HCl.

2. Analysis

Analysis of aspartyl β-methyl esters in the hydrolyzate can be performed by either automatic amino acid analyzer, paper chromatography, or electrophoresis in a similar manner as described for glutamyl γ-methyl ester.

To determine the optical configuration of isolated aspartyl β-methyl ester, the procedure of Manning and Moore[111] has been successfully applied, which involves the synthesis of dipeptide leucyl diasteromers.[108,111]

C. Isoaspartyl α-Methyl Ester

The unusual aspartyl α-methyl ester is another recently identified product catalyzed by mammalian protein-carboxyl *O*-methyltransferase[82] and is also implicated as an intermediate in the repair process of age-damaged protein.[112] It has been reported that deamidation of asparagine-25 in adrenocorticotropin leads to the formation of an atypical isopeptide bond in which the resulting aspartyl residue is linked to the adjacent glycine-26 via its side-chain β-carboxyl group. Thus, the free α-carboxyl of aspartyl residue is now a methyl acceptor site for protein-carboxyl *O*-methyltransferase, yielding isoaspartyl α-methyl ester (L configuration). Indirect evidence, i.e., by identifying the weak alkali hydrolyzable methyl group, indicates that this methylation of α-carboxyl group of aspartyl isopeptide β-linkage is a stoichiometric reaction. Readers should refer to the recent publication by Aswad.[82]

D. Reduction of Protein-Methyl Ester by Calcium Borohydride

As an alternate means to identify the labile protein-carboxyl methyl esters, protein-methyl esters can be reduced to corresponding stable alcohols with (^{3}H)-labeled calcium borohydride.[113] Aspartyl β-methyl ester would yield homoserine, whereas glutamyl γ-methyl ester yields α-amino-6-hydroxyvaleric acid. The resulting products are now stable to treatment with 6 *N* NCl.

$Ca_2B(^3H)_4$ is prepared by mixing 0.02 mℓ of 0.1 *M* $NaB(^3H)_4$ (8.8 Ci/mmol) in 0.05 *M* NaOH, 0.01 mℓ of 0.1 *M* $CaCl_2$ in dimethylsulfoxide (DMSO), and 6 μℓ of 1 *M* HCl. This mixture is then incubated with 0.4 mℓ of the methylesterified erythrocyte membrane proteins in DMSO for 6 hr at room temperature under a stream of N_2 gas. The reaction is terminated by the addition of 0.1 mℓ of cold acetone. The membranes are washed four times by resuspension in a large volume of acetone, twice in 15% trichloroacetic acid, and once in ethanol.

The membrane protein is pelletted by centrifugation and hydrolyzed in 6 *N* HCl *in vacuo* for 36 hr at 110°C. The hydrolyzate is dried *in vacuo,* taken up in 0.5 mℓ of water, and dried completely to remove any trace of exchangeable tritium. This is now analyzed for the hydroxy amino acid in a standard amino acid analyzer. It has been shown that the radioactivity eluted only with homoserine, which had been derived from aspartyl β-methyl ester.[113]

ACKNOWLEDGMENTS

This work was supported by the research grants AMO9602 from the National Institute of Arthritis, Diabetes, Digestive, and Kidney Diseases, CA12226 from the National Cancer Institute, and RG1765-A-1 from the National Multiple Sclerosis Society.

REFERENCES

1. **Uy, R. and Wold, F.,** Posttranslational covalent modification of proteins, *Science,* 198, 890, 1977.
2. **Vickery, H. B.,** The history of the discovery of the amino acids. II. A review of amino acids described since 1931 as components of native proteins, in *Advances in Protein Chemistry,* Vol. 26, Anfinsen, C. B., Edsall, J. T., and Richards, F. M., Eds., Academic Press, New York, 1972, 81.
3. **Wold, F.,** In vivo chemical modification of proteins (Posttranslational modification), *Annu. Rev. Biochem.,* 50, 783, 1981.
4. **Horakova, M. and Deyl, Z.,** Chromatographic and electrophoretic behaviour of amino acids arising from post-translational reactions in proteins, *J. Chromatogr.,* 159, 227, 1978.

5. **Wold, F. and Moldave, K., Eds.,** *Methods in Enzymology,* Vol. 106, Academic Press, New York, 1984.
6. **Wold, F. and Moldave, K., Eds.,** *Methods in Enzymology,* Vol. 107, Academic Press, New York, 1984.
7. **Paik, W. K. and Kim, S.,** Protein methylation, *Science,* 174, 114, 1971.
8. **Paik, W. K. and Kim, S.,** Protein methylation: chemical, enzymological, and biological significance, in *Advances in Enzymology,* Meister, A., Ed., Academic Press, New York, 42, 227, 1975.
9. **Paik, W. K. and Kim, S.,** *Protein Methylation,* John Wiley & Sons, New York, 1980.
10. **Benoiton, N. L.,** N^{ω}-Alkyl diamino acids: chemistry and properties, in *Chemistry and Biochemistry of Amino Acids, Peptides and Proteins,* Weinstein, B., Ed., Marcel Dekker, New York, 1978, 163.
11. **Huszar, G.,** Methylated lysine and 3-methylhistidine in myosin: tissue and developmental differences, in *Methods in Enzymology,* Vol. 106, Wold, F. and Moldave, K., Eds., Academic Press, New York, 1984, 287.
12. **Ambler, R. P. and Rees, M. W.,** ε-N-Methyl-lysine in bacterial flagellar protein, *Nature (London),* 184, 56, 1959.
13. **Glazer, A. N., DeLange, R. J., and Martinez, R. J.,** Identification of ε-N-methyllysine in *Spirillum serpens* flagella and of ε-N-dimethyllysine in *Salmonella typhimurium* flagella, *Biochim. Biophys. Acta,* 188, 164, 1969.
14. **Murray, K.,** The occurrence of ε-N-methyllysine in histones, *Biochemistry,* 3, 10, 1964.
15. **Comb, D. G., Sarkar, N., and Pinzino, C. J.,** The methylation of lysine residues in protein, *J. Biol. Chem.,* 241, 1857, 1966.
16. **Fambrough, D. M. and Bonner, J.,** Sequence homology and role of cysteine in plant and animal arginine-rich histones, *J. Biol. Chem.,* 243, 4434, 1968.
17. **Hardy, M. F. and Perry, S. V.,** *In vitro* methylation of muscle proteins, *Nature (London),* 223, 300, 1969.
18. **Huszar, G. and Elzinga, M.,** ε-N-Methyllysine in myosin, *Nature (London),* 223, 834, 1969.
19. **Venkatesan, M., Nachmias, V., and McManus, I. R.,** *In vivo* methylation of contractile proteins in *Physarum polycephalum, Fed. Proc.,* 34, 671, 1975.
20. **Weihing, R. R. and Korn, E. D.,** ε-N-Dimethyllysine in amoeba actin, *Nature (London),* 227, 1263, 1970.
21. **Reporter, M. and Reed, D. W.,** Methylation of bovine rhodopsin, *Nature (London) New Biol.,* 239, 201, 1972.
22. **Kalasz, H., Kovacs, G. H., Nagy, J., Tyihak, E., and Barnes, W. T.,** Identification of N-methylated basic amino acids from human adult teeth, *J. Dent. Res.,* 57, 128, 1978.
23. **Minami, Y., Wakabayashi, S., Wada, K., Matsubara, H., Kerscher, L., and Oesterhelt, D.,** Amino acid sequence of a ferredoxin from thermo acidophilic archaebacterium, *Sulfolobus acidocaldarius.* Presence of an N^6-monomethyl-lysine and phyletic consideration of Archaebacteria, *J. Biochem. (Tokyo),* 97, 745, 1985.
24. **Ames, G. F. and Niakido, K.,** *In vivo* methylation of prokaryotic elongation factor Tu, *J. Biol. Chem.,* 254, 9947, 1979.
25. **Hiatt, W. R., Garcia, R., Merrick, W. C., and Sypherd, P. S.,** Methylation of elongation factor 1α from the fungus *Mucor, Proc. Natl. Acad. Sci. U.S.A.,* 79, 3433, 1982.
26. **Matsuoka, Y., Kumon, A., Nakajima, T., Kakimoto, Y., and Sano, I.,** Identification of ε-N-methyl-L-lysine and δ-N-methylornithine in bovine brain, *Biochem. Biophys. Acta,* 192, 136, 1969.
27. **Perry, T. L., Diamond, S., and Hansen, S.,** ε-N-Methyllysine: an additional amino acid in human plasma, *Nature (London),* 222, 668, 1969.
28. **Weatherall, I. L. and Haden, D. D.,** ε-N-Methyllysine in sheep plasma, *Biochim. Biophys. Acta,* 192, 553, 1969.
29. **Asatoor, A. M.,** The occurrence of ε-N-methyllysine in human urine, *Clin. Chim. Acta,* 26, 147, 1969.
30. **Kim, S. and Paik, W. K.,** Studies on the origin of ε-N-methyl-L-lysine in protein, *J. Biol. Chem.,* 240, 4629, 1965.
31. **Paik, W. K. and Kim, S.,** ε-N-Dimethyllysine in histones, *Biochem. Biophys. Res. Commun.,* 27, 479, 1967.
32. **Kuehl, W. M. and Adelstein, R. S.,** The absence of 3-methylhistidine in red, cardiac and fetal myosins, *Biochem. Biophys. Res. Commun.,* 39, 956, 1970.
33. **Hardy, M. F., Harris, C. I., Perry, S. V., and Stone, D.,** Occurrence and formation of the N^{ϵ}-methyllysines in myosin and the myofibrillar proteins, *Biochem. J.,* 120, 653, 1970.
34. **Chang, F. N., Navickas, I. J., Chang, C. N., and Dancis, B. M.,** Methylation of ribosomal proteins in HeLa cells, *Arch. Biochem. Biophys.,* 172, 627, 1967.
35. **Kakimoto, Y. and Akazawa, S.,** Isolation and identification of N^G,N^G- and N^G,N'^G-dimethylarginine, N^{ϵ}-mono-, di, and trimethyllysine, and glucosyl-galactosyl- and galactosyl-δ-hydroxylysine from human urine, *J. Biol. Chem.,* 245, 5751, 1970.
36. **Hempel, K., Lange, H. W., and Birkofer, L.,** N^{ϵ}-methylated lysine in histones from chicken erythrocytes, *Hoppe-Seyler's Z. Physiol. Chem.,* 349, 603, 1968.

37. **DeLange, R. J., Glazer, A. N., and Smith, E. L.,** Presence and location of an unusual amino acid, ε-N-trimethyllysine, in cytochrome *c* of wheat germ and *Neurospora*, *J. Biol. Chem.*, 244, 1385, 1969.
38. **DeLange, R. J., Glazer, A. N., and Smith, E. L.,** Identification and location of ε-N-trimethyllysine in yeast cytochrome *c*, J. Biol. Chem., 245, 3325, 1970.
39. **Hill, G. C., Chan, S. K., and Smith, L.,** Purification and properties of cytochrome c_{555} from a protozoan, *Crithidia fasciculata*, *Biochim. Biophys. Acta*, 253, 78, 1971.
40. **Brown, R. H. and Boulter, D.,** The amino acid sequence of cytochrome *c* from *Allium porrum* L. (Leek), *Biochem. J.*, 131, 247, 1973.
41. **Alix, J. H. and Hayes, D.,** Properties of ribosomes and RNA synthesized by *Escherichia coli* grown in the presence of ethionine. III. Methylated proteins in 50S ribosomes of *E. coli* EA2, *J. Mol. Biol.*, 86, 139, 1974.
42. **Chang, F. N., Chang, C. N., and Paik, W. K.,** Methylation of ribosomal proteins in *Escherichia coli*, *J. Bacteriol.*, 120, 651, 1974.
43. **Jackson, R. L., Dedman, J. R., Schreiber, W. E., Bhatnagar, P. K., Knapp, R. D., and Means, A. R.,** Identification of ε-N-trimethyllysine in rat testis calcium-dependent regulatory protein of cyclic nucleotide phosphodiesterase, *Biochem. Biophys. Res. Commun.*, 77, 723, 1977.
44. **Matojima, K. and Sakaguchi, K.,** Part of the lysyl residues in wheat α-amylase is methylated as N-ε-trimethyllysine, *Plant Cell Physiol.*, 23, 709, 1982.
45. **Bloxham, D. P., Parmelee, D. C., Kumar, S., Wade, R. D., Ericsson, L. H., Neurath, H., Walsh, K. A., and Titani, K.,** Primary structure of porcine heart citrate synthase, *Proc. Natl. Acad. Sci. U.S.A.*, 78, 5381, 1981.
46. **Wang, C., Lazarides, E., O'Connor, C. M., and Clarke, S.,** Methylation of chicken fibroblast heat shock proteins at lysyl and arginyl residues, *J. Biol. Chem.*, 257, 8356, 1982.
47. **L'Italien, J. J. and Laursen, R. A.,** Location of the site of methylation in elongation factor Tu, *FEBS Lett.*, 107, 359, 1979.
48. **Ohba, M., Koiwai, O., Tanada, S. and Hayashi, H.,** *In vivo* methylation of elongation factor Tu of *Escherichia coli*, *J. Biochem. (Tokyo)*, 86, 1233, 1979.
49. **Takemoto, T., Daigo, K., and Takagi, N.,** Studies on the hypotensive constituents of marine algae. I. A new basic amino acid "Laminine" and the other basic constituents isolated from *Laminaria angustate*, *Yakugaku Zasshi* (in Japanese), 84, 1176, 1964.
50. **Larsen, P. O.,** N^6-Trimethyl-L-lysine betaine from seeds of *Reseda luteola* L., *Acta Chem. Scand.*, 22, 1369, 1968.
51. **Tomita, T. and Nakamura, K.,** Isolation and identification of N^G-monomethyl-, N^G,N^G-dimethylarginine and $N^ε$-trimethyllysine from human placenta. *Hoppe- Seyler's Z. Physiol. Chem.*, 358, 413, 1977.
52. **Nakajima, T. and Volcani, B. E.,** ε-N-Trimethyl-L-δ-hydroxylysine phosphate and its nonphosphorylated compound in diatom cell walls, *Biochem. Biophys. Res. Commun.*, 39, 28, 1970.
53. **Paik, W. K. and Kim, S.,** Enzymatic methylation of protein fractions from calf thymus nuclei, *Biochem. Biophys. Res. Commun.*, 29, 14, 1967.
54. **Friedman, M., Shull, K. H., and Farber, E.,** Highly selective *in vivo* ethylation of rat liver nuclear protein by ethionine, *Biochem. Biophys. Res. Commun.*, 34, 857, 1969.
55. **Baldwin, G. S. and Carnegie, P. R.,** Specific enzymic methylation of an arginine in the experimental allergic encephalomyelitis protein from human myelin, *Science*, 171, 579, 1971.
56. **Tuck, M., Farooqui, J., and Paik, W. K.,** Two histone H1-specific protein-lysine N-methyltransferases from *Euglena gracilis*, *J. Biol. Chem.*, 260, 7114, 1985.
57. **Kasai, T., Sano, M., and Sakamura, S.,** N^G-Methylated arginines in broad bean seed, *Agric. Biol. Chem.*, 20, 2449, 1976.
58. **Reporter, M. and Corbin, J. L.,** N^G,N^G-Dimethylarginine in myosin during muscle development, *Biochem. Biophys. Res. Commun.*, 43, 644, 1971.
59. **Boffa, L. C., Karn, J., Vidali, G., and Allfrey, V. G.,** Distribution of N^G,N^G-dimethylarginine in nuclear protein fractions, *Biochem. Biophys. Res. Commun.*, 74, 969, 1977.
60. **Christensen, M. E., Beyer, A. L., Walker, B., and Lestourgeon, W. M.,** Identification of N^G,N^G-dimethylarginine in a nuclear protein from the lower eukaryote *Physarum polycephalum* homologous to major proteins of mammalian 40S ribonucleoprotein particles, *Biochem. Biophys. Res. Commun.*, 74, 621, 1977.
61. **Boffa, L. C., Sterner, R., Vidali, G., and Allfrey, V. G.,** Post-synthetic modification of nuclear proteins: high mobility group proteins are methylated, *Biochem. Biophys. Res. Commun.*, 89, 1322, 1979.
62. **Lischwe, M. A., Roberts, K. D., Yoeman, L. C., and Busch, H.,** Nuclear specific acidic phosphoprotein C23 is highly methylated, *J. Biol. Chem.*, 257, 14600, 1982.
63. **Lischwe, M. A., Ochs, R. L., Reddy, R., Cook, R. G., Yoeman, L. C., Tan, E. M., Reichlin, M., and Busch, H.,** Purification and partial characterization of a nucleolar scleroderma antigen (mw = 34,000; pI, 8.5) rich in N^G,N^G-dimethylarginine, *J. Biol. Chem.*, 260, 14304, 1985.

64. **Inukai, F., Suyama, Y., Inatome, H., Morita, Y., and Ozawa, M.,** Identification of δ-N-methylarginine, abstract of papers, Annual Meeting of the Agricultural Chemistry Society of Japan, Nagoya, Japan, 1968, 183.
65. **Johnson, P., Harris, C. I., and Perry, S. V.,** 3-Methylhistidine in actin and other muscle proteins, *Biochem. J.*, 105, 361, 1967.
66. **Asatoor, A. M. and Armstrong, M. D.,** 3-Methylhistidine, a component of actin, *Biochem. Biophys. Res. Commun.*, 26, 168, 1967.
67. **Gershey, E. L., Haslett, G. W., Vidali, G., and Allfrey, V. G.,** Chemical studies of histone methylation, *J. Biol. Chem.*, 244, 4871, 1969.
68. **Tallan, H. H., Stein, W. H., and Moore, S.,** 3-Methylhistidine, a new amino acid from human urine, *J. Biol. Chem.*, 206, 825, 1954.
69. **Lhoest, J. and Colson, C.,** Genetics of ribosomal protein methylation in *Escherichia coli, Mol. Gen. Genet.*, 154, 175, 1977.
70. **Pettigrew, G. W. and Smith, G. M.,** Novel N-terminal protein blocking group identified as dimethylproline, *Nature (London)*, 265, 661, 1977.
71. **Martinage, A., Briand, G., Van Dorsselaer, A., Turner, C. H., and Suatiere, P.,** Primary structure of histone H2B from gonads of the starfish *Asterias rubens:* identification of N-dimethylproline residue at the amino-terminal, *Eur. J. Biochem.*, 147, 351, 1985.
72. **Wittman-Liebold, B. and Pannenbecker, R.,** Primary structure of protein L33 from the large subunit of the *Escherichia coli* ribosome, *FEBS Lett.*, 68, 115, 1976.
73. **Lederer, F., Alix, J. H., and Hayes, D.,** N-Trimethylalanine, a novel blocking group found in *E. coli* ribosomal proteins L11, *Biochem. Biophys. Res. Commun.*, 77, 470, 1977.
74. **Henry, G. D., Dalgarno, D. C., Marcus, G., Scott, M., Levine, B. A., and Taryer, I. P.,** The occurrence of α-N-trimethylalanine as the N-terminal amino acid of some myosin light chains, *FEBS Lett.*, 144, 11, 1982.
75. **Brasius, J. and Chen, R.,** The primary structure of protein L16 located at the peptidyltransferase center of *Escherichia coli* ribosome, *FEBS Lett.*, 68, 105, 1976.
76. **Brauer, D. and Wittman-Liebold, B.,** The primary structure of the initiation factor IF-3 from *Escherichia coli, FEBS Lett.*, 79, 269, 1977.
77. **Olsson, M. and Lindahl, T.,** Repair of alkylated DNA in *Escherichia coli, J. Biol. Chem.*, 255, 10569, 1980.
78. **Farooqui, J. Z., Tuck, M. T., and Paik, W. K.,** Purification and characterization of enzymes from *Euglena gracilis* that methylate methionine and arginine residues of cytochrome *c. J. Biol. Chem.*, 260, 537, 1985.
79. **Kleene, S. J., Toews, M. L., and Adler, J.,** Isolation of glutamic acid methyl ester from an *Escherichia coli* membrane protein involved in chemotaxis, *J. Biol. Chem.*, 252, 3214, 1977.
80. **Van DerWerf, P. and Koshland, D. E., Jr.,** Identification of γ-glutamyl methyl ester in bacterial membrane protein involved in chemotaxis, *J. Biol. Chem.*, 252, 2793, 1977.
81. **Janson, C. A. and Clarke, S.,** Identification of aspartic acid as a site of methylation in human erythrocyte membrane proteins, *J. Biol. Chem.*, 255, 11640, 1980.
82. **Aswad, D. W.,** Stoichiometric methylation of porcine adrenocorticotropin by protein carboxyl methyltransferase requires deamidation of asparagine 25, *J. Biol. Chem.*, 259, 10714, 1984.
83. **Kakimoto, Y., Matsuoka, Y., and Konishi, H.,** Methylated amino acid residues of proteins of brain and other organs, *J. Neurochem.*, 24, 893, 1975.
84. **Matsuoka, Y.,** N-Methylated lysine and N^G-methylated arginines, *Seikagaku* (in Japanese), 44, 252, 1972.
85. **Hempel, V. K. and Lange, H. W.,** Traktionierung und eigenschaten N-methyl-ierter lysine, *Hoppe-Seyler's Z. Physiol. Chem.*, 350, 966, 1969.
86. **Markin, R. T.,** Isolation of N-methylated basic amino acids from physiological fluids and protein hydrolyzates, *Biochem. Med.*, 13, 23, 1975.
87. **Chang, C. N. and Chang, F. N.,** Methylation of the ribosomal proteins in *Escherichia coli.* Nature and stoichiometry of the methylated amino acids in 50S ribosomal proteins, *Biochemistry*, 14, 468, 1975.
88. **Rebouche, C. J. and Broquist, H. P.,** Carnitine biosynthesis in *Neurospora crassa:* enzymatic conversion of lysine to ε-N-trimethyllysine, *J. Bacteriol.*, 126, 1207, 1976.
89. **Tyihak, E., Ferenczi, S., Hazai, I., Zoltan, S., and Patthy, A.,** Combined application of ion-exchange chromatographic methods for the study of minor basic amino acids, *J. Chromatogr.*, 102, 257, 1974.
90. **Klagsbrun, M. and Furano, A. V.,** Methylated amino acids in the proteins of bacterial and mammalian cells, *Arch. Biochem. Biophys.*, 169, 529, 1975.
91. **Crampton, C. F., Stein, W. H., and Moore, S.,** Comparative studies on chromatographically purified histones, *J. Biol. Chem.*, 225, 363, 1957.
92. **Kakimoto, Y.,** Methylation of arginine and lysine residues of cerebral proteins, *Biochim. Biophys. Acta*, 243, 31, 1971.

93. **Deibler, G. E. and Martenson, R. E.,** Determination of methylated basic amino acids with amino acid analyzer, *J. Biol. Chem.*, 248, 2387, 1973.
94. **Zarkadas, C. G.,** A simple chromatographic method for the determination of the methylated basic amino acids in proteins, *Can. J. Biochem.*, 53, 96, 1975.
95. **Zarkadas, C. G.,** A rapid chromatographic method for the determination of ω-N-methylarginines in protein and muscle tissues, *Can. J. Biochem.*, 56, 952, 1978.
96. **Helm, R., Vancikova, O., Macek, K., and Deyl, Z.,** Separation of methylated basic amino acids, *J. Chromatogr.*, 133, 390, 1977.
97. **Lange, H. W. and Hempel, K.,** Saeulenchromatographische & trennung von N^G,N^G-dimethylarginin, N^G,N'^G-dimethylarginin, N^G-monomethylarginin, N^ϵ-monomethyllsin, N^ϵ-dimethyllysin and N^ϵ-trimethyllysin. *J. Chromatogr.*, 107, 389, 1975.
98. **Seely, J. H., Edattel, R., and Benoiton, N. L.,** A quick separation of ε-N-methyllysine using the amino acid analyzer, *J. Chromatogr.*, 44, 618, 1969.
99. **Lange, H. W., Loewer, R., and Hempel, K.,** Verbesserte saeulenchromatographische bestimmung N^ϵ-methylierter lysin in physiologischen fluessigkeiten, *J. Chromatogr.*, 76, 252, 1973.
100. **Lou, M. F. and Siena, M.,** Quantitation of methylated basic amino acids in biological fluid, *Biochem. Med.*, 25, 309, 1981.
101. **Paik, W. K., DiMaria, P., Pearson, E., and Kim, S.,** Preparation of radioactive ε-N-methylated lysine, not involving elaborate organic synthesis, *Anal. Biochem.*, 90, 262, 1978.
102. **Davis, A. T., Ingalls, S. T., and Hoppel, C. L.,** Determination of free trimethyllysine in plasma and tissue specimens by high performance liquid chromatography, *J. Chromatogr.*, 306, 79, 1984.
103. **Kim, S. and Paik, W. K.,** Labile protein-methylester: comparison between chemically and enzymatically synthesized, *Experientia*, 32, 982, 1976.
104. **Kim, S. and Paik, W. K.,** New assay method for protein methylase II activity, *Anal. Biochem.*, 42, 255, 1971.
105. **Ahlgren, J. A. and Ordal, G. W.,** Methyl esterification of glutamic acid residues of methyl-accepting chemotaxis proteins in *Bacillus subtilis*, *Biochem. J.*, 213, 759, 1983.
106. **Springer, W. R. and Koshland, D. E., Jr.,** Identification of a protein methyltransferase as the *che* R gene product in the bacterial sensing system, *Proc. Natl. Acad. Sci., U.S.A.*, 74, 533, 1977.
107. **Kort, E. N., Goy, M. F., Larsen, S. H., and Adler, J.,** Methylation of a membrane protein involved in bacterial chemotaxis, *Proc. Natl. Acad. Sci. U.S.A.*, 72, 3939, 1975.
108. **McFadden, P. N. and Clarke, S.,** Methylation of D-aspartyl residues in erythrocytes: possible step in the repair of aged membrane proteins, *Proc. Natl. Acad. Sci. U.S.A.*, 79, 2460, 1982.
109. **Galletti, P., Paik, W. K., and Kim, S.,** Methyl acceptors for protein methylase II from human erythrocyte membrane, *Eur. J. Biochem.*, 97, 221, 1979.
110. **Frietag, C. and Clarke, S.,** Reversible methylation of cytoskeletal and membrane proteins in intact human erythrocytes, *J. Biol. Chem.*, 256, 6102, 1981.
111. **Manning, J. M. and Moore, S.,** Determination of D- and L-amino acds by ion exchange chromatography as L-D and L-L dipeptides, *J. Biol. Chem.*, 243, 5591, 1968.
112. **O'Connor, C. M., Aswad, D. W., and Clarke, S.,** Mammalian brain and erythrocyte carboxyl methyltransferase are similar enzymes that recognize both D-aspartyl and L-isoaspartyl residues in structurally altered protein substrates, *Proc. Natl. Acad. Sci. U.S.A.*, 81, 7757, 1984.
113. **Kwon, D. S., Jun, G.-J., and Kim, S.,** Calcium borohydride method for identification of a site of methylation in human erythrocyte membrane proteins, *J. Korean Res. Inst. Better Living*, 34, 29, 1984.

Chapter 12

IDENTIFICATION OF PHENYLTHIOHYDANTOIN DERIVATIVES OF AMINO ACIDS

A. S. Bhown and J. C. Bennett

TABLE OF CONTENTS

I. INTRODUCTION

Technological advancements in protein sequence analysis methodologies have made possible the acquisition of reliable amino acid sequence data from subnanomole to subpicomole quantities of polypeptides. The breakthrough which has led to this advancement is the development of an identification and detection system for phenythiohydantoin (PTH) derivatives of amino acids by reverse-phase high-pressure liquid chromatography (RP-HPLC). The superb resolution and excellent reproducibility of RP-HPLC has made this system the method of choice for identification of PTH amino acids generated as the end product of sequential Edman degradation. A number of reports have appeared from this and other laboratories[1-4] demonstrating the use of different solvents, columns, and programs for the separation of PTH amino acids. Use of a computer for base line correction followed by peak weight expansion has enabled Hunkapiller and Hood[5] to separate as little as 5 pmol of 19 standard PTH amino acids. A computer-controlled system has been developed by Bhown et al.[6] to separate and collect each peak of PTH amino acid as it elutes from the HPLC column directly into the scintillation vial for the purpose of radioactive microsequencing. In order to achieve shorter run time and better resolution on smaller samples, research has been directed to evaluate the effects of changing the RP-HPLC colume dimensions. Bhown and Bennett[7] and Issac[8] have employed short RP-HPLC columns for the separation of PTH amino acids, while Silver et al.[9] and Cunico et al.[10] employed microbore RP-HPLC columns. In addition, there are a number of reports[2-4,11-15] dealing with different buffer systems, gradients, and flow rates to achieve the separation of PTH amino acids in the quickest possible time with maximum efficiency.

Separation of colored PTH amino acids has also been reported.[16] While reviewing the literature on this subject, one gets the feeling that there are as many methods available for PTH amino acid separation as there are protein sequencing laboratories. It would be out of the scope of this book to describe in detail the experimental procedures for all different methods of PTH amino acid separation; nonetheless, a few of them will be described in detail.

In most cases the analyst prefers to work under less stringent conditions. In the RP-HPLC system for PTH amino acid separation, the most commonly employed organic modifiers are acetonitrile and/or methanol, although use of other modifiers such as ethanol has been reported.[3] In this chapter we will describe in detail those methods which employ methanol and acetonitrile as an organic modifier. In addition, a detailed procedure to separate colored derivatives of PTH amino acid will be included.

II. EXPERIMENTAL

The separation of PTH amino acids by RP-HPLC has been found to be dependent largely on the buffer system, flow rate, and type of the reverse-phase column, and to a lesser extent on the instrumentation. This section, therefore, will be divided into two parts, dealing with the chemical and instrumental aspects of PTH amino acid separation by RP-HPLC.

A. Chemicals

Under this heading different buffer systems, flow rates, and columns will be described.

1. Buffers

Most of the methods for an efficient and reproducible separation of PTH amino acids by RP-HPLC employ a two-buffer (expressed as "A" and "B" buffer) gradient system.

"A" buffer is usually an aqueous system with a pH range of 3.5 to 5.0, depending upon the method employed. On the other hand, "B" buffer is an organic solvent which should be completely miscible with the "A" buffer. It is also important that both "A" and "B"

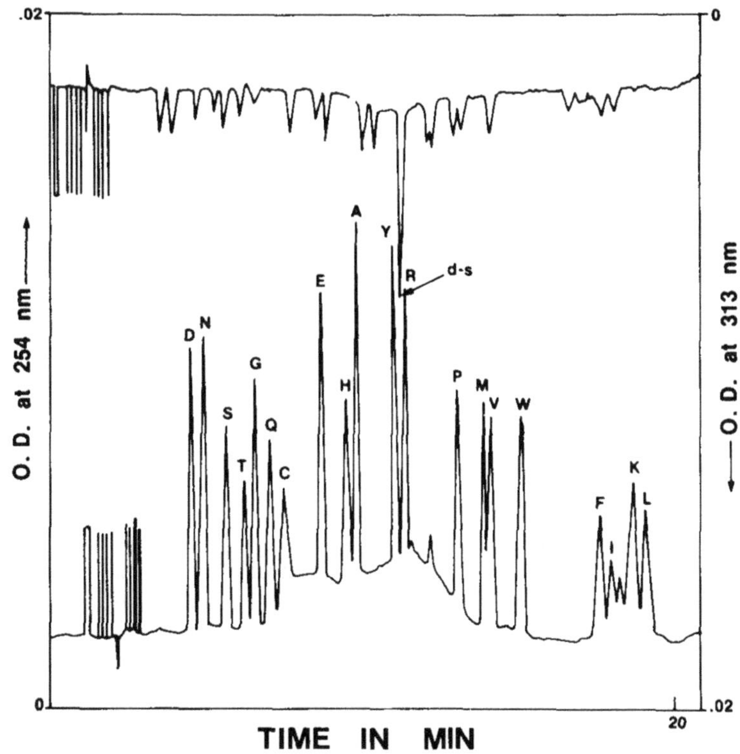

FIGURE 1. Rise in base line caused by impure water. When water purified with milli-Q system is employed, the big hump in the center of the run is abolished. Compare with Figure 2.

buffers and their constituents should not absorb at 254 and 313 nm, the wave lengths at which PTH amino acids are commonly monitored. While preparing these buffers, it is necessary that the following precautions be strictly observed.

The glasswares used should be absolutely clean and rinsed just before use in order to remove any solid or dust particles. Water for the aqueous buffers should be of highest purity. We recommend the use of Milli-Q System available from Millipore Corporation (Cat. #ZD20 230 74). The deionized water is circulated and passed through four cartridges arranged in the order shown below:

Cartridge	Cat. #
1. Carbon	CDFC 01204
2. ION-X	CDMB 01202
3. ION-X	CDMB 01202
4. Organex-Q	CDEX 01201

The water is then passed through a 0.2-$\mu\Omega$ filter manufactured by Gelman Sciences, Inc. (Michigan, Product #12122). It is advisable that water should read 15 to 18 MΩ and should be used fresh and not sotred for later use. And, finally, only glass containers should be used for collecting the water and preparing buffers. The aqueous buffer is filtered through a 0.2-μm filter just before use. It is important to mention that prity of water plays a major role in minimizing the rise in base line (Figure 1) during separation of PTH amino acids at a higher sensitivity by RP-HPLC. The different organic solvents used as "B" buffer should also be of highest purity grade. Most of these, especially methanol and acetonitrile, are now commercially available as "HPLC grade" solvents, some prefiltered, and can be directly used without any treatment. Irrespective of the type of buffers (A and B) used, we strongly recommend degassing these by sonication before use and, also, priming the HPLC pumps before starting the instrument.

2. Flow Rate

An efficient and reproducible separation of PTH amino acids by RP-HPLC is to some extent dependent upon the flow rate of the mobile phase. Separation of all common PTH amino acids has been achieved by Tarr[11] under isocratic condition by increasing the flow rate during separation. Since column pressure is directly related to the flow rate, it is advisable not to increase the flow rate, for this would cause a back pressure, more than what is recommended for that particular column by the manufacturer. It is advantageous for the column life to keep the pressure below the recommended value.

3. Column

Although separation of PTH amino acids by HPLC is dependent upon the column used, basically it has been achieved on reverse-phase columns, of different dimensions and packing. The time and order of elution is essentially a function of reversed-phase packing, although other parameters, such as the pH of "A" buffer, do influence these functions.

III. INSTRUMENTATION

There are a number of manufactueres who marked HPLC instruments. Basically, there are two types of HPLC equipment available, a two-pump and a one-pump system. Both types are efficient, automated, microprocessor controlled with data storage capabilities, and can run elution of essentially any gradient, with flow rates as low as 0.01 mℓ/min. However, minor changes in the separation parameters may be needed to obtain identical separation of PTH amino acids if two different types of HPLC or the same type of HPLC from two different manufacturers is employed. Irrespective of the type of HPLC used, proper care and maintenance of the system is highly warranted.

In order to get reproducible results, it is important that not too many variables be introduced into the separation system. The most commonly introduced variables are (1) changing the HPLC, (2) using different columns of the same manufactuere or similar columns from different manufacturers, and (3) buffers. To minimize the HPLC and column problems, it is strongly recommended that the samples for PTH amino acid identification be analyzed on one type of HPLC if possible. As far as the column is concerned, it should not be used for any other purpose, and when not in use it should be stored following the manufacturer's instructions. Similarly, if the HPLC is not in use, it should be primed first with water and then with methanol.

Preparation of buffers is the single most important source of variations which will adversely affect the separation of PTH amino acids. It is, therefore, strongly suggested that buffers be prepared from the highly purified water (Milli-Q) and HPLC grade organic modifiers. Unless specified by the manufacturer as prefiltered solvents, buffers must be filtered through a 0.2-μm filter after they are prepared; this is most important for aqueous buffers. Chemicals used to prepare "A" and "B" buffers should be kept separate to be used solely for HPLC purposes, and the same holds true for glasswares used to prepare and store these solvents.

IV. PROCEDURAL

As mentioned earlier, two methods of identification for PTH amino acids will be described in detail in this section. The first method employs methanol, and the second, acetonitrile, as the organic modifier, while "A" is an aqueous buffer.

Table 1
PARAMETERS FOR THE SEPARATION OF PTH AMINO ACIDS

Instrument	Flow rate (mℓ / min)	Initial condition		Gradient		Final condition		Chart speed (cm/min)
		% B	Time (min)	Type	Time (min)	% B	Hold (min)	
Waters	1.5	23	2	Linear	5	48	10	1
Beckman	1.5	23	2	Linear	7	46.5	10	1
Perkin-Elmer	1.5	23	2	Linear	5	42.6	11	1

A. Methanol System

The use of methanol as an organic solvent for the separation of PTH amino acids was described by Bhown et al.[6] and has been modified recently by Bhown and Bennett.[17] This procedire, which separates all the known PTH derivatives of 20 amino acids, will be described in detail. The system takes advantage of the computer-controlled precise mixing of the solvents "A" and "B" to achieve accurate pH and, thus avoid the necessity of pH adjustment of "A" buffer. The procedure is simple and highly reproducible.

1. Chemicals

HPLC grade methanol (OmniSolv®; MX0488-1) and glass-distilled acetone (OmniSolv®; AX0116-1) are the products of EM Science (Cherry Hill, N. J.); triethylamine (#25108), acetic acid, and standard PTH amino acids are obtained from Pierce Chemicals (Rockford, Ill.), and 0.2-μm Nylon-66 filters (#38-111) are purchased from Rainin Instrument Co., Inc. (Woburn, Mass.).

2. Instrumentation

The method has been developed on two different types of HPLC. The one-pump system, manufactured by Perkin-Elmer, is their Series 4 microprocessor-controlled solvent delivery system equipped with Waters Autoinjector Model WISP 710B and a dual-channel absorbance detector, Model 440. The two-pump HPLC System employed is supplied by Waters Associates® (Milford, Mass.) and consists of two modified 6000-A solvent delivery pumps, Model 720 System Controller, WISP 710B Autoinjector, and a Model 440 Dual-Channel Absorbance Detector. More recently, we have replaced the Waters two-pump system with a Beckman two-pump system comprised of Model 110B Solvent Delivery Module, 421A Controller, Waters Autoinjector (WISP 710B), and a dual-channel absorbance detector (Model 440). For all these three different types of HPLC systems, an ultrasphere ODS 5-μm column (0.46 × 15.0 cm) manufactured by Altex Scientific, Inc. and marketed by Beckman Instruments (part #235330) is employed as stationary phase to separate the PTH derivatives of amino acids.

Solvents "A" and "B" are constituted as follows: to make buffer "A", 750 to 1000 μℓ of acetic acid and 50 μℓ of acetone are added to 1 ℓ of water purified through a Milli-Q water purification system. It is filtered through a 0.2-μm filter and stored in a colored glass bottle rinsed with the same buffer. Solvent "B" is pure methanol containing 900 μℓ of triethylamine per liter and is used without filtration. Parameters to achieve separation of all the 20 PTH amino acids for different types of HPLC employed are given in Table 1. The effluent is monitored simultaneously at 254 and 313 nm.

3. Sample Preparations

Samples for the analysis are obtained as anilinothiazolinone (ATZ) if the sequencer is not equipped with an autoconversion device (Beckman Model 890-C), or as phenylthiohydantoin

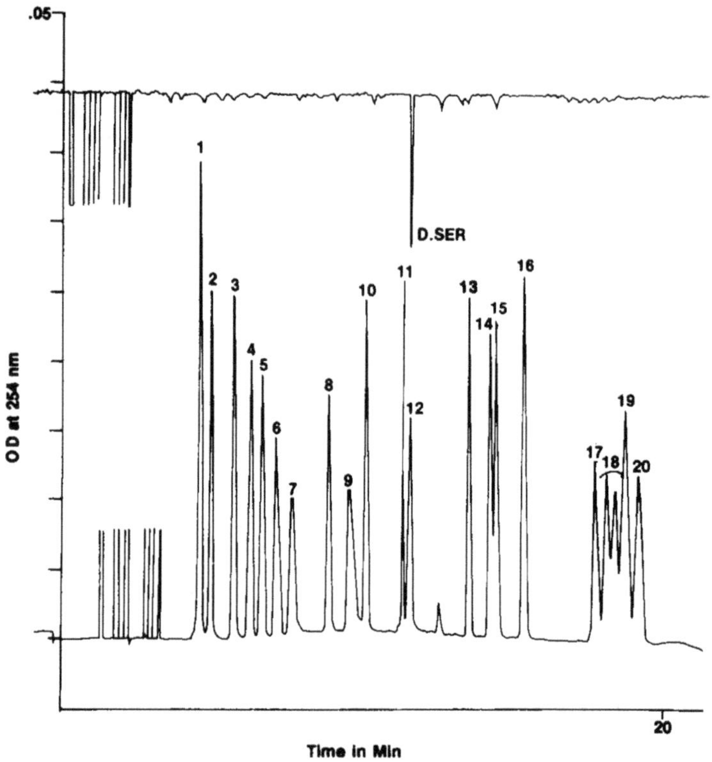

FIGURE 2. Separation of a mixture of all the 20 PTH derivatives of amino acids (200 p*M* each). 1, Asp; 2, Asn; 3, Ser; 4, Thr 5, Gly; 6, Gln; 7, CMC; 8, Glu; 9, His; 10, Ala; 11, Tyr; 12, Arg; 13, Pro; 14, Met; 15, Val; 16, Trp; 17, Phe; 18, Ile + AlloIle; 19, Lys; 20, Leu. For details see text. (From Bhown, A. S. and Bennett, J. C., *Anal. Biochem.*, 150, 457, 1985. With permission.)

derivatives (PTH) if equipped with autoconversion device as in the Beckman 890M, the Applied Biosystem Gas Phase Sequencer, or Porton Protein Sequencer. If obtained as ATZ derivative, then the samples need to be converted to their respective PTH forms as described in Chapter 3 of this book. In either case, they are dried and should be redissolved into the initial buffer (23% B + 77% A) as follows.

The dry sequencer sample is first dissolved in 46 μℓ of "B", vortexed, and 154 μℓ of Solvent "A" is then added. The sample is now in 200 μℓ of initial buffer. Appropriate volumes are then loaded into the autoinjector or injected manually.

4. Analysis

Figure 2 shows the separation of a mixture of all the 20 PTH amino acids, and Figure 3 shows the results of the analysis of samples obtained from amino acid sequence analysis.

B. Acetonitrile System

Although there is more than one method for the separation of PTH amino acids by RP-HPLC which employ acetronitrile as the organic eluent, the most commonly used is that of Hunkapiller and Hood.[18] It is this method which will be described in this section. The method separates a mixture of 19 PTH amino acids by RP-HPLC at an elevated temperature and has an advantage in being completely automated. For complete automation details, the reader is advised to consult the original publication.[18] This section will deal with the basics of this method.

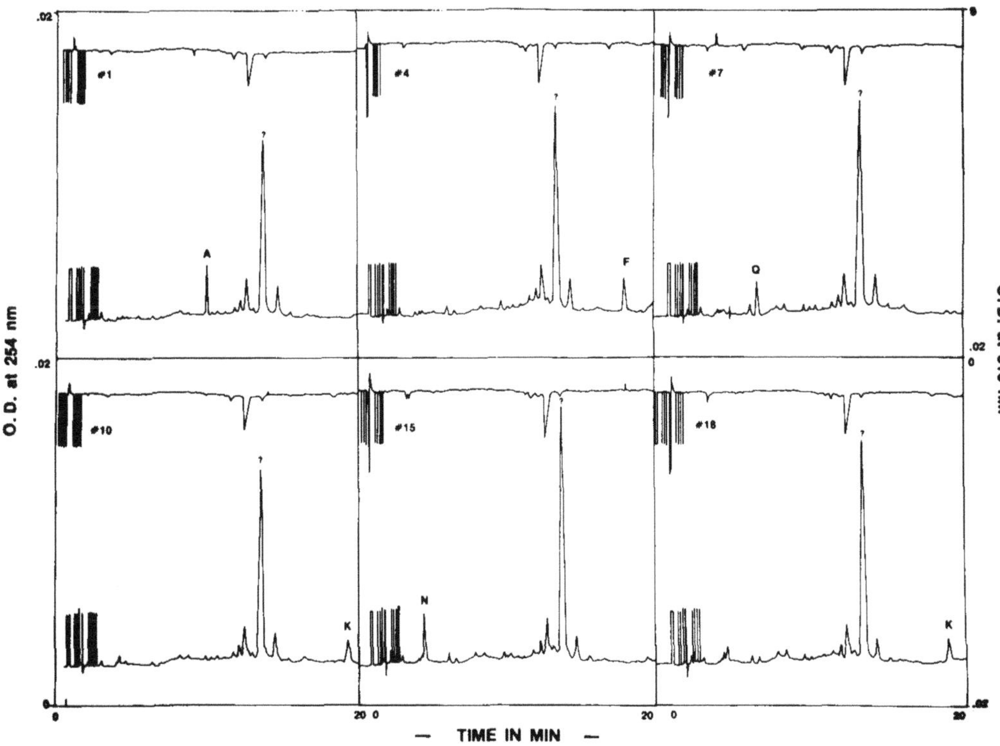

FIGURE 3. Amino acid sequence analysis of 1 n*M* of viral protein.

1. Chemicals

Acetic acid, acetylchloride, and HPLC grade methanol are supplied by J. T. Baker (Phillipsburg, N. J.); sodium acetate (reagent grade), from Mallinckrodt (St. Louis, Mo.), and acetonitrile is obtained from Burdick and Jackson (Muskegon, Mich.). Standard mixture of PTH amino acid is constituted from PTH derivatives obtained from Pierce Chemical Co., (Rockford, Ill.), and SDS is the product of Bio-Rad (Richmond, Calif.). Glass-distilled water is used to prepare solution "A". Authors have recommended[18] further purification of some of the reagents used.

2. Instrumentation

A two-pump system equipped with an autoinjector, dual-channel absorbance detector, Hewlett-Packard (Palo Alto, Calif.) Series 3354 Laboratory Automation Computer, and a constant-temperature forced-air column oven is employed. Separation is achieved at 32 ± 1°C on a Cyano HPLC Analytical Column (0.48 × 25 cm) manufactured by IBM Instruments (Wallington, Conn.).

"A" and "B" buffers are prepared as follows: Solvent "A" is a 0.015- to 0.045-*M* solution of sodium acetate, pH 5.8, and "B" is a mixture of acetonitrile-methanol (4:1). Separation of PTH amino acids is achieved by employing the conditions tabulated in Table 2.

3. Sample Preparation

As described earlier, if the samples are obtained as an ATZ derivative, they are first converted to their PTH form and then dissolved in purified methanol and loaded into the autoinjector.

Table 2
PARAMETERS FOR THE SEPARATION OF PTH AMINO ACIDS

Time (min after injection)	Function	Value	Duration (min)
0.0	%B	15%	—
0.0	Flow/rate	1.0 mℓ/min	—
0.0	%B	35%	1.0
1.0	%B	45%	10.2
11.2	%B	55%	1.9
13.1	%B	60%	3.7
16.8	%B	15%	3.0
35.0	%B	100%	4.0
59.0	Flow/rate	0.0 mℓ/min	1.0

From Hunkapiller, M. W. and Hood, L. E., *Methods in Enzymology*, Vol. 91, Part 1, Hirs, C. H. W. and Timasheff, S. N., Eds., Academic Press, New York, 1983, 486. With permission.

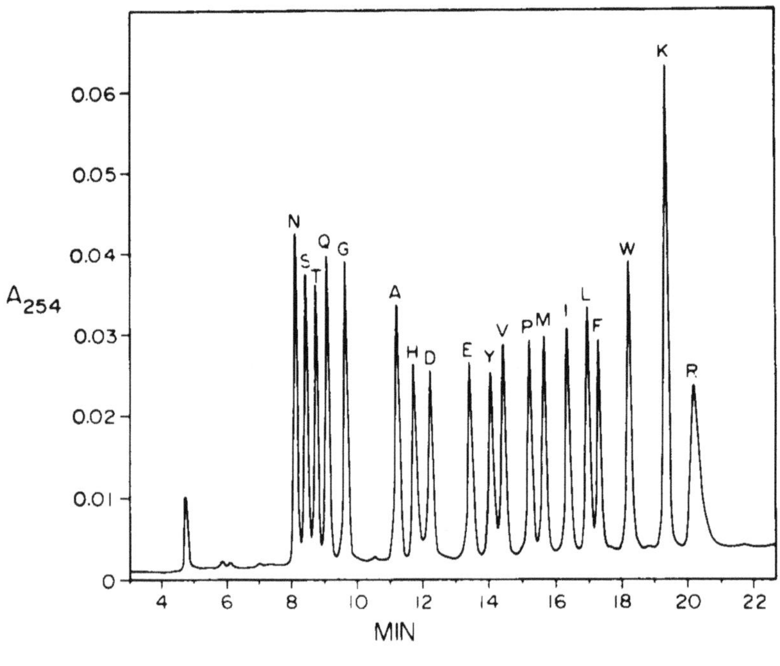

FIGURE 4. Separation of a mixture of standard PTH amino acids. (From Hunkapiller, M. W. and Hood, L E., *Methods in Enzymology*, Vol. 91, Part I, Hirs, C. H. W. and Timasheff, S. N., Eds., Academic Press, New York, 1983, 486. With permission.)

4. Analysis

Figures 4 and 5 show the separation of 19 PTH amino acids and the samples obtained from a sequencer run, respectively.

C. Analysis of Colored Amino Acid Thiohydantoin Derivatives

Chang et al.[19] and Chang and Creaser[20] have introduced a new coupling reagent, 4-*N*,*N*-dimethylaminobenzene-4′-isothiocyanate (DABITC), which has been employed for micro-

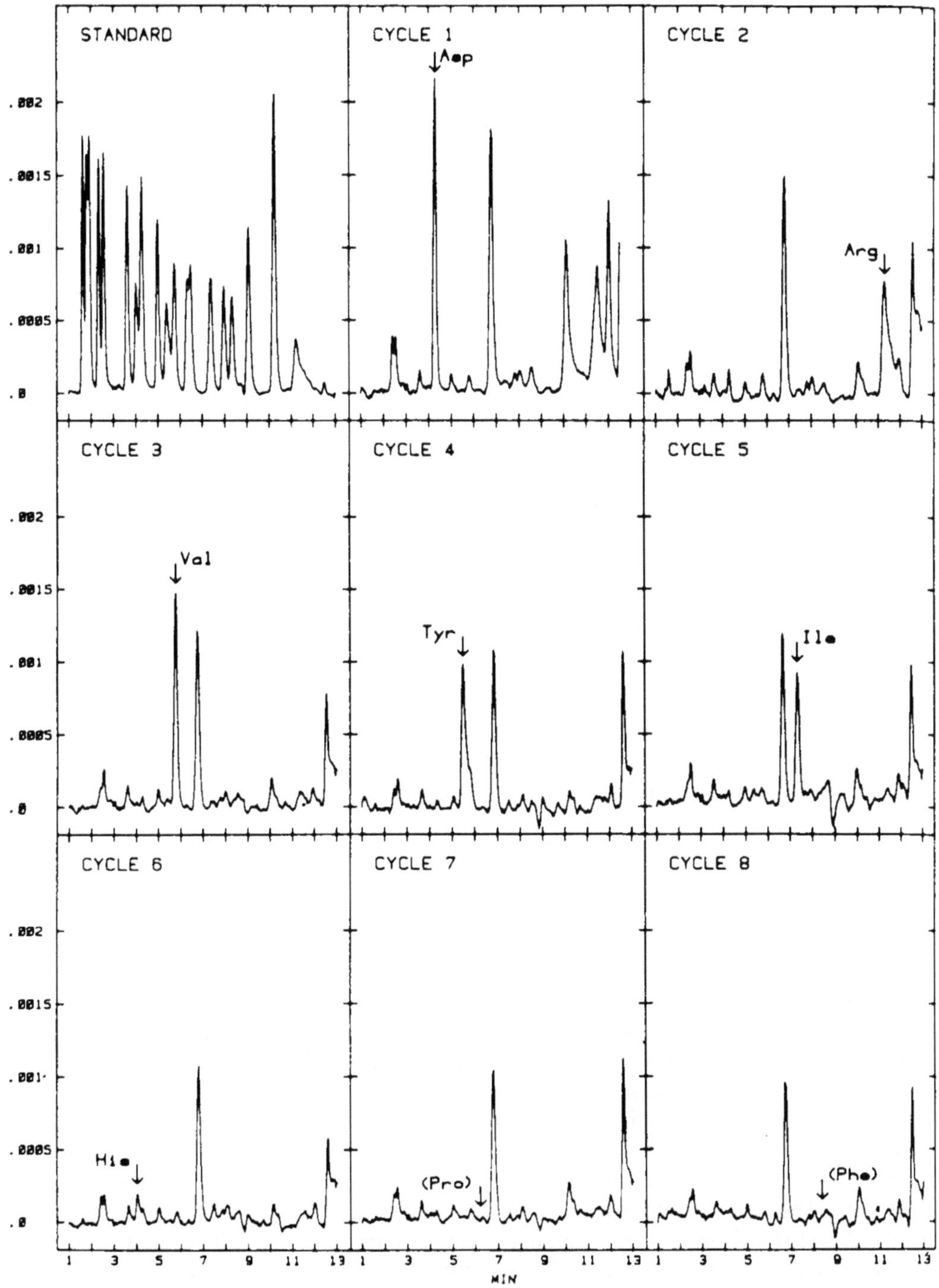

FIGURE 5. Amino acid sequence analysis results obtained on 50 pmol of sample. (From Hunkapiller, M. W. and Hood, L. E., *Methods in Enzymology,* Part I, Hirs, C. H. W. and Timasheff, S. N., Eds., Academic Press, New York, 1983, 486. With permission.)

sequencing.[16] The end product of sequential degradation under these coupling conditions is not a PTH derivative; instead, it is a dimethylaminobenzene thiohydantoin (DABTH), which is a colored product and can be separated and identified by conventional RP-HPLC methods. Chapter 6 describes in detail the methodological aspects of using DABITC for amino acid sequence analysis; this section will describe the procedural details of identification of these colored DABTH amino acid derivatives by RP-HPLC as developed by Lehmann and Wittmann-Liebold.[16]

1. Chemicals

Acetonitrile (LiChrosolv grade) and 1,2-dichloroethane (Uvarol grade) are purchased from Merck and used without purification.

2. Instrumentation

Any of the HPLC systems described earlier can be used for the identification of the DABTH amino acids, but the effluent is monitored at 436 nm, instead of 254 and 313 nm as in the case of PTH derivatives.

The separation method reported by Lehmann and Wittmann-Liebold[16] employs isocratic separation. Only one buffer system is described which is constituted as follows: 12 m*M* sodium acetate (pH 5.0) to acetonitrile to dichloroethane (50:50:0.5). Volume-to-volume ratio of sodium acetate to acetonitrile may have to be changed between 48 and 49% and 52 and 51%, respectively, to obtain a reproducible elution profile. This ratio is dependent upon the column batch.

3. Preparation of Standard DABTH Amino Acids

Commercial unavailability of standard DABTH derivatives of amino acids necessitates that these derivatives by synthesized and purified in the laboratory. The procedure to accomplish this is described as follows.

200 μ*M* of each amino acid is dissolved in 10 mℓ of water to acetone to acetic acid to triethylamine (50:50:0.6:0.4 by volume) and adjusted to 10.4 pH with triethylamine; 50 μ*M* DABITC (Pierce Chemical Co., Rockford, Ill., recrystallized from boiling acetone) in 10 mℓ of acetone is added to the amino acid solution. The tube is flushed with nitrogen, vortexed, and incubated at 60°C for 60 min with stirring. After this the solvent is removed in a rotary evaporator, and finally over KOH under vacuum. To the dried residue, 6 mℓ of 40% trifluoroacetic acid (TFA) is added. The residue is then flushed with nitrogen and again incubated at 50°C for 60 min (asparagine and glutamine at 45°C). The mixture is dried over KOH under vacuum and dissolved in 5 to 15 mℓ of mixture of acetonitrile to water, as shown in Table 3. The crude DABTH derivative is purified by HPLC using Hypersil MOS, 5 μm analytical column (0.4 × 25.0 cm); 1.0-mℓ injection is made and column developed at 2.0 mℓ/min flow using the solvent system given in Table 3, and the effluent is monitored at 436 nm. The fractions are pooled, diluted with 0.5 mℓ of water, lyophilized, and stored at −20°C. DABTH derivatives of arginine and histidine are repeatedly dissolved in water and lyophilized. To obtain the DABTH derivatives of serine, threonine, and lysine, natural or synthetic peptides containing these amino acids are degraded and DABTH derivatives at the desired cycle stored. Separation of the mixture of standard DABTH derivatives is achieved on a Hypersil MOS (C_8) 5-μm column (0.4 × 25.0 cm) kept at 45°C by a thermostated aluminum block and developed at 1.2 mℓ/min flow of a mixture of 12 mM sodium acetate (pH 5.0) to acetonitrile to dichloroethane (50:50:0.5 by volume). The effluent is monitored at 436 nm.

4. Sample Preparation

The procedure described earlier in this chapter may be used for the sample preparation.

Table 3
PARAMETERS FOR THE PURIFICATION OF STANDARD DERIVATIVES OF DABTH AMINO ACIDS

DABTH derivative of	Dissolved in water:CH_3CN	Solvent for the purification by HPLC water:CH_3CN:DCE
Asp	40:60	40:60:0.5
Glu	40:60	40:60:0.5
Asn	40:60	40:60:0.5
Gln	40:60	40:60:0.5
Ala	50:50	52:48:0.5
Ser[a]	50:50	40:60:0.5
Thr[a]	50:50	40:60:0.5
Gly	50:50	40:60:0.5
Tyr	50:50	40:60:0.5
Met	50:50	40:60:0.5
Val	50:50	30:70:0.5
Pro	50:50	30:70:0.5
Pho	50:50	30:70:0.5
Cys	50:50	30:70:0.5
Ile	50:50	30:70:0.5
Leu	50:50	30:70:0.5
Typ	25:75	30:70:0.5
His[b]	20:80	20:80:0.5
Arg[b]	20:80	20:80:0.5

[a] Includes the degradation products.
[b] Water is replaced by 8 m*M* sodium acetate, pH 5.0.

From Lehmann, A. and Wittman-Liebold, B., *FEBS Lett.*, 176, 380, 1984. With permission.

5. Analysis

Figures 6 and 7 show the separation of a mixture of standard DABTH amino acids and some degradation cycles of natural peptides,[16] respectively.

D. Additional Methods

As mentioned earlier in the chapter, there are a number of methods which allow one to achieve the separation of PTH amino acids by RP-HPLC. Table 4 lists some of those systems as a reference guide to the reader.

ACKNOWLEDGMENTS

The authors wish to express their thanks to Robert W. Herrick for science writing, to James Wayland and Lisa Kallman for their expert technical assistance, as well as to Sandra Reid for typing this manuscript. The work was supported by Grants AM-03555 and AM-20614.

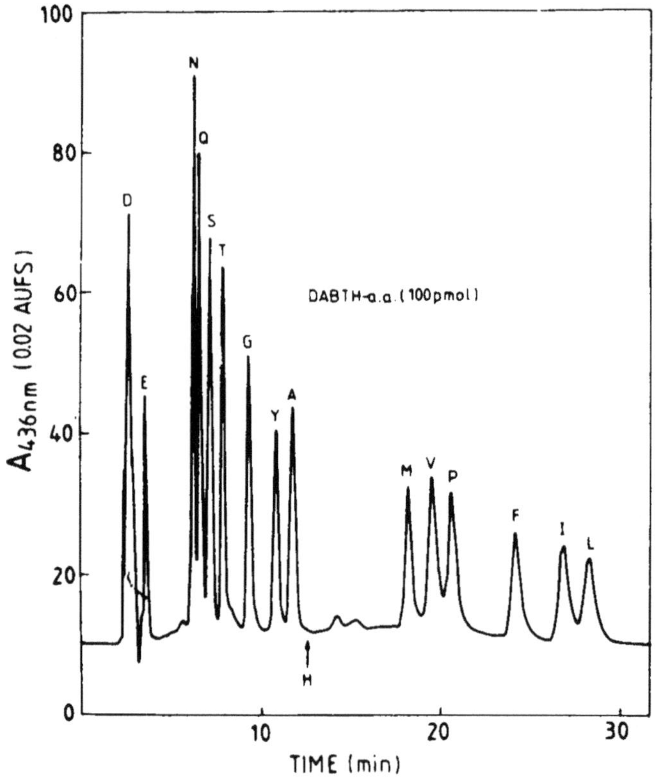

FIGURE 6. Separation of standard DABTH derivatives of amino acids. (From Lehmann, A. and Wittmann-Liebold, B., *FEBS Lett.*, 176, 380, 1984. With permission.)

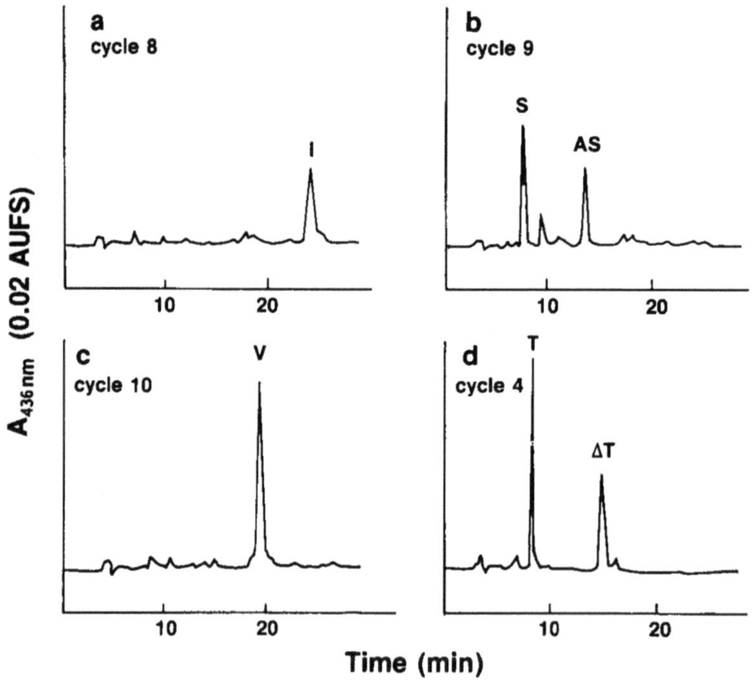

FIGURE 7. Results of the degradation cycles of natural peptide. (From Lehmann, A. and Wittmann-Liebold, B., *FEBS Lett.*, 176, 380, 1984. With permission.)

Table 4
ADDITIONAL METHODS FOR SEPARATION OF PTH AMINO ACIDS BY RP-HPLC

Solvent system A/B	Column	Ref.
NaAc/MeOH + CH_3CN	Zorbax CN	2
Isocratic: acetonitrile + THF + NaAc	Ultrasphere ODS	17
NaAc/EtOH	Waters C_{18}	3
NaPro/MeOH	μ Bondapak® phenylalkyl	4
NH_4Ac/CH_3CN	Ultrasphere ODS	11
$NaAc/CH_3CN$	Ultrasphere ODS	18
$NaAc/CH_3CN$ + THF	Ultrasphere ODS	19
NH_4Ac/CH_3CN	Ultrasphere ODS	20
TEA:AC/CH_3CN		

REFERENCES

1. **Bhown, A. S., Mole, J. E., and Bennett, J. C.,** An improved procedure for high sensitivity microsequencing: use of aminoethyl aminopropyl glass beads in the Beckman sequencer and the ultrasphere ODS column for PTH amino acid identification, *Anal. Biochem.*, 110, 355, 1981.
2. **Johnson, N. D., Hunkapiller, M. W., and Hood, L. E.,** Analysis of phenylthiohydantoin amino acids by high performance liquid chromatography on DuPont Zorbax Cyanopropylsilane Columns, *Anal. Biochem.*, 100, 335, 1979.
3. **Fohlman, J., Rask, L., and Peterson, P. A.,** High pressure liquid chromatographic identification of phenylthiohydantoin derivatives of all twenty common amino acids, *Anal. Biochem.*, 106, 22, 1980.
4. **Henderson, L. E., Copeland, T. D., and Oroszlan, S.,** Separation of amino acid phenylthiohydantoins by high performance liquid chromatography on phenylalkyl support, *Anal. Biochem.*, 102, 1, 1980.
5. **Hunkapiller, M. W. and Hood, L. E.,** New protein sequenator with increased sensitivity, *Science*, 207, 523, 1980.
6. **Bhown, A. S., Mole, J. E., Hollaway, W. L., and Bennett, J. C.,** Computer assisted high pressure liquid chromatography of radio-labelled phenylthiohydantoin amino acids, *J. Chromatogr.*, 156, 35, 1978.
7. **Bhown, A. S. and Bennett, J. C.,** Use of mini-octadecylsilane ultrasphere column in high pressure liquid chromatography for protein structural studies, *Anal. Biochem.*, 137, 256, 1984.
8. **Issac, H. J.,** High performance liquid chromatography separation using short columns packed with spherical ODS particles. II. Effect of mobile phase composition on resolution, *J. Liq. Chromatogr.*, 7, 475, 1984.
9. **Silver, M. R., Trosper, T. D., Gould, M. R., Dickinson, J. E., and Desotelle, G. A.,** The use of large volume injections for the isocratic separation of phenylthiohydantoin amino acids by microbore liquid chromatography, *J. Liq. Chromatogr.*, 7, 559, 1984.
10. **Cunico, R. L., Simpson, R., Correia, L., and Wehr, C. T.,** High sensitivity phenylthiohydantoin amino acid analysis using conventional and microbore chromatography, *J. Chromatogr.*, 336, 105, 1984.
11. **Tarr, G. E.,** Rapid separation of amino acid phenylthiohydantoin by isocratic high performance liquid chromatography, *Anal. Biochem.*, 111, 27, 1981.
12. **Kim, S. M.,** Isocratic separation of phenylthiohydantoin amino acids by reversed phase high performance liquid chromatography, *J. Chromotogr.*, 247, 103, 1982.
13. **Pucci, P., Sannia, S., and Marino, G.,** Separation of phenylthiohydantoin amino acids by high performance liquid chromatography, *J. Chromatogr.*, 270, 371, 1983.
14. **Somack, R.,** Complete phenylthiohydantoin amino acid analysis by high performance liquid chromatography on ultrasphere octadecyltrimethyloxysilane, *J. Chromatogr.*, 104, 464, 1980.
15. **Black, S. and Coon, M. J.,** A simple, rapid and highly efficient separation of amino acid phenylthiohydantoin by reversed phase high performance liquid chromatography, *Anal. Biochem.*, 121, 281, 1982.

16. **Lehmann, A. and Wittmann-Liebold, B.,** Complete separation and quantitative determination of DABTH amino acid derivatives by isocratic reversed phase high performance liquid chromatography, *FEBS Lett.*, 176, 380, 1984.
17. **Bhown, A. S. and Bennett, J. C.,** A modified system for thiazolinone conversion to thiohydantoin derivatives and their separation by high pressure liquid chromatography, *Anal. Biochem.*, 150, 457, 1985.
18. **Hunkapiller, M. W. and Hood, L. E.,** Analysis of phenylthiohydantoins by ultrasensitive gradient high-performance liquid chromatography, in *Methods in Enzymology*, Vol. 91, Part I, Hirs, C. H. W. and Timasheff, S. N., Eds., Academic Press, New York, 1983, 486.
19. **Chang, J. Y., Creaser, E. H., and Bentley, K. W.,** 4-NN-Dimethylaminoazobenzene 4'-isothiocyanate, a new chromophoric reagent for protein sequence analysis, *Biochem. J.*, 153, 607, 1976.
20. **Chang, J. Y. and Creaser, E. H.,** A novel manual method for protein sequence analysis, *Biochem. J.*, 157, 77, 1976.

APPENDIX 1

LIST OF ABBREVIATIONS

Å	Angstrom
APG	Aminopropyl glass
APS	Aminopolystyrene
ATZ	2-Anilino-5-thiazolinone
BAMPITC	Boc-aminoethyl phenylisothiocyanate
CNBr	Cyanogen bromide
CI	Chemical ionization mode
DAP	Dipeptidyl aminopeptidase
DABITC	4-*N*,*N*-Dimethylaminoazo-benzene-4′ isothiocyanate
DABTH	Dimethylaminoazobenzene thiohydantoin
DCP	Dipeptidyl carboxypeptidase
DFP	Diisopropylfluorophosphate
DITC	Diphenyl isothiocyanate
DMAA	Dimethylallylamine
DMBA	Dimethylbutylamine
DMF	Dimethylformamide
DMPTU	*N*,*N*-dimethyl-*N*′-phenylthiourea
DMSO	Dimethyl sulfoxide
DNSAPITC	Dimethylaminophytylsulfonylamino-phenylisothiocyanate
DNSAPTH	Dimethylaminophtylsulfonylamino phenyl thiohydantoin
DP	Dipeptidyl peptidase
DPTU	Diphenylthiourea
DTT	Dithiothreatol
EA	Ethyl acetate
EDTA	Ethylenediamine tetraacetate
EI	Electron bombardment ionic mode
EtOH	Ethanol
FAB	Fast atom bombardment
FD	Field desorbtion
FITC	Fluoresceinisothiocyanate
G.C.	Gas chromatography
GC-MS	Gas Chromatography-mass spectrometery
HF	Hydrofluoric acid
HFA	Hexafluoroacetone trihydrate
HFBA	Heptafluorobutyric acid
HIC/HI-HPLC	Hydrophobic interaction high pressure/performance liquid chromatography
HPLC	High pressure/performance liquid chromatography
IEC/IE-HPLC	Ion exchange high pressure/performance liquid chromatography
Ig	Immunoglobulin
LC-MS	Liquid chromatography-mass spectrometry
MCP	Methyl accepting chemotactic proteins
MeOH	Methanol
MeCN	Acetonitrile
MITC	Methylisothiocyanate
MS	Mass spectrophotometry
MW	Molecular weight

NCS	*N*-Chlorosuccinimide
NTCB	2-Nitro-5-thiocyanobenzoic acid
PAGE	Poly acrylamide gel electrophoresis
PCMB	*p*-Chloromercuribenzoate
pI	Isoelectric point
PMSF	Phenylmethylsulfonyl fluoride
PTC	Phenylthiocarbamyl
PTH	Phenythiohydantoin
PTU	Phenylthiourea
RPC/RP/HPLC	Reverse phase high pressure/performance liquid chromatography
R.Y.	Repetitive yield
SAS	Sample application subroutine
SDA	Sodium dodecyl sulfate
SEC/SE-HPLC	Size exclusion high pressure/performance liquid chromatography
SIMS	Secondary ion mass spectrometry
TATG	Thioacetylthioglycolic acid
TEA	Triethylamine
TETA	Tetraethylenetetraminepolystyrene
TMA	Trimethylamine
TFA	Trifluoro acetic acid
UV	Ultraviolet

APPENDIX 2

ABBREVIATIONS FOR AMINO ACIDS

Amino acid	Molecular weight	Three letter	One letter
Alanine	89	Ala	A
Arginine	174	Arg	R
Asparagine	132	Asn	N
Aspartic acid	133	Asp	D
Cysteine	121	Cys	C
Glutamic acid	147	Glu	E
Glutamine	146	Gln	Q
Glycine	75	Gly	G
Histidine	155	His	H
Isoleucine	131	Ile	I
Leucine	131	Leu	L
Lysine	146	Lys	K
Methionine	149	Met	M
Phenylalanin	165	Phe	F
Proline	115	Pro	P
Serine	105	Ser	S
Threonine	119	Thr	T
Tryprophane	204	Trp	W
Tyrosine	181	Tyr	Y
Valine	117	Val	V

APPENDIX 3

STRUCTURES OF AMINO ACIDS

Ala

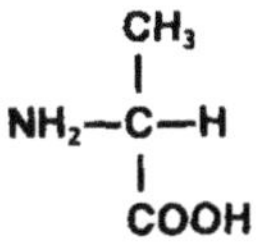

Arg

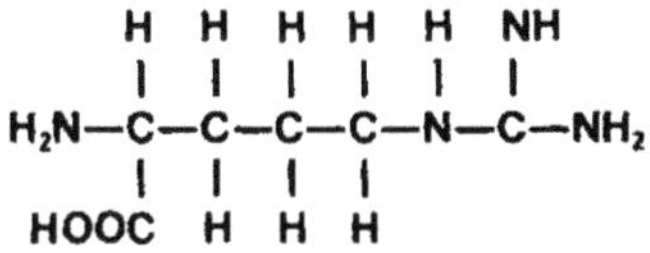

Asn

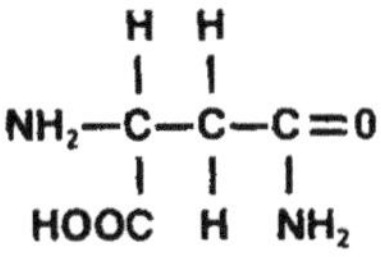

Asp

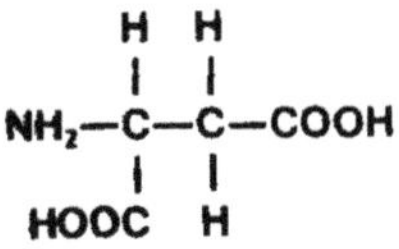

Cys

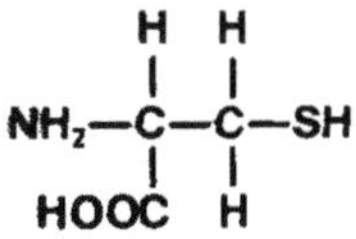

Gln

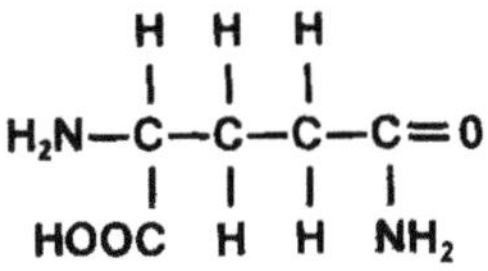

Glu

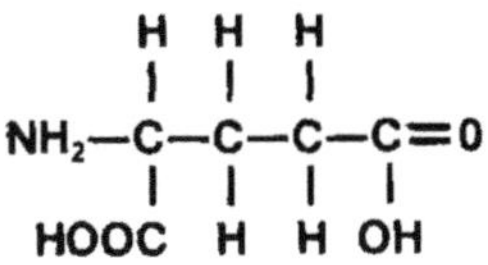

Gly

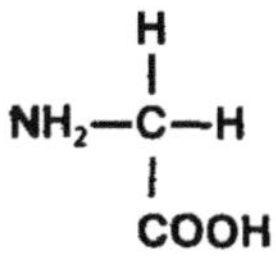

His

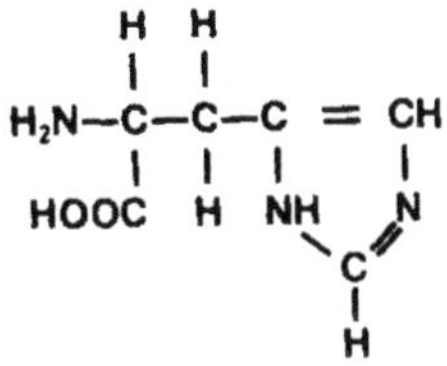

Ile

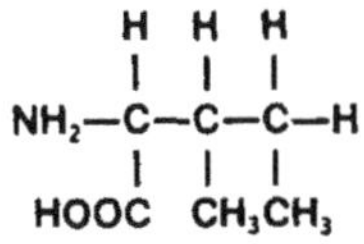

Leu

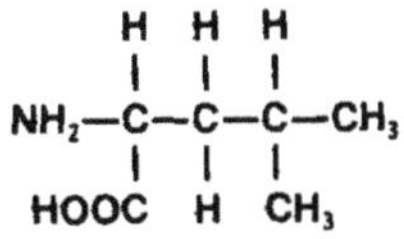

Lys

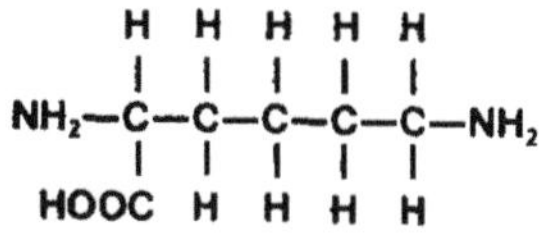

Met

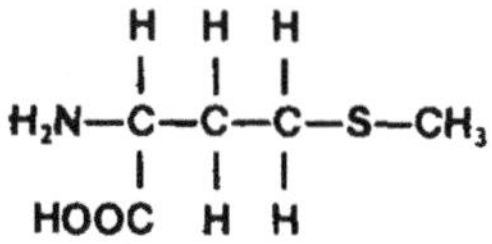

Phe

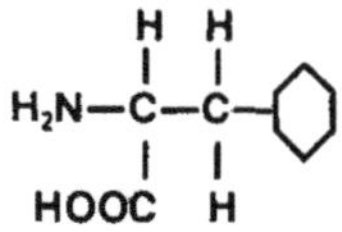

Pro

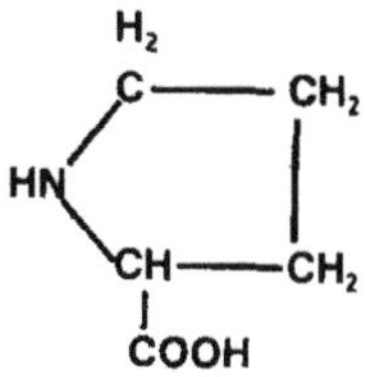

Ser

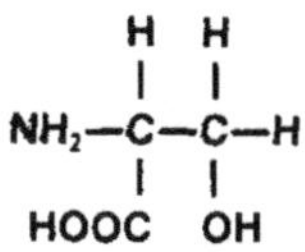

Thr

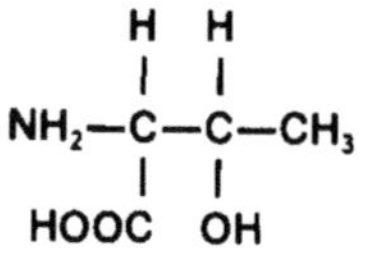

Try

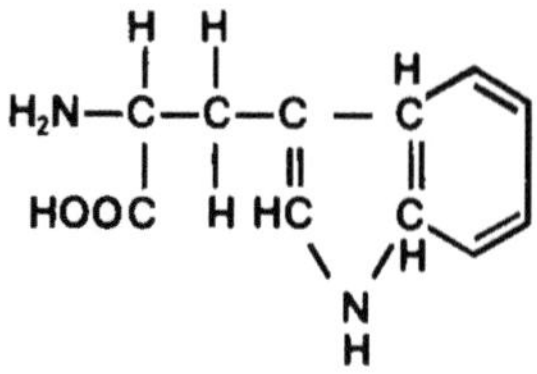

Tyr

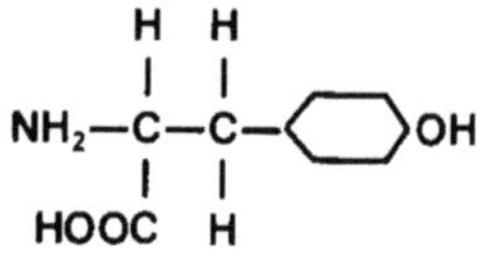

Val

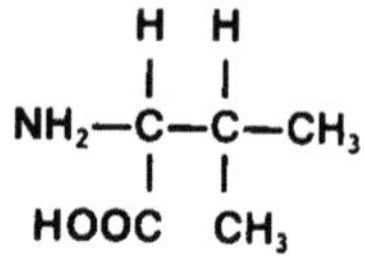

APPENDIX 4

LIST OF VENDORS

Alltech Associates, Inc.
2051 Waukegan Road
Deerfield, IL 60015
(312) 948-8600

Altex
1716 Fourth Street
Berkeley, CA 94710
(415) 527-5900

Applied Biosystems, Inc.
250 Lincoln Center Drive
Foster City, CA 94404
(415) 570-6667

Beckman Instruments, Inc.
1050 Page Mill Road
Palo Alto, CA 94304
(415) 857-1150

DuPont Co.
Biotechnology System
Barley Mill Plaza
Building 24
Wilmington, DE 19898
(302) 992-4743

EM Science
111 Woodcrest Road
Cherry Hill, NJ 08034
(609) 354-9200

Hamilton, Company
P. O. Box 10030
Reno, NV 89510
(702) 786-7077

Hewlett Packard
P. O. Box 10301
Palo Alto, CA 94303
(415) 857-5731

IBM Instruments, Inc.
Orchard Park
P. O. Box 332
Danbury, CT 06810
(203) 796-2500

Interaction Chemicals, Inc.
1615 Plymouth Street
Mountain View, CA 94043
(415) 969-2400

ISCO, Inc.
P. O. Box 5347
Lincoln, NE 68505
(402) 464-0231

Lazar Scientific
920 N. Formosa Avenue
Los Angeles, CA 90046
(213) 384-6195

LKB Instruments, Inc.
9319 Gaither Road
Gaithersburg, MD 20877
(301) 763-3200

MG Industries
Gas Products
2460 Boulevard of the Generals
P. O. Box 945
Valley Forge, PA 19482
(215) 630-5400

Millipore
Bedford, MA 01730
(617) 275-9200

Peptides International, Inc.
10101 Linn Station Road
Suite 445
Louisville, KY 40223
(502) 425-8765

Perkin-Elmer Corp.
Main Avenue (MS-12)
Norwalk, CT 06856
(203) 762-1000

Phenomenex
6100 Palos Verdes Drive South
Rancho Palos Verdes, CA 90274
(213) 541-0606

Polymer Laboratories, Inc.
P. O. Box 1581
Stow, OH 44224
(216) 688-7339

Porton Instruments
5445 Balboa Boulevard
Suite 115
Encino, CA 91316
(218) 995-2471

Rainin Instrument Co., Inc.
Mack Road
Woburn, MA 01801
(617) 935-3050

SCI-CON
631 Executive Drive
Winter, Park, FL 32789
(305) 647-4900

Scientific Glass Engineering (SGE)
2007 Kramer Lane
Austin, TX 78758
(512) 837-7190

Spectra Physics
AutoLab Division
3333 North First Street
San Jose, CA 95134
(408) 946-9682

SynChrom, Inc.
P. O. Box 110
Linden, IN 47955
(317) 423-4694

Waters
34 Maple Street
Milford, MA 01757
(617) 478-2000

Whatman Lab Sales, Inc.
5285 N.E. Elam Young Parkway
Suite A-400
Hillsboro, OR 97124
(503) 648-0762

INDEX

A

B

D

E

F

G

H

I

L

M

N

O

P

Q

R

S

For Product Safety Concerns and Information please contact our EU
representative GPSR@taylorandfrancis.com
Taylor & Francis Verlag GmbH, Kaufingerstraße 24, 80331 München, Germany

www.ingramcontent.com/pod-product-compliance
Ingram Content Group UK Ltd.
Pitfield, Milton Keynes, MK11 3LW, UK
UKHW052052230926
475702UK00025B/1125
* 9 7 8 1 3 1 5 8 9 6 9 8 4 *